Bioinformatics Methods

The past three decades have witnessed an explosion of what is now referred to as high-dimensional "omics" data. ***Bioinformatics Methods: From Omics to Next Generation Sequencing*** describes the statistical methods and analytic frameworks that are best equipped to interpret these complex data and how they apply to health-related research. Covering the technologies that generate data, subtleties of various data types, and statistical underpinnings of methods, this book identifies a suite of potential analytic tools and highlights commonalities among statistical methods that have been developed.

An ideal reference for biostatisticians and data analysts that work in collaboration with scientists and clinical investigators looking to ensure the rigorous application of available methodologies.

Key Features:

- Survey of a variety of omics data types and their unique features
- Summary of statistical underpinnings for widely used omics data analysis methods
- Description of software resources for performing omics data analyses

Shili Lin, PhD is a Professor in the Department of Statistics and a faculty member in the Translational Data Analytics Institute at the Ohio State University. Her research interests are in statistical methodologies for high-dimensional and big data, with a focus on their applications in biomedical research, statistical genetics and genomics, and integration of multiple omics data.

Denise Scholtens, PhD is a Professor and Chief of the Division of Biostatistics in the Department of Preventive Medicine at Northwestern University Feinberg School of Medicine. She is interested in the design and conduct of large-scale multi-center prospective health research studies, and in the integration of high-dimensional omics data analyses into these settings.

Sujay Datta, PhD is an Associate Professor and the Graduate Program Coordinator in the Department of Statistics at the University of Akron. His research interests include statistical analyses of high-dimensional and high-throughput data, graphical and network-based models, statistical models and methods for cancer data, as well as sequential/multistage sampling designs.

Bioinformatics Methods
From Omics to Next Generation Sequencing

Shili Lin
Denise Scholtens
Sujay Datta

CRC Press
Taylor & Francis Group
Boca Raton London New York

CRC Press is an imprint of the
Taylor & Francis Group, an **informa** business

First edition published 2023
by CRC Press
6000 Broken Sound Parkway NW, Suite 300, Boca Raton, FL 33487-2742

and by CRC Press
4 Park Square, Milton Park, Abingdon, Oxon, OX14 4RN

CRC Press is an imprint of Taylor & Francis Group, LLC

© 2023 Taylor & Francis Group, LLC

ISBN: 978-1-498-76515-2 (hbk)
ISBN: 978-1-032-34183-5 (pbk)
ISBN: 978-1-315-15372-8 (ebk)

DOI: 10.1201/9781315153728

Typeset in CMR10 font
by KnowledgeWorks Global Ltd.

Publisher's note: This book has been prepared from a camera-ready copy provided by the authors.

The authors dedicate this book to their families.

Contents

Preface

The past three decades have witnessed an explosion of what we now often refer to as high-dimensional "omics" data. Initial investigations into the sequence of the human genome and variation in gene expression as evidenced by the assays of messenger RNA abundance, i.e. genomics, ushered in a paradigm of high-throughput assay development and data generation for molecular and cellular biology that has revolutionized the way scientists think about human health. Emergence of these varied, complex and often related data types demands development of statistical methodologies and analytic paradigms to support meaningful interpretation of these data. This book is written to describe a sampling of statistical methods and analytic frameworks that have been developed to address the unique features of several types of high-dimensional molecular omics data. The content describes the technologies that generate data, subtleties of various data types, basic statistical underpinnings of methods that are used for their analysis, available software for implementing analyses and discussions of various applications in health related research. The intended audience is collaborative biostatisticians and data analysts who are looking to ensure rigorous application of available methodologies as they partner with basic scientists and clinical investigators.

The introductory Chapter 1, The Biology of a Living Organism, summarizes the basic cellular components and molecular biology that the various omics technologies discussed in this book are designed to assay. This description is included to give context for the resultant data and highlight how, in many cases, different data types provide multiple points of view on a highly coordinated physiologic system. An Appendix provides similar foundational content, but from a statistical point of view. It is expected that most readers of this book will be familiar with basics of probability, random variables, probability distributions, basics of stochastic processes, hidden Markov models, and both frequentist and Bayesian inference procedures. Nevertheless, appendix content is provided to help fill in gaps where needed.

The primary content of the book focuses on various types of omics data. Chapters 2 and 3 are dedicated to Protein-Protein Interactions and Protein-Protein Interaction Network Analyses. Chapter 4 summarizes Detection of Imprinting and Maternal Effects through epigenetic mechanisms. Sequencing data are the focus of both Chapter 5 Modeling and Analysis of Next-Generation Sequencing Data and Chapter 6 Sequencing-Based DNA Methylation Data. Chapter 7 Modeling and Analysis of Spatial Chromatin Interactions and Chapter 8 Digital Improvement of Single Cell Hi-C Data describe some of the most recently emerging data types and related methods. Finally, Chapter 9 Metabolomics Data Preprocessing and Chapter 10 Metabolomics Data Analysis are dedicated to a survey of metabolomics data. A strong common theme in each chapter is the importance of specifying statistical models that closely reflect the underlying biology that is being queried. Themes of preprocessing to handle for technical variation and artifacts, accounting for testing multiplicity and cross-feature dependencies, and computational efficiency also arise.

The content herein is by no means a comprehensive survey of the state-of-the-art. Given the pace of technology development and data availability, new techniques, descriptive and inferential procedures, analysis packages and potential applications emerge almost daily. It is our hope that the survey offered here will identify a suite of potential analytic tools, highlight commonalities among statistical methods that have been developed, and motivate continued statistical methods development for meaningful, integrated analyses of high-dimensional molecular omics data.

1

The Biology of a Living Organism

CONTENTS

There is no unique definition of bioinformatics in the scientific literature. Generally speaking, it is a rapidly evolving field of science that harnesses the powers of mathematics, statistics and information science to analyze very large biological datasets using modern-day high-performance computing. It is all about extracting knowledge from the deluge of information generated by cutting-edge high-throughput technologies in the biomedical sciences. A great deal of research in the current century has gone into developing innovative tools and techniques enabling us to handle the information overload that the scientific community is currently experiencing. In this book, one of our goals is to shed some light on those tools and techniques.

Among the natural sciences, biology is the one that studies highly complex systems known as living organisms. Before the era of modern technology, it was primarily a descriptive science that involved careful observation and detailed documentation of various aspects of a living being. (e.g. its appearance, behavior, interaction with the surrounding environment, etc.). These led to a reasonably accurate classification of all visible forms of life on the earth (the *binomial nomenclature* by Carolas Linnaeus) and to a theory of how the diversity of life-forms we see all around us came into being over billions of years (the theory of *evolution* by Charles Darwin). However, the technology available in those days was not good enough for probing the internal mechanisms that made life possible and sustained it – the complex biochemistry of metabolism, growth and self-replication. So the biological datasets that were available for statistical analysis in those days (including clinical, epidemiological and taxonomic data) were relatively small and manageable. Standard classical statistical procedures such as two-sample t-tests, ANOVA and least-squares regression were adequate to handle them. But it all

DOI: 10.1201/9781315153728-1

began to change around the middle of the twentieth century with a series of key breakthroughs in the biomedical sciences and rapid technological development. These breakthroughs enabled us to probe the inner sanctum of a living organism at the molecular level and brought increased sophistication to our concept of medicine. A series of new discoveries gave us unprecedented insight into the *modus operandi* of living systems, the most famous of which was the Franklin-Watson-Creek *double helix* model for DNA. Building on these foundations, an emerging goal of medical scientists became personalized medicine. New experiments powered by advanced technologies were now generating enormous amounts of data on various aspects of life, and increasingly efficient computers made it possible to create and query gigantic databases. Statisticians and information scientists are now inundated with an incredible amount of data and the demand for methods that can handle huge datasets that often violate some of the basic assumptions of classical statistics. Enhanced data-storage capability enables an abundance of prior information on unknown parameters, and the lack of closed-form analytical solutions matters less given the capability for approximate numerical solutions. In summary, modern-day statisticians almost invariably find themselves navigating an ocean of information where traditional statistical methods and their underlying assumptions often are often inappropriate.

Understanding the intricate chemistry by which life survives, propagates, responds and communicates is the overarching goal of molecular biology – a field which relies increasingly on automated technologies for performing biological assays and on computers for analyzing the results. While the complexity and diversity of life on earth are truly awe-inspiring, at the molecular level there are many commonalities between different forms of life. In the next section, we briefly review some essential concepts in cellular and molecular biology, which will provide the necessary background and motivation for the bioinformatics problems discussed throughout the book. First, we discuss cells and organelles, followed by an introduction to DNA (the genetic material) and the chromosomes' 3D structure. Proteins are considered next. Finally, biochemical networks are discussed, with a focus on cell metabolism and control of proliferation. For further details on any of these topics, the reader is referred to any modern textbook on molecular and cellular biology.

1.1   Cells

The basic structural and functional unit of a living organism is a *cell*. Robert Hooke, a British scientist of the seventeenth century, made a chance discovery of them when he looked at a piece of cork under the crude microscope that was available at that time. Soon it was hypothesized, and ultimately verified that every living organism is made of cells. In more than three and a half centuries since their discovery, thanks to increasingly powerful microscopes and other sophisticated technologies, astonishing details are available regarding all aspects of cells, including their internal structure, day-to-day function, proliferation and death. Primitive organisms such as *bacteria* (e.g. *Proteobacteria* that include pathogens such as *Escherichia coli* and *Salmonella*, *Cyanobacteria* that are capable of converting the sun's electromagnetic energy to chemical energy through photosynthesis and *Archaebacteria* that survive in extreme environments such as the deep-ocean hydrothermal vents) are single-cell organisms known as *prokaryotes*. They lack an organelle, called a nucleus, that is present in the cells of all higher organisms known as *eukaryotes*. Eukaryotes, which include everything from a bath sponge to an elephant, may be uni-cellular or multi-cellular. In multi-cellular organisms, the cells are organized into *tissues* or groups of cells that collectively perform specific tasks, such as transport of oxygen or absorption of nutrients. Several such tissues collectively form an *organ*, such as the leaves of a plant or the lungs in our body. All the different types of cells in a multi-cellular organism originate from the same "mother cells" in the embryo, known as *embryonic stem cells*, which are *pluripotent* (i.e. capable of transforming into various types of specialized cells through a process called *differentiation*). This is why embryonic stem cells are considered so promising in the biomedical sciences as a possible cure for certain debilitating and degenerative diseases. The process of differentiation is still not fully understood, and bioinformatics may play a key role in helping us understand it.

The human body consists of trillions of eukaryotic cells. There are at least two hundred different types of cells in our body, and they may be as little as a few thousandths of a millimeter in diameter or up to a meter in length (e.g. neurons with long axons). Each cell has an outer membrane which encloses the *cytoplasm* and various components called *organelles*. They include the *nucleus*, the *endoplasmic reticulum*, the *Golgi apparatus*, the *mitochondria*, the *ribosomes* and the *lysosomes*. A plant cell also has large cavities called *vacuoles* and different types of *plastids* (*chloroplasts* containing chlorophyll for photosynthesis, *leucoplasts* and *chromoplasts*). Each of these organelles has some specific role to play in the life of a cell. Because mitochondria and chloroplasts can multiply within a cell and bear structural similarities to certain unicellular prokaryotic organisms, it is hypothesized that a long time ago, some prokaryotic cells started living inside others, forming a *symbiotic* or mutually beneficial relationship with their hosts. This process, called *endosymbiosis*, is believed to have led ultimately to the development of eukaryotic cells.

There are four major classes of small organic (carbon-containing) molecules in cells: *sugars*, *lipids*, *amino acids* and *nucleotides*. These molecules may exist on their own in the cell but may also be bound into *macromolecules* or *polymers*, large molecules comprising specific patterns of different types of small organic molecules. For example, simple sugars such as glucose act as a source of energy for the cell and as a means for storing energy. Glucose can also be assembled into *cellulose*, a structural polymer that forms the cell walls of plant cells. Lipids are a crucial component of cellular membranes, including the outer cell membrane and the membranes of organelles. They are also used for the storage of energy and act as *hormones*, transmitting signals between different cells. Amino acids are the building blocks of *proteins* which perform numerous functions in a living organism. The nucleotide adenosine triphosphate (ATP) is used to transfer energy needed in a variety of reactions, while long chains of nucleotides form the *deoxyribonucleic acids* (DNA) and *ribonucleic acids* (RNA) that act as the basis of heredity and cellular control. In addition to these organic molecules, there are a variety of inorganic molecules inside the cell. By mass, a cell for the most part is water. A few other important inorganic compounds are oxygen and carbon monoxide (which are usually considered inorganic despite containing carbon), various ions (the medium for transmitting electrical signals in neurons and heart cells) and phosphate groups (crucial in intracellular signaling).

The cell membrane is primarily made of *phospholipid*, with various *membrane proteins* and other substances embedded in it. A phospholipid molecule has a *hydrophobic* (water-repelling) end and a *hydrophilic* (water-attracting) end. A cell membrane consists of a phospholipid bi-layer, with the hydrophobic "heads" of the molecules tucked inside and the hydrophilic "tails" forming surfaces that touch water. In addition to serving as an enclosure for the organelles, the membrane controls the passage of substances into and out of the cell, and various receptors embedded in it play a crucial role in *signal transduction*. The endoplasmic reticulum primarily serves as a transport network and storage area for various cellular substances. It is continuous with the membrane of the nucleus. Various nuclear products (e.g. the messenger RNA produced by transcription) are transported to places where they are needed through the endoplasmic network. This reticulum has smooth surfaces and rough surfaces, the rough appearance being the result of numerous ribosomes attached to it. The ribosomes are the cell's protein-manufacturing plants where the macro-molecules of proteins are synthesized component by component according to the instructions in the messenger RNA. The proteins produced at the ribosomes are transported to the Golgi apparatus (named after the Italian scientist Camillo Golgi who first noticed them) where they are further modified as needed and then sent off in *vesicles* (little bubble-like sacks) to other organelles or to the cell membrane for secretion outside the cell. This Golgi apparatus has a *cis* end that is nearer to the endoplasmic reticulum and a *trans* end that is farther from it. The incoming proteins are received at the *cis* end and the outgoing vesicles bud off from the *trans* end. The lysosomes are the cell's scavengers or garbage cleaners, facilitating the disintegration of unwanted or harmful substances in the cell. The mitochondria are the energy-producing plants where the cell's energy currency (ATP) is synthesized. A mitochondrion has a membrane quite similar to that of a prokaryotic bacterium and inside it, there are protruded folds of the membrane called *christae*. The rest of the space inside it is called the *lumen* where ATP synthesis takes place. In a plant cell, a chloroplast also has an outer membrane resembling that of a prokaryotic bacterium and inside it, the lumen contains stacks of flat disc-like structures called *thylakoids*. This is where chlorophyll, a magnesium-containing protein, is found that plays a key role in photosynthesis – the plant's food production process. Chlorophyll is also the reason why leaves are green. Chromoplasts contain other pigments such as carotene and xanthophylls instead of chlorophyll. These are the reason behind the bright display of yellow, orange and brown in autumn. The remaining major organelle in a eukaryotic cell is the nucleus. It is roughly spherical with an outer membrane that has pores in order to let nuclear products out into the cytoplasm or let cytoplasmic substances in. The central, denser region of a nucleus is called the nucleolus. Most importantly, a living organism's blueprint of life (i.e. its DNA) is stored inside the nucleus.

1.2 Genes, DNA and RNA

Deoxyribonucleic acid or DNA is present in all living organisms and is of central importance to regulating cellular function and conveying hereditary information. It does so primarily through the *genes* it encodes. All cells in our body contain DNA in their nuclei (except a few such as the erythrocytes or red blood cells). As mentioned earlier, DNA is a polymer of nucleotides, conceptually arranged like a ladder. The functionally interesting component of a nucleotide is its *base*. There are four different bases: *adenine, guanine, thymine* and *cytosine*, often referred to by their single-letter abbreviations A, G, T and C. Adenine and guanine are called *purines*, while the other two are called *pyrimidines*. Pairs of bases, one from each strand of the double-stranded DNA, bind together to form the steps of the ladder. These base-pairs are held in place by two backbones of sugar (deoxyribose) and phosphate molecules, which form the sides of the ladder. Each purine binds with only one kind of pyrimidine – A with T and G with C. This is known as *complementary base pairing*.

In reality, DNA molecules are not shaped like a straight ladder but have a complex three-dimensional (3D) structure. The two strands of the DNA (i.e. the two sides of the ladder) are twisted into the well-known *double helix* shape, famously discovered in 1953 by James Watson and Francis Crick with the help of X-ray crystallographic studies done by Rosalind Franklin. At a larger scale, the DNA double-helix is wrapped around barrel-shaped protein molecules called *histones*. The DNA and histones are, in turn, wrapped into coils and other structures, depending on the state of the cell.

In some cells, most or all of the DNA occurs in the form of a single molecule. In prokaryotes such as bacteria, most of the DNA resides in a single *chromosome*, a long loop of DNA without beginning or end. Bacteria also often contain *plasmids*, which are much shorter loops of DNA that can be exchanged between bacteria as a means of sharing beneficial genes. In eukaryotes, the DNA is divided into several chromosomes, each a linear stretch of DNA with a definite beginning and end. Many eukaryotes are *polyploid,* carrying more than one copy of each chromosome. For example, humans are diploid as most human cells carry two copies of each chromosome – one inherited from each parent. Some plants are tetraploid, carrying four copies of each chromosome. The total length of DNA and the number of chromosomes into which it is divided vary from one organism to another. The single chromosome of the bacterium *Escherichia coli* has about 4.6 million base pairs, while humans have over 3 billion base pairs divided into 23 pairs of chromosomes. Fruit flies (*Drosophila melanogaster* have less than 200 million base pairs divided into 8 pairs of chromosomes, whereas rice has roughly 400 million base pairs divided into 24 pairs of chromosomes.

When a cell divides to produce two daughter cells, as is necessary for organism growth and for replacing old or damaged tissue, the DNA in the parent cell must be *replicated* to provide each daughter cell with its own copy. Complementary base-pairing is key to this process. During replication, a group of proteins including *DNA polymerase* travels along the DNA strands. The two strands of DNA are separated, much like a zipper being unzipped, and complementary base pairs are filled in along both strands, resulting in the two needed copies.

Less than 5% of the DNA is believed to encode useful information for protein synthesis. The other 95%, the so-called "junk" DNA, was believed to have no apparent function until recently. However, cutting-edge research in the last few years has shed more light on the relevance of this non-protein-coding DNA in terms of gene regulation and other important things. Of the 5% functional DNA, the majority is part of some gene. Genes are regions of the DNA comprising two parts – a regulatory region or *promoter* and a coding region. The regulatory region is partly responsible for specifying the conditions under which the gene-product is produced, or the extent to which it is produced (i.e. the *expression* of the gene). The coding region specifies the functional molecular product or products – often a protein but sometimes an RNA (ribonucleic acid). Proteins, as mentioned earlier, are sequences of amino acids. For protein-coding genes, the coding region specifies the amino-acid sequence of the protein. However, the protein is not directly constructed from the DNA. Instead, the DNA is *transcribed* into an RNA intermediate, called a messenger RNA (mRNA), which is then *translated* into protein at the ribosome. For RNA-coding genes, the RNA coded by it is itself the final product and serves some important function instead of being translated into protein.

RNA consists of a single sequence, or strand, of nucleotides. These nucleotides are similar to the ones in DNA, except that the backbone uses the sugar *ribose,* and the base *uracil* (U) is used instead of thymine (T). Transcription (i.e. the construction of an RNA chain from the DNA) is similar to DNA replication. A group of proteins, collectively known as RNA polymerases, open up the DNA at the start of the coding region and move along the DNA until reaching the end of the coding region. The *transcript* is produced one nucleotide at a time by complementary base-pairing (A with U, C with G, G with C and T with A). The resulting RNA molecule does not stay bound to the DNA template that generated it. As it is constructed, the RNA nucleotide chain separates from the DNA and the two DNA strands then re-join, going back to the same state as they were in before transcription. Instead of forming a stable and neat geometric shape (such as the DNA double-helix), the nucleotides in an RNA strand bind to each-other forming a complex 3D shape unique to each RNA sequence. For RNA-coding genes, features of this 3D shape determine its functional properties (for example, the ability to bind to specific parts of certain protein molecules). For protein-coding genes, the creation of the RNA is only the first step in protein production.

Proteins are composed of 20 naturally occurring amino acids. The RNA nucleotide sequence encodes an amino acid sequence in a relatively straight-forward way. Each triplet of nucleotides, called a *codon*, specifies one amino acid. For example, the RNA nucleotide triplet AAA is the codon for the amino acid lysine, while the triplet AAC codes for asparagine. Since there are 4 nucleotides, there are $4^3 = 64$ distinct possible codons. However, as there are only 20 distinct amino acids, there is quite a bit of redundancy among the codons. For example, the amino acid leucine is encoded by six different codons: CUU, CUG, CUC, CUA, UUG and UUA. The entire length of the RNA strand does not code for amino acids. A small amount of RNA at the start and the end of the transcript (called untranslated regions or UTRs) are not translated into amino acids. For some genes, especially in eukaryotes, there are untranslated gaps in the middle of an RNA transcript, called *introns*. The portions of the transcript that are translated are known as *exons*. In eukaryotes, a process called *splicing* removes the introns before the transcript leaves the nucleus for translation. Splicing can result in the elimination of some exons as well, leading to different *splice-variants* of a protein. This is referred to as *alternative splicing*, which allows a cell to produce different versions of a protein from the same DNA. Ribosomes translate the RNA transcript into the corresponding amino acid sequence. In prokaryotes, free-floating ribosomes can begin translation as soon as the transcript is created, or even while it is being created. In eukaryotes, the transcripts must be transported outside the nucleus through the nucleopores and reach a ribosome attached to the rough endoplasmic reticulum.

Transcription, splicing and translation are regulated in various ways. Characteristic patterns of nucleotides in the DNA indicate where transcription should begin and end. For example, in many bacteria, transcription begins just after a stretch of DNA with the sequence TATAAT. However, some variations on this sequence are allowed and, conversely, not every occurrence of this sequence in the DNA is immediately followed by the coding region of a gene. A different pattern indicates the end of the coding region. In eukaryotes, there is much more variation in these patterns. In either case, even when the DNA of an organism is completely sequenced, there remains uncertainty about the exact locations and number of the genes.

Variations in transcription initiation patterns have apparently evolved in order to control the rate or frequency with which the RNA polymerase binds to the DNA beginning the process of transcription. However, transcription is primarily regulated by *transcription factors.* These proteins bind to the regulatory region of the gene on the DNA, usually located within a few hundreds or thousands of base pairs of the transcription initiation site, and influence the rate or frequency of transcription by one or more of the following mechanisms: blocking the binding of RNA polymerase, changing the shape of the DNA to expose the transcription initiation site and thereby increasing RNA polymerase binding, attracting RNA polymerase to the region of transcription initiation, or acting as a dock or blocker for other proteins which themselves have similar effects. Since these transcription factors are proteins themselves and thus generated from genes, the logical conclusion is that some genes regulate others, giving rise to *gene regulatory networks.* Transcription factors bind to the DNA at specific sites known as *transcription factor binding sites* by virtue of their 3D shapes. These sites are typically between 8 and 20 nucleotides long. An organism may have hundreds of different proteins acting as transcription factors, each of which binds to different characteristic patterns of nucleotides. However, there are variations in these patterns, as was the case for transcription initiation sites. Some of these variations may serve the purpose of influencing the frequency or strength with which the transcription factor binds. Because of the short lengths of these sites, the variability seen in their patterns, and the virtual "haystack" of DNA in which they can be situated, it is not easy to identify transcription factor binding sites. Further, the binding of a transcription factor to a site does not guarantee any influence on the transcription process. Thus, identifying *functional* binding sites is an additional layer of difficulty.

In eukaryotes, and for a few genes in prokaryotes, splicing and alternative splicing follow transcription. The signals that regulate splicing are partly in the transcribed RNA sequence itself, although clear signals have not been identified. RNA from RNA-coding genes also plays a role, especially in alternative splicing. However, this depends on a complex interplay between the 3D structures of the regulatory RNA, the transcript and the splicing machinery.

Translation is not as heavily regulated as transcription, and its regulation is better understood. A ribosome assembles the sequence of amino acids specified by the transcript in the order in which they are encountered. A type of RNA, known as *transfer RNA* or tRNA, plays an important role in this assembly process. Any of the three codons UAA, UAG and UGA indicate the end of the protein and the termination of translation. In eukaryotes, ribosomes bind to the start of the transcript and move along it until encountering the first AUG codon, when translation begins. Sometimes the translation machinery will skip over the first (or even the second) occurrence of AUG and begin at the next occurrence. This skipping is influenced by the adjacent nucleotides, and is another mechanism by which different variants of a protein are created from the same DNA coding region. In bacteria, the ribosomes do not bind to the start of the transcript. Instead, there is a longer nucleotide sequence (including an AUG codon) to which they bind. The ribosomes can bind to that nucleotide sequence anywhere it occurs in the transcript and begin translation. As a result, it is possible for a single transcript to code for several proteins, each separated by a region that is not translated. Such a gene is called an *operon* and is a common means by which proteins that have related functions are co-regulated. For example, the well-known *lac* operon in the bacterium *Escherichia coli* includes three genes, one of which helps the organism to metabolize the sugar lactose and another transports lactose into the cell from the extracellular environment. Translation of an RNA transcript can also be regulated by other proteins or RNAs (such as micro-RNAs), which can, for example, bind with the RNA transcript and block the translation mechanism.

1.3   Proteins

The reason why proteins are called the "work-horses" of a cell is that they are involved in virtually every biological process in cells, including sensing of the cell's environment and communication between cells. Proteins are polymer chains with amino acids as their building blocks. A backbone comprising one nitrogen and two carbon atoms is bound to various hydrogen and oxygen atoms. The central carbon is also bound to a unit (which can be a single atom or a group of atoms) that distinguishes among different amino acids. In *glycine*, the simplest amino acid, that unit is just one hydrogen atom. In *methionine*, that unit contains three carbon, seven hydrogen and one sulfur atoms. In different amino acids, that unit differs in its chemical properties, most relevantly in size, acidity and polarity. When amino acid molecules bind together to form polymers, water molecules are released, and the parts of the amino acids that still remain are called *residues*. Such a bond formation is known as peptide bonding. Shorter amino acid sequences are *peptides*, while longer ones are *proteins* (typically containing hundreds or thousands of amino-acid residues).

As a protein is generated by translation from a transcript, it folds into a complex 3D shape, which depends on its amino acid sequence as well as the chemical environment in which it is folding. The structure of a protein can be described at four different levels. The *primary* structure of a protein is simply its amino acid sequence. The *secondary* and *tertiary* structures refer to the 3D folded shape of the protein. The tertiary structure is a specification of 3D coordinates for every atom in the protein in an arbitrary coordinate system, as well as that of which atoms are chemically bound to each other. Tertiary structures contain commonly occurring patterns, such as *alpha-helices* and *beta-sheets* Alpha-helices are stretches of the protein that wind into a helical form. A beta-sheet is a set of *beta-strands* that align with each other lengthwise, forming a sheet-like shape. The secondary structure of a protein assigns each amino acid to participating in an alpha-helix, participating in a beta-sheet, participating in a bend between two alpha-helices, and so on. So the secondary structure of a protein is an abstraction of its tertiary structure components. Many proteins participate in complexes (groups of proteins and other molecules that are weakly bound together). A complex may contain proteins of different types. Conversely, one type of protein can participate in different complexes. A specification of how proteins and other molecules organize into complexes is called the *quaternary* structure.

In addition to serving as the building material for various parts of an organism's body, proteins play a number of important roles related to DNA and RNA. Transcription factors regulate transcription rates, RNA polymerase creates the transcripts, DNA polymerase opens up the DNA double-strand like a zipper and then replicates it, histones (little packs of proteins in the chromatin wrapped by the DNA double-strand) affect the 3D structure of the DNA and influence transcription. Proteins are also responsible for repairing damage to DNA (caused by exposure to radiation or certain toxic chemicals, among other things) and for untangling DNA. Proteins also act as *enzymes*, molecules that facilitate the occurrence of chemical reactions (most often very specific reactions). For instance, the enzyme *lactase* facilitates the conversion of the sugar lactose to two other simpler sugars, glucose and galactose. The roles that DNA polymerase and RNA polymerase play can also be viewed as enzymatic. The specificity of an enzyme to the reactions it *catalyzes* (i.e. accelerates) depends on its *active sites* (i.e. portions of the tertiary structure of the enzyme that bind to particular reactants.

Cell membranes, such as the outer membrane of a cell or the nuclear membrane, are impermeable to many types of molecules. *Trans-membrane* proteins control the flow of molecules across membranes and convey signals from one side of the cell to the other. These proteins reside within the membrane and protrude on either side. *Channels* allow the transport of molecules (mostly very specific molecules). For example, ion channels in cardiac muscles or neurons control the flow of ions (such as sodium, potassium, calcium and chlorine) by opening to allow the flow or closing to stop it, depending on triggers such as electrical activity. The dynamical properties of these channels (opening and closing), and the resulting changes in ion flows, drive larger-scale electrophysiological phenomena, such as the beating of the heart and the transmission of action potentials down the axon of a nerve cell. Nuclear pore proteins form complexes that allow molecules such as RNA transcripts and transcription factors across the nuclear membrane. *Signaling* is a related task but does not necessarily involve the transport of a molecule from one side of a membrane to the other. Often, the binding of a specific molecule to a trans-membrane protein causes a structural change to the part of the protein on the other side of the membrane. This structural change then typically sets off a chain of reactions which convey the signal (about the presence of the molecule on the other side of the membrane) to its appropriate destination.

Structural proteins provide rigidity to cells and perform various mechanical functions. The *cytoskeleton* is made of long, filamentous protein complexes and maintains the shape of cells and organelles by forming a network beneath the cellular or organellar membrane. The cytoskeleton is also involved in changes to cell shape, as in the formation of pseudopods. *Microtubules*, a part of the cytoskeleton, act as conduits or roads, directing the movement of molecules and organelles within the cell. Structural proteins are also involved in movements at the cellular and whole-organism levels. The flagella of bacteria and that of sperms are composed of microtubules. The movement of the protein myosin along actin filaments (part of the cytoskeleton) generates the contractile force in animal muscle cells.

Proteins and peptides are found in many other locations and are involved in many other processes (e.g. *antibodies* in the immune system, *hemoglobin* in the red blood cells, *hormones*, *neuro-transmitters*, etc.).

1.4 The epigenome

The epigenome is the answer to questions such as "Why did only one of a pair of identical twins develop cancer and the other remain healthy?" or "How did a muscle cell end up being so different from a nerve cell, given that all cells in the human body have the same genetic blueprint?" The short answer is that it happens due to DNA modifications that do not change the DNA sequence but still affect gene activities. An example of such modification is the addition of a chemical compound to an individual gene. The epigenome is the totality of all the chemical compounds that have been added to the entirety of one's DNA (genome) as a way of regulating the expressions of all the genes in that genome. Although these chemical compounds are not part of the DNA sequence, they remain in position as the cell divides and, in some cases, can be inherited through the generations. This is the mechanism by which environmental influences such as the experience of an acute famine or the exposure to toxic chemicals are sometimes stored as genetic "memory" and handed down to subsequent generations.

Epigenetic changes can help determine whether genes are turned on or off and can, thereby, have an impact on protein production, ensuring that only necessary proteins are produced. For example, proteins that are needed for muscle growth are not produced in nerve cells. Patterns of epigenomic modification vary among individuals. Among different tissue types within an individual and even sometimes among different cells within a tissue. A commonly encountered epigenomic modification is *methylation*, which involves the attachment of small molecules called methyl groups (CH_3) to DNA segments. When methyl groups are added to an individual gene, that gene is turned off or silenced (i.e. its transcription and translation don't happen). Other types of epigenomic modification are *acetylation, ubiquitination*, etc.

The process of epigenomic modification is not error-free. Silencing a wrong gene or failing to silence a gene that needs to be suppressed can lead to abnormalities and cause genetic disorders (including cancers).

1.5   Metabolism

The totality of all biochemical reactions that occur in a cell is known as metabolism. It involves all the four types of fundamental biochemical molecules mentioned earlier, plus the cooperative and coordinated action of various catalysts called enzymes (which are actually proteins). Often a number of biochemical reactions are linked to one another through cofactors forming what are known as pathways. A metabolic pathway is, in some sense, like a roadmap showing the steps in the conversion process of one biochemical compound to another. The important "landmarks" or points of interest on this map are, for example, enzymes that may be targets for drugs or toxins (i.e., poisons), enzymes that are affected by diseases and other regulating factors. Learning a metabolic pathway basically means being able to identify these crucial "landmarks," understanding how metabolic intermediates are related to each other and how perturbations of one system may affect other systems.

Metabolic pathways can be broadly categorized into three groups: catabolic, anabolic and central. Catabolism means disassembly of complex molecules to form simpler products, and its main objectives are to produce energy or provide raw materials to synthesize other molecules. The energy produced is temporarily stored in high-energy phosphate molecules (adenosine triphosphate or ATP) and high-energy electrons (NADH). Anabolism means synthesis of more complex compounds from simpler ingredients, and it usually needs the energy derived from catabolic reactions. Central pathways are usually involved in interconversions of substrates (i.e., substances on which enzymes act), and these can be regarded as both catabolic and anabolic. An example is the citric acid cycle. Catabolic pathways are convergent because through them, a great diversity of complex molecules is converted to a relatively small number of simpler molecules and energy-storing molecules. Anabolic pathways are divergent since through them, a small number of simple molecules synthesize a variety of complex molecules. Another way of classifying metabolic pathways is to categorize them as linear, branched, looped or cyclic. A linear pathway can be is the simplest of all; it is a single sequence of reactions in which a specific initial input is ultimately converted to a specific end-product with no possibility of alternative reactions or digressions in the pathway. This kind of pathway is not found very often. A more common type is a branched pathway in which an intermediate compound can proceed down one branch or another, leading to possibly different end-products. Typically at each branching point there are several enzymes competing for the same substrate, and the "winner" determines which branch will be followed. A looped pathway is one that involves many repetitions of a series of similar reactions. A cyclic pathway is one whose end-product is the same as the initial substance it started with. An example is the urea synthesis cycle in humans.

Enzymes are proteins that act as biological catalysts, accelerating the rates of various reactions, but they themselves are not irreversibly altered during the reactions. They are usually required in small amounts and have no effect on the thermodynamics of the reactions they catalyze. They differ from inorganic (i.e., non-biological) catalysts in that the latter may speed up reactions hundreds or thousands of times, whereas enzymes often speed up reactions a billion or a trillion times. There are other important differences such as the high specificity of an enzyme for its substrate, lack of unwanted products or harmful side-reactions that might possibly interfere with the main reaction, ability to function in the physiological environment and temperature of an organism's interior, etc. To understand how an enzyme accomplishes its task, one needs to know the different types of forces that are at play in a chemical compound. Inside a molecule, there are ionic or covalent bonds that hold the atoms together and between molecules, there are weaker forces such as hydrogen bonds and Van der Waals force. During a chemical reaction, existing covalent bonds are broken and new ones are formed. Breaking covalent bonds requires an energy input in some form, such as heat, light, or radiation. This is known as the activation energy of the reaction. This energy excites the electrons participating in a stable covalent bond and shifts them temporarily to orbitals further from the atomic nucleus, thereby breaking the bond. These excited electrons then might adopt a different stable configuration by interacting with electrons from other atoms and molecules, thereby forming new covalent bonds and releasing energy. This energy output may be exactly the same as, higher than or lower than the initial activation energy. In the first case, the reaction is called energetically neutral, in the second case, it is called exothermic and in the last case, endothermic. It is important to know that a reaction usually does not proceed in one direction only. At least in principle, if two compounds C1 and C2 can react with each other to form two other compounds C3 and C4, the products C3 and C4 can also react to form C1 and C2. In practice, starting with only C1 and C2, first the forward reaction alone will occur producing C3 and C4, but with increasing accumulation of the latter, the reverse reaction will also start taking place forming C1 and C2. Continuing in this manner, a stage will be reached when the rate of the forward reaction will be identical to that of the reverse reaction. This is known as equilibrium and at this stage, the ratio of the total amount of C1 and C2 to the total amount of C3 and C4 will be constant for a given temperature. Any addition or removal of any of the four compounds will temporarily disturb the equilibrium, and the reaction will then proceed to restore it. With this background, now we are in a position to understand how enzymes do what they do. An enzyme lowers the activation energy of a reaction and increases the rate at which a reaction comes to equilibrium. It does so primarily in the following three ways: (a) by providing a surface on which the molecules participating in a reaction can come together in higher concentrations than in a free solution, so that they are more likely to collide and interact; (b) by providing a microenvironment for the participating molecules that is different from the free-solution environment (e.g., a non-aqueous environment in a watery solution) and (c) by taking up electrons from or donating electrons to covalent bonds. The "lock and key" type binding of an enzyme to its substrate involves interaction between the latter and the reactive groups of the enzyme's amino acid side-chains that are part of its active site. These reactive side-chains may be far apart in the primary amino acid sequence of the enzyme but come closer together when the enzyme molecule folds and assumes its characteristic 3-D shape. This shows why the 3-D structure of a protein is important for its functions, and it also gives the reason behind the extreme "choosiness" or specificity of an enzyme for its substrate. The interaction between an enzyme, and its substrate is usually non-covalent (i.e., ionic bonds, hydrogen bonds, hydrophobic interactions, etc.), although occasionally transient covalent bonds may be formed. Some key factors affecting the activity of an enzyme are (a) temperature, (b) the pH (or negative logarithm of the hydrogen-ion concentration) of the solution environment, (c) concentration of the substrate and (d) presence or absence of inhibitors (e.g., a medical drug or toxic substance that interferes with the enzyme-substrate interaction). Depending on the type of reaction they catalyze, enzymes can be classified into categories such as hydrolases (involved in hydrolysis of covalent bonds), oxido-reductases (involved in oxidation and reduction), transferases (involved in transferring a reactive group from one substrate to another) and so forth.

1.6   Biological regulation and cancer

In order for a highly complex system such as a living organism to survive and function, it is critical that the variety of biochemical pathways that sustain the organism and the countless biomolecules that participate in them be *regulated*. There are many different levels and forms of biological regulation. It can happen at the genetic level (e.g. transcription regulation via the binding of transcription factors to the DNA, translation regulation via the degradation or inactivation of mRNAs by micro-RNAs, etc.) or at the proteomic or metabolomic level through enzymes, hormones and other regulatory agents.

Also, there can be many different control mechanisms. For example, control on the quantity of a metabolite can be achieved through a *supply-demand pathway* (where two other metabolites serve as its "source" and "sink" simultaneously) or through *feedback inhibition* (where a sufficient concentration of the end-product of a metabolic pathway inhibits the pathway itself). Feedback inhibition can be further classified into *sequential* feedback, *concerted nested* feedback, *cumulative nested* feedback and so on. In an earlier subsection, we discussed genetic regulation. Here we shed some light on the regulation of *cell proliferation* and describe the consequences of uncontrolled growth.

Cells in almost all parts of our body are constantly proliferating, although most often it goes unnoticed because it is a slow process and usually does not result in any visible growth. The primary purpose of this ongoing proliferation is to replenish the cells lost or damaged through daily wear and tear. For example, cells in the outermost layer of our skin (epidermis) and those in the lining (or epithelium) of our intestine are subject to frequent wear and tear and need replenishment. Another purpose is to respond to a trauma or injury, where cell proliferation has to accelerate in order to expedite wound healing. Importantly, as soon as the proliferating cells fill the incision created by an injury, they stop their growth "overdrive" and return to the normal "wear and tear" rate of proliferation. Occasionally, the cells in a wound proliferate a little bit beyond what is needed for complete healing, thereby creating a *hypertrophied keloid* (or heaped-up scar), but even this is considered normal.

Two processes control cell proliferation. One involves substances called *growth factors* and *growth inhibition factors*. Growth factors stimulate cells to grow and multiply. They are produced all the time for the sake of daily replenishment, but in greater amounts in the case of an injury. It is exactly the opposite for growth inhibition factors, whose production is reduced during a trauma and goes back to the everyday level once the healing is complete. The other process is *apoptosis* or programmed cell-death. It allows individual cells within a group to die, thereby leaving the group at the same size in spite of new cells produced by proliferation. Some substances enhance apoptosis in certain kinds of tissue.

When cells do not respond to the body's built-in control mechanisms for proliferation, the result is uncontrolled growth and a *tumor* is produced. Sometimes a tumor crosses the normal boundaries of the tissue that it originally belonged to and invades surrounding tissues. Even worse, sometimes tumor cells can get into blood vessels or lymph vessels, travel to body-parts that are distant from their place of origin and spread the phenomenon of uncontrolled growth to those areas. In addition, some of them may produce substances that interfere with the normal functioning of various systems in the body, such as the musculoskeletal system or the nervous system. When they do all these, we call the resulting condition *cancer*. The mechanism by which cancer spreads from one body-part to another is called *metastasis*. Upon reaching a distant region in the body, a small lump of metastatic cancer cells break out of the blood-capillary wall, establish themselves there and continue their uncontrolled growth to produce a new tumor. To crown it all, once the new tumor has grown to a certain size, new blood-vessels grow to supply it with more oxygen and nutrients (a process known as *angiogenesis.*

There are many different varieties of cancer, but as a whole, it remains one of the deadliest diseases worldwide. Despite years of intense research, there is no universal cure for cancer. Some types of cancer can be cured, or at least temporarily remedied, depending on the stage at which they are diagnosed. The traditional methods used to combat the disease include *radiation therapy* (destroying tumor cells by irradiating them), *chemotherapy* (destroying all proliferating cells, including tumor cells, by administering a combination of chemicals into the body) and surgery. But all of them induce significant collateral damage (i.e. destruction of nearby healthy tissue) and have side effects, not to mention the risk of a relapse (i.e. re-occurrence of the disease). Recently, more "directed" methods with a higher precision for destroying cancer cells and fewer side effects have been developed, such as *proton beam therapy* and *drug-induced angiogenesis inhibition.* Another promising approach that has received a lot of attention lately is *immunotherapy* (training the cells in the body's own defense mechanism, the immune system, to recognize tumor cells as alien invaders and to destroy them). But none of these is still widely available.

In cancer research, the central question is *why* some cells in an organism defy the built-in regulatory mechanisms for proliferation and show uncontrolled growth. Scientists do not have a definitive answer yet, but progress has been made in answering this question in the genomic era. Certain genes have been identified as *oncogenes* and *tumor suppressor genes* that play a key role in the development of cancer. In many cases, a mutation in one of those genes or some other kind of damage to it will trigger the uncontrolled growth. There are a variety of causes (*mutagens*) for such mutations, including exposure to radiation and certain types of chemicals. Sometimes, if a cell is subjected to oxidative stress (i.e. exposed to highly reactive free oxygen-radicals due to an abundance of them in the bloodstream), it will undergo DNA damage. It has also been observed that chronic inflammation increases the risk of cancer.

1.7   Data generating technologies

Much of the material discussed in the previous section was learned using traditional "low-throughput" laboratory techniques. However, one of the driving forces behind bioinformatics is the advent of "high-throughput" technologies for detecting and quantifying the abundances of a variety of biomolecules, as well as interactions among them. In this section, we describe some of the traditional low-throughput (but still quite accurate) techniques of molecular biology, along with their more recent and high-throughput counterparts.

Perhaps the best known area of application for bioinformatics is in the analysis of DNA sequences. Traditional methods for sequencing DNA (i.e. determining the sequence of the four nucleotides, A, C, G and T, comprising a gene, a chromosome or even the entire genome of an organism) were painstaking, hands-on procedures that could only produce sequences of limited length and at a high cost. DNA sequencing was revolutionized during the 1980's and 1990's by the introduction of machines or robotic systems that could carry out the labwork automatically (or semi-automatically), along with computers for storing and analyzing the data produced thereby. Today, the genomes of many species (including humans) have been completely sequenced. While this trend continues, the emphasis of sequencing has broadened to include the sequencing of each individual's genome, in order to detect the differences between individuals, which are mostly single-letter changes in the sequence (called *single nucleotide polymorphisms* or SNP) that are the basis for much of the observable differences between individuals.

Other technologies focus not on static features of an organism, but rather on properties that may change over time or that may differ in different parts of an organism (e.g. the concentrations of various RNAs and proteins). *Gel electrophoresis* is a traditional method for detecting the presence of DNA, RNA or proteins in tissue samples. In one-dimensional gel electrophoresis (often called 1D PAGE due to the polyacrylamide gels typically used), a sample of molecules from a tissue is inserted at one end of a rectangle-shaped plate of gel. In one variant, an electric field causes the molecules to migrate towards the other side of the gel. DNA and RNA molecules are naturally negatively charged and, therefore, are accelerated by the field. Proteins are usually bound with substances that make them negatively charged. While the electric field accelerates the molecules, friction decelerates them. Friction is greater for larger (and hence heavier) molecules, and so the molecules separate by size, with the smallest molecules traveling farthest through the gel. The separated molecules can be visualized by staining, fluorescence or radiation and show up as bands at different lengths along the gel. If the gel is calibrated with molecules of known sizes, and if the sample contains relatively few types of molecules, the presence or absence of bands can be directly interpreted as the presence or absence of particular molecules. This can be used to detect if a specific gene is being transcribed by, for example, looking for RNAs of the appropriate size or finding out if the gene's protein products are present in the sample. Often several samples are run side-by-side in a single gel. enabling us to compare the molecules present in each one. If the molecules cannot be identified, due to lack of calibration or other reasons, they can be extracted from the gel to be subjected to a separate identification process. Another version of gel electrophoresis for proteins involves a gel with a pH gradient and an electric field. In this case, the protein molecules move to a position in the gel corresponding to their *isoelectric point*, where the pH balances the protein's electric charge.

In *2D gel electrophoresis* (2D PAGE), typically used with proteins, the sample is inserted at one corner of the gel and sorted in one direction first (often by isoelectric point) and then the electric field is shifted 90 degrees and the proteins are additionally separated by size. After staining, what we get is a gel with spots corresponding to proteins of different isoelectric points and sizes. By comparing gels, one can look for proteins that are present under one condition and not under another. On a 2D gel, it is possible to separate hundreds or even thousands of different kinds of proteins, which makes 2D PAGE a high-throughput measurement technology. 1D and 2D gel electrophoresis techniques are typically used to determine the presence or absence of a kind of molecule, rather than to quantify how much of it is present (although the darkness or spread of a band or spot can sometimes be an indicator of the quantity.

The *Southern blot*, the *Northern blot* and the *Western blot* are extensions of gel electrophoresis that are used to determine the identity of DNA, RNA or protein molecules in the gel respectively. After running the gel, the molecules are transferred onto and bound to a film. A solution of *labeled probes* is washed over the film. For example, if one were interested in testing for the presence of single-stranded DNA for a specific gene, the probes could be DNA strands with bases that are complementary to (i.e. binding partners for) some portion of that gene's DNA and not complementary to any portion of any other gene's DNA. When washed over the film, these probes would bind only to the DNA from the gene of interest. Complementary DNA probes are also used to detect specific RNAs. Antibodies are used as probes to detect specific proteins. Once the probes bind to their target molecules, the solution is washed off to remove unbound probes. The locations of the probes are determined on the basis of their fluorescent or radioactive labels, which are easily imaged.

Gene expression *microarrays*, which revolutionized the monitoring of gene expression, are an adaptation of the blotting idea to a massively parallel scale. Microarrays, which primarily measure messenger RNA levels, are small glass or silicon slides with many thousands of spots on them. In each spot, there are probes for a specific gene, so a single slide can contain probes for virtually all genes in the entire genome of an organism. There are two main types of microarrays: *one-channel* or *one-color* arrays (also known as *oligonucleotide* arrays) and *two-channel* or *two-color* arrays (also known as *cDNA* arrays). In one-channel arrays, RNAs are extracted from a tissue sample, bound with *biotin* and then washed over the array. They then *hybridize* (i.e. bind with) the probes. Ideally, an RNA molecule will hybridize with only the probes that correspond to the gene that generated the RNA, but in reality, there is some cross-hybridization (i.e. RNA molecules binding with probes corresponding to other genes which only have partial complementarity with them). Fluorescent molecules are applied that bind to the biotin attached to the RNAs. The fluorescent molecules are then excited by a laser and imaged. The higher the fluorescence intensity from a spot, the greater the number of RNAs bound to the probes therein, which is an indicator of the expression level of the corresponding gene. For two-channel arrays, RNA is extracted from two different samples. The RNA from each sample is converted by *reverse transcription* into single-stranded *complementary DNAs* (cDNAs) and labeled with fluorescent molecules. Each sample is labeled with molecules that fluoresce at different wavelengths (commonly, red and green). The labeled cDNAs from the two samples are then mixed and washed over the microarray, where they bind to probes. The relative fluorescence in each wavelength of each spot indicates the relative expression of the corresponding gene in the two samples.

Another massively parallel means of measuring gene expression was *Serial Analysis of Gene Expression* (SAGE). In this technology, RNAs are extracted from a sample and a small stretch from one end of each RNA molecule is converted into cDNA. These cDNAs are then bound together in long chains and sequenced by a DNA sequencing machine. These sequences can then be examined to identify and count the source cDNAs, effectively counting the number of RNA molecules that had their ends converted to those cDNAs earlier.

Along the lines of SAGE, another digital technology for transcription profiling was introduced by Lynx Therapeutics, Inc. of California. *Massively Parallel Signature Sequencing* (MPSS) uses the Lynx Megaclone technology and measures gene expression by transcript counting. A 17-nucleotide sequence is generated for each mRNA at a specific site upstream from its poly-Adenine tail. These short sequences, generated by sequencing cDNA fragments, are called *identification signatures* for the corresponding mRNAs. Next, they are cloned into a library of *tags*, attached to nylon *micro-beads* and exposed to the tissue sample. Then the total number of signature-tag conjugates for the mRNAs corresponding to each gene is counted and used as an indicator of the gene's expression. This counting is carried out with 2 - 4 replications. This technology is claimed to have greater accuracy than SAGE and a greater dynamic range (i.e. the ability to measure the expressions of genes that are more than 100 times up- or down-regulated). It is designed to capture the whole transcriptome of an organism and can be used with any organism (as opposed to microarrays that are mostly limited to specific organisms for which, arrays are commercially available). However, the downside of MPSS is its prohibitive cost compared to microarrays.

It is undeniable that microarrays and SAGE were revolutionary in their ability to quantitatively measure the expression of thousands of genes simultaneously. However, they have their own drawbacks. The measurements are notorious for being noisy and having significant variability due to inevitable variations in the complex measurement procedure as well as differences in experimental conditions, equipment used and technicians carrying out the experiment. Also, while the expression of thousands of genes may be measured in each sample, most studies involve only a handful of samples (a few tens to a few hundreds), giving rise to the well-known "p >> n" problem in the subsequent statistical analysis. The statistical and machine-learning issues engendered by such data are significant and have been a major area of study in the last couple of decades.

Labeled probes are also used in living or recently-living tissue. In *in-situ hybridization*, labeled probes for DNA or RNA sequences are inserted into cells and imaged. This reveals the spatial distribution of the target within the cell or tissue and, if observations are taken over a period of time, the temporal distribution as well. *Immunohistochemistry* or *immunostaining* follows the same idea, but the probes inserted are the antibodies and the target molecules are proteins. As some proteins are markers for (i.e. indicative of) specific organelles in a cell or tissue-types in the body, immunostaining is often used to determine the locations of such structures. While one probe can reveal the spatio-temporal distribution of a single type of molecule, two different probes with different labels can be used to study the differences or similarities in the distributions of different molecules. For example, this technique is used to determine whether two different proteins collocate (which could indicate a functional protein-protein interaction) or under what conditions they collocate. It is technically difficult to introduce and image more than a few different kinds of probes. As a result, these methods are very limited in the number of different types of molecules that can be studied simultaneously.

A widely used technique for detecting and quantifying proteins in a tissue-sample is *mass spectrometry*. Mass spectrometers separate and quantitate ions with various *mass-to-charge ratios*. The basic underlying principle is similar to that of gel electrophoresis. Electric or magnetic fields accelerate the ions differently, until they reach a detector. Usually, proteins are enzymatically digested into much smaller fragments (peptides) which are then fed into the mass spectrometer. The outcome is a measured distribution of ions with different mass-to-charge ratios. In some cases, this peptide *mass fingerprint* is sufficient to identify which proteins are present in the sample. Such identification, however, is not always possible. In *tandem mass spectrometry*, ions traveling through a first mass analyzer (which separates them according to mass-to-charge ratios) can be selectively sent through one or more additional mass analyzers This allows the successive selection and measurement of specific ranges of peptides, allowing for more definite identification. The most recent techniques even allow the enzymatic digestion step to be skipped. Instead, they introduce entire protein molecules to the first stage of a tandem mass spectrometer. However, ionizing the proteins without breaking them down first is a much more delicate process than it is for smaller peptide fragments. It requires more sophisticated and expensive equipment.

As mentioned earlier, many proteins act together in complexes. A traditional technique for determining the complexes which a given protein belongs to, and the conditions under which they do so, is *co-immunoprecipitation*. The method of immunoprecipitation extracts a specific protein from a solution by binding it with an antibody specific to that protein. The antibodies are then bound to insoluble proteins or other constructs (such as agarose beads) which are easily separated out of the solution. If the target protein is in a complex with other proteins, then these complex-sharing proteins will also be extracted in this process and can subsequently be identified by mass spectrometry.

Another method of determining protein-protein interactions is the *yeast two-hybrid* screen. This procedure relies on the following important feature of eukaryotic transcriptional regulation: While many transcription factors contain two domains (a DNA-binding domain and an activation domain that stimulates transcription), these two domains need not be in a specific position or orientation with respect to each other. A yeast two-hybrid screen for the interaction between two proteins (say, P_1 and P_2) works by inserting 3 new genes in a yeast cell. One gene (call it $1'$) codes for the protein P_1 fused with a transcription factor binding domain. Another gene (call it $2'$) codes for the protein P_2 fused with an activation domain. The third gene (call it $3'$) is a so-called *reporter gene* having the following two important features: its regulatory region contains a binding site for the $1'$ protein and its expression is easily measurable (one way of ensuring which is to choose a $3'$ gene that produces a fluorescent protein). If the proteins P_1 and P_2 interact and form a complex, then the expectation is that the proteins produced by $1'$ and $2'$ will also interact and form a complex. Also, the $1'$ portion of this complex will bind to the promoter region of $3'$ and the activation domain of $2'$ will stimulate the expression of $3'$. The expression of $3'$ or lack thereof is then an indicator of whether P_1 and P_2 interact or not.

A variant of immunoprecipitation, *chromatin immunoprecipitation*, is used to study where transcription factors bind to the DNA. In a given tissue-sample, depending on the conditions of the experiment, certain transcription factors will be expressed and bind to the DNA at various locations. The transcription factors are first *cross-linked* (i.e. bound) to proteins comprising the DNA's chromatin. Next, the cells are broken apart and the DNA broken into fragments by *sonication* (i.e. bombardment by sound waves). A chosen transcription factor can then be immunoprecipitated, bringing with it the chromatin to which it is bound and the DNA. Then the DNA is separated from the chromatin and the transcription factor, and can be identified. Ideally, this DNA represents only stretches that were bound by the transcription factor and nothing else. Identification of these DNA segments can be by traditional, low-throughput methods (mentioned earlier) or in a high-throughput manner using a special DNA microarray with probes designed to cover the whole genome of the organism. This is known as *ChIP-chip*, the first 'ChIP' being an acronym for the chromatin immunoprecipitation procedure and the second 'chip' referring to the microarray. Repeating this procedure for every known transcription factor in an organism will produce information on the entire set of transcription factor binding sites on the DNA. However, many sites may be missed, as the transcription factor may not be expressed or may not bind under the conditions of the experiment. In addition, transcription factors may bind to many locations on the DNA without affecting transcription. Such non-functional binding sites are usually not the ones researchers are looking for.

Recent developments in high-throughput sequencing technologies have opened another avenue for measuring mRNA levels in individual samples. While the detailed steps of the sequencing process depend on the specific technology being used, typically the input DNA to be sequenced (mRNA converted into cDNA in the case of mRNA sequencing) is fragmented into smaller pieces and a random sample of these fragments is selected to be sequenced. Such a random sample can have tens of millions of fragments. For each fragment selected, a portion of it is sequenced. Usually the portion sequenced is at the end of the fragment (or at both ends of it in the case of a *paired-end* sequencing). *Strobe reads* sequence many large portions interspersed across the input DNA-fragment. A steady advancement of the sequencing technology in the last few years has made it possible to increase both the length and the number of reads from a single run. As a result, millions of reads of hundreds or even thousands of base-pairs is now possible. The resulting sequences (or reads) are then analyzed to determine where the input fragment originated from by *aligning* or *mapping* the genes to the genome. For mRNA sequencing, the sequence has to be related back to the individual mRNA transcripts. In the presence of splicing, a read may span non-contiguous regions of the genome and the possible set of such splice junctions is generally not completely known in advance. Consequently, the process of aligning reads to the genome is sure to miss some valid fields. One remedy is to provide a set of possible splice junctions based on known annotation and restrict further analysis to those reads that match the known annotation. More sophisticated methods try to determine gapped alignments for reads that did not map to the genome.

RNAseq methods are derived from generational changes in sequencing technology. *First generation* high-throughput sequencing typically refers to Sanger dideoxy sequencing. Second generation or *next generation* sequencing refers to methods that use sequencing by synthesis chemistry of individual nucleotides, performed in a massively parallel format, so that the number of sequencing reactions in a single run can be in the millions. *Third generation* sequencing refers to methods that are also massively parallel and use sequencing by synthesis chemistry, but have individual molecules of DNA or RNA as templates. RNAs are typically isolated from freshly dissected or frozen tissue-samples using commercially available kits such as RNAEasy (Qiagen, Hilden, Germany), TRIZOL (Life Technologies, Carlsbad, California) or RiboPure (Ambion, Austin, Texas). Commonly, the RNA thus isolated will be contaminated by genomic DNA. Therefore it is treated with DNase to digest the contaminating DNA prior to library preparation. It is highly recommended for best results that RNAs be quality-checked for degradation, purity and quantity before library preparation. Several platforms are available for doing this, such as Nanodrop, QubitFluorometer (Life Technologies) and Agilent Bioanalyzer. After quality checking and prior to sequencing, the RNAs in a sample are converted into a cDNA library representing all of the RNA molecules in the sample. This needs to be done, as the RNA molecules themselves are unstable as opposed to the cDNA molecules which are not only chemically stable but also more amenable to the protocols of each sequencing platform. The library preparation protocols for each commercial platform come with its kits or are available at the manufacturer's website. However, third-party library preparation kits are available as well, and it is also possible for a user to come up with his/her own kit using commonly available molecular biology components.

We conclude this section by introducing the major RNAseq platforms that are available in the market. The Illumina platform is one of the most widely used ones. Illumina provides a wide range of instruments with different throughputs. The Hi-Seq 2500 machine produces up to six billion paired-end reads in a single run. If it sounds like an overkill, Illumina also provides a smaller sequencer with lower throughput, called the MiSeq system. It can produce thirty million reads (PE250 or paired-end reads with 125 nucleotides at each end) within a two-day runtime. The SOLID platform made available by Applied Biosystems (Carlsbad, California) is an acronym for sequencing by oligonucleotide ligation and detection. A library of cDNA fragments (originally obtained from the RNA molecules under study) is attached to magnetic beads – one molecule per bead. The DNA on each bead is them amplified in an emulsion, so that the amplified products remain with the bead. The resulting amplified products are then covalently bound to a glass slide. Using several primers that hybridize to a universal primer, di-base probes with fluorescent labels are competitively ligated to the primer. If the bases in the first two positions of the di-base probe are complementary to the sequence, then the ligation reaction will occur and the fluorescent label will generate a signal. This unique ligation chemistry allows for two checks of a nucleotide position and thus provides greater sequencing accuracy (up to 99.99%). The latest machines in this platform (such as the 5500 W) avoids the bead amplification step and use flow chips instead of amplifying templates. The Roche 454 platform is also based on adaptor-ligated double-stranded DNA library sequencing by synthesis chemistry. After double-stranded DNA is attached to beads and amplified in a water-oil emulsion, the beads are placed into pico-titer plates that have massive numbers of wells. That is where the sequencing reactions take place. The detection method differs from other platforms, as the synthesis chemistry involves the detection of an added nucleotide by means of a three-step reaction involving several enzymes. This method, known as *pyrosequencing*, allows for longer reads than other platforms. Read lengths of up to 1000 bases are possible. Based on the Roche 454 platform are both the bigger GS FLX+ system and the smaller GS Junior system. The Ion Torrent platform is more recent and, while it uses the adaptor-ligated library and sequencing-by-synthesis chemistry of other platforms, it distinguishes itself from other platforms by detecting changes in the pH of the solution in a well when a nucleotide is added and hydrogen ions are produced (instead of detecting fluorescent signals). Overall, this platform produces fewer reads than the others in a single run. However, the run time is only 2–4 hours instead of days or weeks on other platforms. The Pacific Biosciences platform (a third-generation platform) utilizes chemistry that is similar to second-generation instruments, but the important distinguishing feature is that it needs only a *single molecule* and is able to read the added nucleotides in *real time* (hence the acronym SMRT). As a result of this speed advantage, the run time can be very short (in the order of 1–2 hours) even with an avrage read-length of 5000 nucleotides. The state-of-the-art version of the system (the PacBio RS II) is capable of producing up to 250 megabase of sequence in a single run. Another useful aspect of this platform is that, due to its direct DNA sequencing of single molecules, it can provide sequencing of DNA modifications (up to 25 base modifications with the current version). Another third-generation, single-molecule technology that is still a work in progress but has great future potential is *nanopore sequencing*. A single enzyme is used to separate a DNA strand which is guided through a protein pore embedded in a membrane. Many ions passing through the pore at the same time generates an electric current, which is measured. The current is sensitive to specific nucleotides passing through the pore, each of which impedes the current flow in its own way. This generates a signal that is measured in the pore. Both Oxford Nanopore and Illumina have their own nanopore sequencing platforms under development. It is expected that, if successful, this technology will shrink the sequencing device size to something that fits into one's pocket.

2

Protein-Protein Interactions

CONTENTS

Protein-protein interaction data capture relationships among pairs of proteins that are present in a cellular system under study. The specific definition of 'interaction' depends on the technology being used, but in general pertains to physical binding either directly between protein pairs or in a complex with other proteins. Several other types of molecular interactions exist as well that summarize functional rather than physical relationships. In this chapter, we will describe technologies that are commonly used to assay protein interactions and summarize node-and-edge graph representations of protein-protein interaction data. We will explain subtle sampling issues that are relevant to protein interaction networks and describe statistical approaches for diagnosing and summarizing systematic and stochastic measurement error.

Two data sets will be used to illustrate these concepts, one resulting from tandem affinity purification (TAP) technology and the other from yeast two hybrid (Y2H) experiments.

2.1 Data sets

The primary data set that will be used to illustrate protein interaction data and statistical analyses in this chapter is a tandem affinity purification (TAP) data set referred to here as "Gavin2002." The Gavin2002 data set is a series of protein complex co-memberships observed for *Saccharomyces cerevisiae* using bait-prey TAP technology, and is one of the earliest published, comprehensive data sets of its kind. TAP data are also often known as Co-immunoprecipitation (CoIP) or Affinity Purification-Mass Spectrometry (AP-MS) data. These data were downloaded from the Biological General

DOI: 10.1201/9781315153728-2

Repository for Interaction Datasets (BioGRID) (build 4.4.201) [322] and were originally published by Gavin et al. (2002) [127]. A total of 1352 proteins and 3399 interactions are represented in Gavin2002. Data were originally reported by Gavin et al. (2002) as a series of pairwise 'interactions' with designations of bait or prey for each protein involved in the observed interaction. Several other widely used TAP (or CoIP or AP-MS) data sets for *Saccharomyces cerevisiae* are also publicly available [127, 128, 174, 238, 239].

The second data set, referred to here as "Ito2000," is one of the earliest published, comprehensive data sets that used yeast-two-hybrid (Y2H) technology to identify direct physical interactions among protein pairs in *Saccharomyces cerevisiae*. A total of 795 proteins and 839 interactions are represented in Ito2000. Y2H screening is also known more simply as two-hybrid screening. The Ito2001 data were also downloaded from BioGRID (build 4.4.201) [322] and were originally published by Ito et al. (2001) [196].

While these two TAP and Y2H data sets in *Saccharomyces cerevisiae* will be used for illustrative purposes, an ever-growing collection of interaction data of various types in a variety of organisms are available in several databases, including STRING [462], BioGRID [47, 412], IntACT [171], MINT [46], APID [6] and HuRI [276]. There are also efforts to unite data among these resources [102, 321]. Standardized schema for reporting experimental interaction data are under development, for example the PSI-MI XML and PSI-MI TAB molecular interaction data exchange formats supported by the HUPO Proteomics Standards Initiative [170]. Table 2.1 lists several highly referenced databases that contain protein interaction data for a wide variety of organisms, along with their websites and references.

2.2   Technologies and data types

TAP or CoIP technology assays co-membership in protein complexes on behalf of protein pairs. If two pairs are members of at least one protein complex in a cell, TAP will detect an "interaction" for the two proteins. Direct physical binding is not requisite for a detected TAP interaction. A pair of proteins that are directly bound to a third or additional proteins in a large multiprotein complex would still be identified as interacting pairs according to TAP technology. For example, in the simple illustration in Figure 2.1, proteins A and C are a part of the same protein complex. They do not bind directly to each other, but are physically bound to other proteins in the same complex. TAP technology would identify an interaction between proteins A and C due to their membership in the same protein complex. There are several helpful reviews describing specifics of TAP and CoIP technology in addition to original publications of highly utilized data sets [127, 128, 174, 238, 239, 247, 349].

In contrast to TAP technology, Y2H technology assays direct physical binding among protein pairs. In Figure 2.1, proteins A and C would not be identified as interacting pairs. Even though they are part of the same protein

TABLE 2.1

Molecular interaction databases

Database	Reference	URL
APID	Alonso-López et al. 2019 [6]	http://apid.dep.usal.es/
BioGRID	Chatr-Aryamontri et al. 2017 [47]	https://thebiogrid.org/
DIP	Salwinski et al. 2004 [372]	http://dip.doe-mbi.ucla.edu/dip/
HPRD	Prasad et al. 2009 [348]	http://www.hprd.org/
HuRI	Luck et al. 2020 [276]	http://www.interactime-atlas.org/
IntACT	Hermjakob et al. 2004 [171]	http://www.ebi.ac.uk/intact/
MINT	Chatraryamontri et al. 2007 [46]	http://mint.bio.uniroma2.it/
PDB	Rose et al., 2017 [368]	http://www.rcsb.org
Reactome	Fabregat et al. 2016 [101]	http://www.reactome.org/
STRING	Szklarczyk et al. 2015	http://string-db.org/

complex, they do not directly physically bind to each other. Y2H interactions would be detected for proteins A and B, and for proteins B and C. Several helpful reviews and publications of original Y2H data sets explain subtleties of Y2H technology [38, 113, 163, 196, 197, 438, 446].

TAP and Y2H are both considered "bait-prey" technologies. In these settings, for a designated "bait" protein, all proteins that interact with the bait are identified as its "prey." In concept, all cellular proteins are eligible for

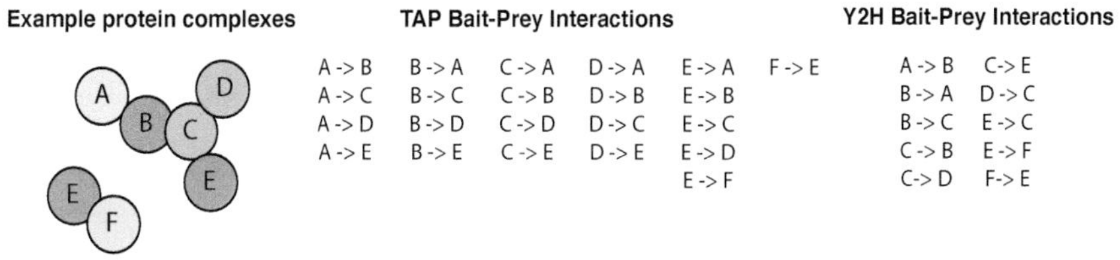

Example protein complexes

TAP Bait-Prey Interactions

A -> B	B -> A	C -> A	D -> A	E -> A	F -> E
A -> C	B -> C	C -> B	D -> B	E -> B	
A -> D	B -> D	C -> D	D -> C	E -> C	
A -> E	B -> E	C -> E	D -> E	E -> D	
				E -> F	

Y2H Bait-Prey Interactions

A -> B	C -> E
B -> A	D -> C
B -> C	E -> C
C -> B	E -> F
C -> D	F -> E

FIGURE 2.1

Two simple, hypothetical protein complexes comprised of proteins labeled A through F. Protein complex co-memberships are assayed by TAP technology and direct physical interactions are assayed by Y2H technology. The lists of directed bait-prey relationships that would be detectable by perfectly sensitive and specific TAP and Y2H technologies are listed.

Gavin2002   Ito2001

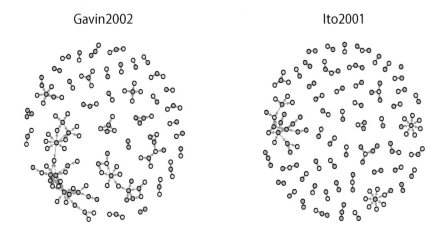

FIGURE 2.2
Plots of graphs for Gavin2002 TAP data and Ito2001 Y2H data for the set of proteins in common for both experiments. Nodes without adjacent edges have been removed. Bait proteins are blue and proteins that are only found as prey and never used as baits are yellow. Gray directed edges extend from baits to prey.

detection as interacting prey for a given bait, assuming they are present in the cell under the given experimental conditions and are amenable to the technology. A set of baits is typically pre-specified at the outset of the experiment, with the set ranging in size for a focused series of investigations [239] to all viable proteins to the extent possible for comprehensive characterization of a given system [128, 238].

Figure 2.2 depicts Gavin2002 TAP and Ito2001 Y2H networks for the set of proteins that were reported in both experiments. Nodes without adjacent edges in this set of common nodes were removed before plotting. Bait proteins are colored blue and proteins that were detected as prey but never used as bait are colored yellow. Visual inspection suggests roughly similar topologies for the two networks, but with increased sparsity for the Ito2001 data. This is to be expected since Y2H technology assays a direct subset of the edges assayed by TAP technology. Data analytic implications according to bait-prey designations will be discussed in later sections.

Protein complex co-memberships (such as those detected by TAP) and direct physical binding interactions (such as those detected by Y2H) are frequently used protein interaction data types. It should be noted, however, that several other types of "interaction" data do exist, involving proteins as well as other molecules. For example, synthetic lethal interactions are functional interactions that are defined by functional disruption when both genes are perturbed in an experiment but a normal functional phenotype when either of the genes are mutated individually [214]. As another example, protein-DNA interactions have also been reported, identifying proteins that bind directly to a DNA sequence that may affect downstream gene expression [81]. Most recently, these types of protein-DNA interactions are identifying using ChIP-seq technology [208]. Compound-protein interaction data have also been catalogued in recent drug development efforts [265,273]. Understanding the nature of the interactions being assayed and any subtleties among the measured relationships is paramount to informative analysis. Analysis teams should work closely with multidisciplinary collaborators to understand the subtleties of interaction types and relevant interpretations of the data.

2.3 Graph representations of protein-protein interaction data

Protein-protein interaction data are often modeled in node-and-edge graphs in which nodes represent proteins and edges represent interactions. Two of the earliest versions of this type of graph modeling for protein-protein interactions were the "spoke" and "matrix" models for bait-prey data [17]. These models designate protein interactions using undirected edges among protein pairs. The spoke model is perhaps the most intuitive; all pairwise interactions that are directly observed among baits and prey in an experiment are represented by edges between these baits and prey. The matrix model extends spoke modeling and attempts to capture edges that are not directly observed. In the matrix model, all spoke model edges are included, and edges are also included among all pairs of prey that were detected as interactors for at least one common bait. The spoke model can be viewed as a direct representation of experimental data, and the matrix model as an effort to impute additional edges among pairs of prey that were not directly tested for interaction. Spoke and matrix models are depicted in Figure 2.3 for a subset of the Gavin2002 TAP data.

When undirected edges are used for spoke and matrix modeling, reciprocity in detecting interactions or lack thereof is not evident. For example, in Figure 2.3, the protein Apl6 detects protein Apm3 as prey when used as a bait, but when Apm3 is used as a bait, it does not detect Apl6 as prey. Strictly speaking, bait-prey data are most accurately represented by directed graphs in which directed edges extend from baits to their interacting prey. Including direction summarizes the nature of the study design since edges will only originate at proteins that are pre-specified as baits. A simple summary of bait-prey data can be to list all "viable baits" and "viable prey." Viable baits can be defined as proteins for which an interaction with at least one prey was observed in the experiment. Viable prey can be defined as proteins that

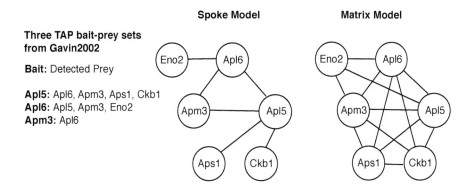

FIGURE 2.3
Spoke and matrix models for three sets of TAP bait-prey edges for the Gavin2002 data set. Spoke modeling includes all detected edges while matrix modeling additionally imputes edges among all pairs of prey that share at least one common bait.

were identified by at least one bait in an interaction. Summarizing viable baits and prey can help describe experimental design and intent, and in some cases identify the extent to which certain proteins were amenable to the specific assay [53, 54]. Recording bait-prey data in a directed graph is also crucial for for understanding sampling subtleties in protein interaction data, identifying systematic biases potentially due to technological factors and for estimating measurement error. These concepts will be developed in detail in the next sections.

Other types of node-and-edge graphs are often utilized for protein interaction data modeling as well. For example, bipartite graphs have been crucial constructs in statistical algorithms that estimate protein complex membership given observed TAP data. Bipartite graphs consist of two node classes and edges among nodes represent affiliation of a node of one type with a node of the other type. An example of a bipartite graph depicting membership of proteins in the simple example protein complexes from Figure 2.1 is depicted in Figure 2.4. A bipartite graph is also often represented as an affiliation matrix in which rows correspond to one node type and columns to the other. In Figure 2.4, for affiliation matrix $\mathbf{A}$, the rows correspond to n proteins and columns to m protein complexes. The co-affiliation matrix $\mathbf{Y} = [\mathbf{A}\mathbf{A}^{\mathbf{T}}] - \mathbf{I}$ where $[x] = 1$ if $x > 0$ identifies protein pairs that are co-members in at least one protein complex. Entries of 1 in $\mathbf{Y}$ represent complex co-membership. In fact, $\mathbf{Y}$ represents the data that TAP and CoIP technologies are designed to assay. Statistical approaches for estimating protein complexes recorded in $\mathbf{A}$ given observations on $\mathbf{Y}$ will be described in the next chapter.

To accommodate the increasing use of node-and-edge graphs for representing protein-protein interaction data, R statistical software

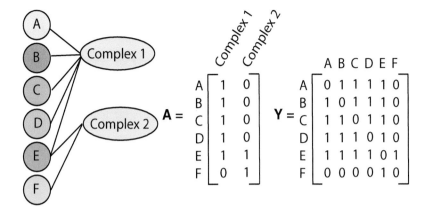

FIGURE 2.4
Bipartite graph and related affiliation (**A**) and co-affiliation ($\mathbf{Y} = [\mathbf{AA^T}] - \mathbf{I}$) matrices for two simple example protein complexes.

(https://cran.r-project.org/) packages including *igraph* [62] provide efficient and informative data structures for storing and representing protein-protein interaction data. As discussed in the next chapter, *igraph* also contains extensive functionality for plotting graphs, and for characterizing both local and global graph features.

2.4 Sampling issues in protein-protein interaction data

As explained previously, directed graph modeling of bait-prey data not only records observed protein-protein interactions, but also summarizes the nature of identified interactions from baits to their prey. Understanding the bait/prey status of a protein is crucial for understanding which edges in a graph have been tested and which have not. In concept, if all proteins in a cell are equally amenable to the bait-prey technology, then given the full set of proteins, all possible edges extending from baits to prey are tested in bait-prey experiments. Edges in a graph model of bait-prey data are included only if the interactions they represent are directly tested and observed according to the technology being used. Importantly, edges that are not included in a graph either represent potential interactions that were not observed when directly assayed, or potential interactions that were never directly tested using the bait-prey technology.

For any pair of proteins that are both used as baits in an experiment, the possible edge between them is tested twice, one time for each bait. If technologies were perfectly sensitive and specific, identical results should be obtained

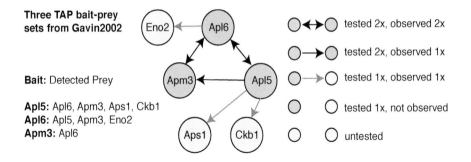

FIGURE 2.5

Directed graph model of TAP data from Gavin et al. (2002) [127]. Baits are represented by blue nodes. Proteins not used as baits but detected as prey are represented by yellow nodes. Edges among pairs of baits that are tested twice are black with reciprocal arrows if observed twice and single arrows if detected once. Edges among baits and prey-only nodes are gray and directed from baits to prey if detected, and otherwise not included in the graph. All edges among prey-only nodes are not tested by the TAP assay and therefore edges are not included in the graph.

with both baits either detecting or failing to detect the pairwise interaction. In reality, observed data for these doubly tested interactions are not always consistent. An interaction may be observed when one protein is used as bait, but not the other. This lack of reciprocity is generally attributed to measurement error, either stochastic or systematic in nature [54]. For a pair of proteins in which only one protein is used as bait but the other is not, the edge is only tested once in the direction from the bait to the prey; reciprocal testing of the edge from the prey to the bait is not performed. Importantly, edges among pairs of proteins that are never used as baits are in fact never directly tested in bait-prey technologies. While some graph modeling approaches, for example the matrix model, induce edges among prey proteins that share a common bait, in fact, there is no direct assay of the edge between two prey-only proteins [382]. Similar to any statistical analysis, it is crucial to understand which data are directly collected and which are not. Directed graph modeling and bait-prey designations are key to distinguishing among edges in a graph that have been tested and observed to exist, tested but no interaction was observed, and never tested [7,379,383]. Figure 2.5 illustrates doubly tested edges among pairs of bait proteins, singly tested edges from baits to prey-only proteins and untested edges among pairs of prey-only proteins for a subset of the Gavin2002 TAP data. Note the simple but critical differences between the spoke model of these same data in Figure 2.3. The directed edges and bait-prey distinctions (denoted by color in Figure 2.5) more accurately represent the experimental design and the frequency with which edges are tested and observed.

TABLE 2.2
Counts of VB, VP and VBP for several
TAP and Y2H data sets

Data set	VB	VP	VBP
TAP data sets			
Gavin2002 [127]	453	1169	270
Ho2002 [174]	492	1308	237
Krogan2004 [239]	28	39	24
Gavin2006 [128]	914	1361	845
Krogan2006 [238]	1605	2011	942
Y2H data sets			
Cagney2001 [38]	19	39	10
Ito2001 [196]	465	506	195
Uetz2000 [446]	484	578	135
Tong2002 [438]	26	134	16

One simple way to describe the sampling scheme underlying bait-prey experiments is to tabulate the numbers of proteins that serve as baits or prey or both in a given experiment. According to Chiang et al. (2007) [54], in bait-prey investigations, "viable baits" (VB) can be defined as proteins for which at least one prey is detected as an interactor. "Viable prey" (VP) can be defined as proteins which are detected as interactors by at least one bait. "Viable bait and prey" (VBP) proteins satisfy both criteria and prove amenable to assay as both baits and prey by the technology being used. Table 2.2 reports the numbers of VB, VP and VBP for several TAP and Y2H data sets in *Saccharomyces cerevisiae*. From these quantities, the numbers of doubly tested, singly tested and untested edges among all pairs of proteins can be directly calculated. These simple descriptive statistics demonstrate the wide range of scope and comprehensiveness of bait-prey experiments.

Identifying tested and untested edges is similar in spirit to understanding the sampling space of a graph. Tested edges are sampled for assay and untested edges are not. The effects of sampling on global graph topology have been summarized. Global summaries often treat untested edges in the exact same way as edges that are tested but not observed. In many cases observed phenomena can be explained by failure to identify the sampling structure when calculating summaries. For example, global topological graph characteristics that identify graphs as "scale-free" can be replicated simply by partial sampling of fuller networks with completely different topologies [155]. Other statistics including overall node degree distributions, clustering coefficients, node betweenness summary statistics and motif frequency can be greatly affected by failure to account for sampling [71]. Specific statistical summaries describing both global and local topological feature of graphs will be developed in detail in the next chapter, but we note here that their typical calculation assumes that all edges in a graph have been tested to support direct observation of edge existence or lack thereof.

Recognizing incomplete sampling in experimental settings, some attempts have been made at compiling data from various sources to increase the quality of protein interaction reporting. Some strategies acknowledge variations in direct assay of edges and incorporate this in their global rankings of hypothesized interactions [444]. Other applications recognize the potential for bias due to sampling and adjust for potential bias when identifying network characteristics of proteins pertaining to disease, for example cancer [378]. There are also several experimental design strategies that explicitly acknowledge the sampling structure of bait-prey technologies and refine sampling according to previous observations in a graph to ensure exploration of untested edges as additional rounds of experimentation are conducted [64, 243, 383, 387].

2.5   Systematic and stochastic measurement errors

Directed graph modeling of bait-prey interactions can also be key for identifying systematic measurement error and for estimating stochastic false positive and false negative probabilities for the set of observed edges. False positive interactions are those detected by the technology that in truth do not exist. False negative interactions are those that are tested but not detected by the technology, when in fact they do exist. Here we define systematic measurement error as errors that occur when edges are either observed or unobserved due to differences in sensitivity and specificity for individual proteins for the technology in use. Proteins maybe be "sticky" and identify more interacting partners than are truly involved in cellular protein interactions. Proteins may also be less amenable to the assay and either fail in large part to detect interacting partners when used as baits or fail to be detected as prey by most baits. While the cause for systematic error cannot always be diagnosed, fairly straightforward statistical approaches can identify systematic bias in observed data. Here, stochastic measurement errors are defined as false positive and false negative observations that occur at random and do not reflect underlying amenability of the proteins or the experimental conditions to the assay.

A simple test based on a binomial distribution can be used for diagnosing bait proteins prone to systematic error by examining a subgraph consisting only of nodes that are viable baits [54, 129]. Edges in this viable bait subgraph are all tested twice, once in each direction. If the bait-prey technology was perfectly sensitive and perfectly specific, then reciprocal interactions or lack thereof should be observed for all protein pairs. However, because of measurement error, unreciprocated edges often do exist such that an interaction is observed if one protein is used as bait, but not the other. If measurement

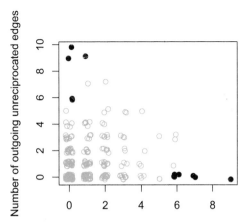

FIGURE 2.6

Plot of outdegree vs. indegree for the unreciprocated edges among the VBP subset of data from the TAP data set published by Gavin et al. (2002). Proteins that show high indegree relative to outdegree or vice versa for unreciprocated edges may be prone to systematic bias. In this figure, light blue points correspond to proteins with p>0.05 when a binomial test is applied to assess extremes in the direction of detection. These proteins are likely not prone to systematic bias. Dark blue points correspond to proteins with p<0.05 using the binomial test and may be subject to systematic bias.

error is purely stochastic, the number of unreciprocated edges directed toward a bait node should roughly equal the number of unreciprocated edges pointing out of a bait node. Given the total number of unreciprocated edges for a bait n_U, the number of outgoing unreciprocated edges n_O follows a binomial distribution with probability $p = 0.5$, or $n_O \sim Binomial(n_U, 0.5)$. A simple two-sided p-value calculation according to the binomial distribution can be used to evaluate whether the number of outgoing unreciprocated edges is as or more extreme than what would be expected under a purely stochastic error model. Nodes with very low binomial p-values may be prone to systematic error and potentially excluded from further analyses. Figure 2.6 illustrates nodes with extreme numbers of incoming and outgoing unreciprocated edges in the VBP subset of TAP data published by Gavin et al. (2002).

Once nodes potentially prone to systematic bias are removed, estimates of false positive and false negative probabilities can also be made using the subgraph of viable bait proteins in a graph. For a graph with N baits, a total of $N(N+1)/2$ possible edges exist and all edges are tested twice. Assuming n of the edges represent true positive interactions and m represent true negative

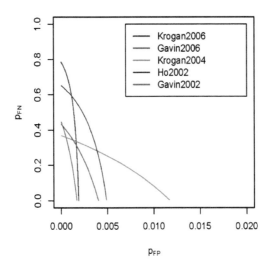

FIGURE 2.7
Plots of solution manifolds that satisfy the systems of equations for p_{FP} and p_{FN} for several published TAP data sets.

interactions such that $n + m = N(N + 1)/2$, multinomial modeling can be used to calculate the expected number of reciprocated and unreciprocated edges in a graph, as well as the number of edges with no observed interaction, assuming measurement errors occur at random. Let X_1, X_2 and X_3 be random variables for the number of reciprocated edges, unreciprocated edges and edges with no observed interaction, and let p_{FP} and p_{FN} represent false positive and false negative probabilities, respectively. The following represent the expected numbers of each type of edge.

$$
\begin{aligned}
E[X_1] &= n(1 - p_{FN})^2 + mp_{FP}^2 \\
E[X_2] &= 2np_{FN}(1 - pFN) + 2mp_{FP}(1 - p_{FP}) \\
E[X_3] &= np_{FN}^2 + m(1 - p_{FP})^2
\end{aligned}
$$

Given $E[X_1] + E[X_2] + E[X_3] = n + m = N(N + 1)/2$ and N is known, any of these two equations imply the third. With algebraic substitution, we therefore have a system of two equations with three unknowns, p_{FP}, p_{FN} and n. Either a solution manifold can be calculated to describe the range of possible true and false positive probabilities given a range of values for n, or a value of n can be determined for example using a gold standard data set culled from the literature [382] and point estimates for p_{FP} and p_{FN} can then be calculated. Figure 2.7 illustrates a solution manifold of false positive and false negative probabilities for several TAP data sets.

Other methods have been proposed for estimating false positive and false negative probabilities. For example, a method specific to Y2H data harnesses raw counts of observed bait-prey clones from Y2H technology to estimate measurement error probabilities rather than dichotomizing to the typical binary approach for recording interaction data [183]. Other methods utilize data resources such as Gene Ontology to reduce false positives included in protein interaction data [281]. Ultimately, is it ideal to formally model measurement error when performing statistical analysis of protein interaction data. Many of the techniques described in the next chapter accommodate measurement error for estimation of graph features.

3

Protein-Protein Interaction Network Analyses

CONTENTS

As described in the previous chapter, node-and-edge graphs are convenient data structures to succinctly describe protein interaction data. Nodes represent proteins and edges represent pairwise interactions among them. Graphical representations of protein-protein interaction data allow summarization of global features to describe overall graph topology. In addition, protein-protein interaction graphs can be queried for local subcomponents that represent underlying biology. Methods are often applied that do not rely on formal statistical modeling, but instead represent helpful descriptive statistics for observed data represented in graphical form. Other approaches build on classic statistical modeling and are used for estimation and/or inference. In this chapter, we will discuss descriptive and statistical estimation strategies for both global network characterization and local feature detection. Examples in which protein-protein interaction data are integrated with other data types are also summarized.

DOI: 10.1201/9781315153728-3

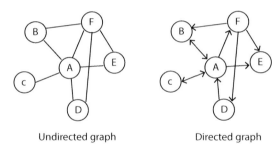

Undirected graph   Directed graph

FIGURE 3.1
Example undirected and directed graphs.

3.1 Node summaries in protein interaction graphs

3.1.1 Node degree

Assume a given protein interaction graph $G(V, E)$ with V representing the set of vertices or nodes ($|V| = n$) and E representing the set of edges. Many of the most commonly used descriptive approaches for protein interaction networks begin by characterizing node degree. In an undirected graph, the *degree* of a node, Δ, is defined as its number of adjacent edges. In a directed graph, *indegree*, Δ_{in} is defined as the number of edges directed toward the node and *outdegree*, Δ_{out} is defined as the number of edges extending from the node to other nodes. Degree is a node-specific descriptor that, in the context of protein interaction networks, identifies the interactivity of the protein under consideration relative to the other proteins represented in the network. In Figure 3.1, the degree of node A in the undirected graph is 5; the indegree and outdegree of node F in the directed graph are 1 and 3, respectively.

For a given protein, node degree is often estimated as a simple sum of adjacent edges as depicted above, and often without regard to directionality of the observed edges. This can be a helpful starting point for understanding the data at hand; however, this approach accounts for neither sampling nor measurement error as described in the previous chapter. In particular, consider node degree for proteins in a bait-prey network that are both used as baits and detectable as prey, i.e. proteins in a VBP network as described in the previous chapter. All edges in this network are assayed twice, once in each direction, while the true underlying graph is an undirected graph with reciprocal edges. For a given protein with node degree Δ, let R_T and U_T be random variables for the number of reciprocated and unreciprocated edges incident on the node for which the edges do truly exist. Let R_F and U_F be random variables for the number of reciprocated and unreciprocated edges incident on the node for which the edge does not truly exist. The naïve estimator of degree for the node, $\hat{\Delta}_{naive} = R_T + R_F + U_T + U_F$, is a simple sum of both reciprocated and unreciprocated edges. This strategy, while simple and commonly used, does not model the frequent false positive and false negative measurements that comprise observed VBP networks.

When estimates of both true and false probabilities, p_{TP} and p_{FP}, for detecting edges in VBP graphs are available, for example using estimation techniques from the previous chapter, a more accurate estimate of node degree is possible using a multinomial modeling strategy [382]. Specifically, given Δ, the joint probability distribution of R_T and U_T is:

$$Pr(R_T = r_T, U_T = u_T | \Delta, p_{TP}) = \frac{\Delta!}{r_T! u_T! (\Delta - r_T - u_T)!} p_{T2}^{r_T} p_{T1}^{u_T} p_{T0}^{(\Delta - r_T - u_T)},$$

where $p_{T2} = p_{TP}^2$, $p_{T1} = 2p_{TP}(1 - p_{TP})$ and $p_{T0} = (1 - p_{TP})^2$ with $0 < p_{TP} < 1$. Similarly, given the number of nodes in the VBP graph or $|VBP|$, the joint probability distribution of R_F and U_F is:

$$Pr(R_F = r_F, U_F = u_F | \Delta, p_{FP}) = \frac{(|VBP| - \Delta)!}{r_F! u_F! ((|VBP| - \Delta) - r_F - u_F)!} \times$$
$$p_{F2}^{r_F} p_{F1}^{u_F} p_{F0}^{(|VBP| - \Delta) - r_F - u_F},$$

where $p_{F2} = p_{FP}^2$, $p_{F1} = 2p_{FP}(1 - p_{FP})$ and $p_{T0} = (1 - p_{FP})^2$ with $0 < p_{FP} < 1$.

In protein interaction network data, R_T, U_T, R_F and U_T are not observed individually since which edges represent true and false detections are unknown. Instead the observed numbers of reciprocated and unreciprocated edges, $R = R_T + R_F$ and $U = U_T + U_F$, are the observed data. The joint distribution of R and U can be specified using the convolution of $Pr(R_T = r_T, U_T = u_T | \Delta, p_{TP})$ and $Pr(R_F = r_F, U_F = u_F | \Delta, p_{FP})$ as specified above. Maximization of the joint distribution for R and U for Δ, given the observed data and estimates for p_{TP} and p_{FP}, results in a maximum likelihood estimator, $\hat{\Delta}_{MLE}$ that is more accurate than $\hat{\Delta}_{naive}$ as described in [382]. The critical difference in this approach and the simple sum is that the MLE uses statistical modeling to account for measurement error and the number of times each edge is tested when estimating node degree.

While degree itself is a node-specific summary, the distribution of degrees for all nodes, or *node degree distribution*, is often referenced as a general descriptor of graph properties. In many cases, a simple empirical distribution of estimated degree for all nodes is used to summarize the graph. However, similar to the problem of estimating individual node degrees that was just described, this simple empirical approach ignores the sampling scheme underlying the observed network data. Zhang et al. (2015) [528] describe a constrained, penalized least squares approach to estimating the node degree distribution for a true underlying network for which only sampled data are observable.

3.1.2   Clustering coefficient

Another common measure of global network topology is the *clustering coefficient*. In an undirected graph $G(V, E)$, for an individual node i with degree d_i, there are $d_i(d_i - 1)/2$ possible edges among the d_i nodes adjacent to node x. The clustering coefficient for node i is defined as the proportion of these $d_i(d_i - 1)/2$ possible edges that exist. More formally, let N_i be the *neighborhood* of nodes adjacent to node i such that $|N_i| = d_i$ and let e_{jk} represent an edge between nodes j and k. Then the clustering coefficient for node i, C_i, is defined as

$$C_i = \frac{|\{e_{jk} : j, k \in N_i, e_{jk} \in E\}|}{d_i(d_i - 1)/2}.$$

Importantly, in an undirected graph e_{jk} and e_{kj} refer to the same edge. Thus, if $e_{jk} = e_{kj} \in E$, this edge is only counted once in the numerator of C_i for an undirected graph written above. For an directed graph, e_{jk} and e_{kj} are distinct from each other. The neighborhood of node i consists of all adjacent edges to i regardless of direction, and the number of possible edges in the neighborhood of node i is $d_i(d_i - 1)$. Therefore, for a directed graph,

$$C_i = \frac{|\{e_{jk} : j, k \in N_i, e_{jk} \in E\}|}{d_i(d_i - 1)}.$$

The mean clustering coefficient for all n nodes in a graph is often viewed as a summary measure of graph topology [474]. This is similar to a global measure of a graph called *transitivity* for an undirected graph which is defined as the number node triples that are completely connected by three edges, divided by the number of node triples that are connected by two or more edges [472]. An alternative to a simple mean is to calculate a weighted average of all node-specific clustering coefficients with weights equal to $d_i(d_i - 1)$; this will result in the measure of transitivity just described. Figure 3.2 depicts a clustering coefficient calculation for the nodes in a simple undirected network, the mean clustering coefficient, and transitivity for the entire network.

3.1.3   Connectivity

In graph theoretic terms, a pair of nodes is *connected* if there is a set of edges, or a path, that connects the two nodes. An entire graph is said to be *connected* if a path exists between all pairs of nodes in the graph. The overall connectivity of a graph is often described both in terms of *node connectivity* and *edge connectivity*. For a given pair of nodes, node connectivity is defined as the number of nodes that need to be removed from the graph so that a path no longer exists that connects the pair of nodes. Node connectivity can also be described for the graph as a whole, and is defined as the minimum node connectivity for each pair of nodes in the graph. Similarly, for a given pair of nodes, edge connectivity is defined as the number of edges that need

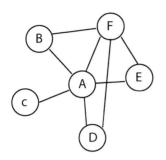

Clustering coefficient (CC)

CC(A) = 3 / [5*4/2] = .30
CC(B) = 1 / [2*1/2] = 1
CC(C) = 1
CC(D) = 1 / [2*1/2] = 1
CC(E) = 1 / [2*1/2] = 1
CC(F) = 2 / [3*2/2] = 0.67
Mean CC = 0.83

Triples connected by 2 edges:
    ABC, ABD, ABE, ACD, ACE, ACF, ADE, BDF, BEF, DEF
Triples connected by 3 edges:
    ABF, ADF, AEF

Transitivity = 3 / 13 = 0.23

FIGURE 3.2
Example clustering coefficient and transitivity calculations.

to be removed to disconnect the node pair. The connectivity of a graph can be interpreted as a measure of robustness of the network to perturbations in its topology.

3.1.4   Betweenness

The *betweenness* of a node in a graph is a measure of its centrality to the graph. For a given node i, betweenness counts the number of shortest paths for all pairs of nodes in a graph that go through i to the number of shortest paths that exist for each pair of nodes. More formally, betweenness, b_i for node i, can be expressed as follows:

$$b_i = \sum_{i \neq j \neq k} \frac{\sigma_{jk}(i)}{\sigma_{jk}}$$

where σ_{jk} is the number of shortest paths from node j to node k and $\sigma_{jk}(i)$ is the number of those paths that include node i along the path.

3.1.5   Applications to protein-protein interaction networks

The graph theoretic measures just described apply to networks in many different settings, but have yielded helpful insights when applied directly to protein-protein interaction networks. In a study of clustering coefficients for protein-protein interaction networks, the bias of this measure under different sampling conditions was demonstrated and its use for ruling out overall network topologies was described [123]. In a study of the yeast protein interaction network, Joy et al. (2005) [212] identified high betweenness for a subset of proteins with low degree in the network. They demonstrated a correlation of these high-betweenness, low-degree proteins with evolutionary age and showed that this subset of proteins is more likely to be essential to cellular function than other proteins. In a study of amyotrophic lateral scerlosis (ALS), network analyses of protein-interaction data were applied to identify essential hub proteins based on node degree that are implicated in causing ALS [284]. Graph-based constructs for modeling protein-protein interaction data and simple graph theoretic measures of network features have been promoted in several discipline-specific contexts including cancer research [376]. While the underlying concepts of node-and-edge graphs are simple, the flexibility of this modeling approach has been fruitful for characterization and discovery in the study of protein-protein interactions.

3.2   Graph models of protein interaction data

Accumulating protein interaction data have motivated several investigations into the overall topological characterization of protein-protein interaction networks. Visual inspection and several summary statistics indicate that protein interaction data are not in general consistent with the classic *Erdös-Renyi random graph*, as described below. Rather, protein interaction data have been described as *scale-free*, with both *hierarchical* and *modular* structure. These graph models are described in what follows.

3.2.1   Erdös-Renyi random graphs

An *Erdös-Renyi random graph* is an undirected graph in which a subset e of all possible $n(n-1)/2$ pairwise edges E among n nodes are randomly represented in the graph such that $Pr(e \in E) = p$ [358]. In a random graph, the node degree distribution is Binomial$(n-1, k)$ since a given node has degree k with probability $Pr(k) = \binom{n-1}{k} p^k (1-p)^{n-1-k}$. In a sparse graph with small p and large n, the node degree distribution for an Erdös-Renyi random graph is Poisson(np). The overall connectivity of the graph depends on the value of p relative to n. If $p > log(n)/n$, close to the entire graph is connected, with the exception of only a few nodes. For lower values of p, the graph often consists of several smaller subcomponents that are not all connected to each other. Figure 3.3 depicts three random graphs with 20 nodes, one each with $p = .05$, $p = .10$ and $p = .20$. The differences in overall connectivity and the degree distributions are obviously quite dependent on p. Note that, as expected, since $.20 > log(20)/20$, the rightmost graph is fully connected. Protein-protein

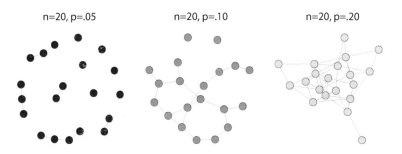

n=20, p=.05   n=20, p=.10   n=20, p=.20

FIGURE 3.3
Example Erdös-Renyi random graphs.

interaction networks do not, in general, resemble Erdös-Renyi random graphs. Erdös-Renyi random graph features are often described to draw a contrast to features that do reflect protein-protein interaction networks.

3.2.2 Scale-free graphs

While Erdös-Renyi random graphs are a helpful construct, real protein interaction graphs that are generated from experimental data tend to demonstrate different features than random graphs. In particular, node degree distributions for protein interaction graphs tend to follow a power-law with $Pr(k) \sim k^{-\lambda}$ for a constant λ and are said to be *scale-free*. For most observed protein-protein interaction networks $2 < \lambda < 3$ [18, 19]. The power law distribution arises from what can be thought of as "preferential attachment," i.e. the tendency of proteins to interact with other proteins that already have a high number of interactions. In addition, clustering coefficients tend to be higher in scale-free networks than in Erdös-Renyi random graphs. For both graph types, the clustering coefficient distribution within a given network tends to be independent of node degree.

As illustrated in Figure 3.4, scale-free and Erdös-Renyi random graphs graphs are visually quite distinct. Both graphs contain 50 nodes and 100 edges, but the overall topology is notably different. Conceptually, it is reasonable that the Erdös-Renyi model is not fitting for a true biological graph since proteins assumedly act in coordinated, rather than random, fashion. Accordingly, the relatively small number of "hub" nodes with high degree in a scale-free network are sometimes ascribed biologically relevant behavior, i.e. hubs have been suggested to represent proteins that are key players in biological processes under study. In contrast, nodes with low node degree have been interpreted as more peripheral to biological function [201]. While these hypotheses may have

Scale-free graph   Erdos-Renyi graph

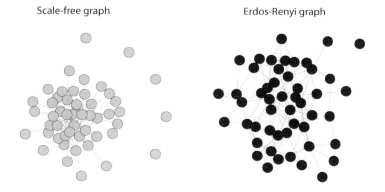

FIGURE 3.4
Example scale-free and Erdös-Renyi random graphs, both with 50 nodes and
100 edges.

merit and in some cases experimental data support these concepts, it is crucial
to understand that descriptive analyses that rely on simple degree calculations
for observed data in general do not model measurement error and frequently
ignore any effects of sampling as described previously. In other words, a node
may be identified as a "hub" simply because it is prone to systematic false
positive edge detections or it may be missed as a hub if it was not used as
bait for a bait-prey technology. Conclusions on degree for individual nodes or
overall degree distributions that do not statistically model sources of variation
should be interpreted with caution.

3.2.3   Hierarchical graphs

Hierarchical graphs are considered to be a certain type of scale-free network,
but with distinguishing characteristics that reflect experimental data, including protein-protein interaction networks. Specifically, as noted previously, the
distribution of clustering coefficients in scale-free graphs tends to be independent of node degree; however, in hierarchical graphs, nodes with higher
degree in general have lower clustering coefficients. In addition, in hierarchical graphs, there is typically no relationship between the average clustering
coefficient and the overall graph size.

A replication model algorithm is typically used to generate hierarchical
graphs. These models start with an initial cluster of nodes. The cluster is
replicated by some factor and attached to the main cluster, and the process repeats. The *replication factor M* in a hierarchical model determines the
overall topology of the network. Hierarchical graphs are considered a subclass
of the scale-free family since, under this replication model, the node degree
distribution follows a power law with $\lambda = 1 + (\ln M)/(\ln(M-1))$ for replication
factor M [316].

The hierarchical nature of protein interaction networks have been confirmed in many studies, including investigations of multiple sources of yeast protein interactions that consistently demonstrate the scale-free, hierarchical nature of protein interaction networks [508]. Hierarchical graph structure was also harnessed for specific insight into proteins that are transcription factors, i.e. proteins that control the transcription of genes and therefore the functional presence of other proteins within a cell. It was noted that transcription factors with varying locations within the hierarchical modular organization of the protein interaction network either serve as "master regulators" or as "middle managers" of specific processes that are critical to cellular function [509]. Hierarchical graph models have also been used to study the self-organization of protein interactions in the context of evolutionary conservation of protein complexes across species. These studies indicate that, from a network topology perspective, the clustering coefficient distribution tends to drive the hierarchical organization of protein-protein interaction networks rather than the overall scale-free nature of the network [126].

3.2.4 Modularity

Consistent in many ways with the notions underlying hierarchical graphs, protein-protein interaction networks are also often described as having high *modularity*. Modularity reflects the extent to which the graph is divided into locally dense communities, i.e. groups of nodes that are more highly connected within the group than outside the group. Conceptually, the modularity statistic quantifies the difference in probabilities that an edge from the observed graph is included within a given module M_i and that it would randomly fall into that module. If $e_{ij} = |\{(u,v) : u \in M_i, v \in M_j, (u,v) \in E\}|/|E|$ and $a_i = |\{(u,v) : u \in M_i, (u,v) \in E\}|/|E|$, then modularity, Q, for a graph with k modules can be defined as

$$Q = \sum_{i=1}^{k}(e_{ii} - a_i^2).$$

Modularity is a key feature of protein-protein interaction networks and has been consistently noted for a wide range of protein-protein interaction data. One of the first publications noting modularity in protein interaction data was in an investigation of the yeast proteome in experiments involving *Saccharomyces cerevisiae*. Han et al. (2004) noted two classes of highly connected "hub" proteins, with some hubs serving as core members of functional protein complexes and other hubs serving as connectors of complexes [154]. These relationships appeared to reflect distinct modularity of the protein-protein interaction network, preserving both functionality of distinct units as well as interactivity among them. This modularity was further confirmed in additional species including *Caenorhabditis elegans* and *Drosophila melanogaster* in a study investigating the utility of protein networks for predicting protein function [397].

3.3 Module detection

3.3.1 Community detection algorithms

Fast-greedy algorithm

As noted, protein interaction networks exhibit topological features consistent with modularity. Naturally, this motivates investigation into specific identification of modules within a network. A variety of module detection algorithms have been developed, many of which are independent of the types of interactions modeled in a network and simply seek to identify sets of nodes within a network that are more densely connected to each other by edges than other nodes in the network. We first briefly describe several of the most popular community detection strategies including the fast-greedy [58], walktrap [346] and label propagation [351] algorithms. All of these module detection strategies are available in the *igraph* R package. In addition, we discuss statistical modeling strategies that are specifically designed to identify functional multiprotein complexes and in many cases depend directly on the types of interaction data that are available.

The *fast-greedy* algorithm [58] is an efficient computational approach that runs in essentially linear time for sparse large-scale networks that demonstrate hierarchical features. This algorithm uses a greedy approach to maximize the modularity score for a given network and thus identify modules. For a network with n nodes and m edges, let d_i represent the degree of node $i(i = 1, ..., n)$ and $\sum_{i=1}^{n} d_i = 2m$. The algorithm first assigns all nodes in the network to their own individual communities. For each node $i = 1, ...n$, $a_i = d_i/(2m)$ is calculated. A potential change in modularity score upon merging communities M_i and M_j (initially equivalent to nodes i and j) is also calculated:

$$\Delta Q_{ij} = \begin{cases} (2m - d_i d_j)/(4m^2) & \text{if } (i,j) \in E \\ 0 & \text{otherwise.} \end{cases}$$

The communities for which ΔQ_{ij} is largest are merged. When a community is merged, a_i is updated to reflect the proportion of edges with at least one node in each community. A simple updating procedure for $\Delta Q'_{jk}$ is then proposed as communities are joined. Assuming communities M_i and M_j are merged into M_j, the update is given as:

$$\Delta Q'_{jk} = \begin{cases} \Delta Q_{ik} + \Delta Q_{jk} & \text{if } k \text{ is connected to both } i \text{ and } j \\ \Delta Q_{ik} - 2a_j a_k & \text{if } k \text{ is connected to } i \text{ but not to } j \\ \Delta Q_{jk} - 2a_i a_k & \text{if } k \text{ is connected to } j \text{ but not to } i. \end{cases}$$

This simple update procedure fosters the computational efficiency of the algorithm. Following the update, the next two communities yielding highest ΔQ_{ij}

are merged and the process repeats until there is one community. The end result is a hierarchical organization of community merges that can be used to identify the most cohesive modules.

Walktrap algorithm

The *walktrap* algorithm [346] is built on the idea of random walks of length t through the network. In a random walk, given the current node, the next step in the walk is determined by a random selection of a node with equal probability among all nodes adjacent to the current node. A walk of length t consists of t steps from a given starting node. Let P_{ik}^t represent the probability that a random walk that starts at node i ends at node k after t steps. The walktrap algorithm is built on two primary concepts:

- If two nodes i and j are members of the same community, then the probability P_{ij}^t will be high.

- If two nodes i and j are members of the same community, then $\forall k$, $P_{ik}^t \simeq P_{jk}^t$. That is, the probability of ending at node k after a random walk of length t will be similar for two nodes in the same community.

Importantly, as t tends toward infinity, P_{ij}^t depends only on the degree of node j and not starting point node i. Specifically, $\lim_{t\to\infty} P_{ij}^t = d_j/\sum_k d_k$. For a given community M, the probability P_{Mj}^t of moving from community M to node j in a random walk of length t is given by

$$P_{Mj} = \frac{1}{|M|} \sum_{i\in M} P_{ij}^t.$$

Using this result, the distance between two communities M_i and M_j, $r_{M_i M_j}$, is then defined as follows:

$$r_{M_i M_j} = \sqrt{\sum_{k=1}^{n} \frac{(P_{M_i k}^t - P_{M_j k}^t)^2}{d_k}}.$$

The walktrap algorithm first assigns all nodes to their own community. The two closest communities are merged based on the calculated value of r, the distances are updated, and the algorithm is repeated until there is one community. Similar to the fast-greedy algorithm, the end result is a hierarchical organization of community merges.

Label propagation algorithm

The label propagation algorithm [351] is a simple iterative algorithm in which community labels are assigned to nodes based on the majority label assignment of their adjacent nodes. Initially, all nodes are labeled as belonging to their own community. For every iteration of the algorithm, each node is labeled according to the label of the maximum number of its adjacent nodes. Ties are broken randomly with equal probability. The algorithm continues until all nodes have a label that matches the maximum number of its adjacent nodes. In contrast to the fast-greedy and walktrap algorithms, the end result of label propagation is not hierarchical. Instead, community membership results without a natural ordering of community merging. Results of label propagation can vary based on the initial configuration, so it is important to run the algorithm several times to inform consensus modules.

Other algorithms

Many other popular community detection algorithms exist in addition to those just described, and have been applied to protein-protein interaction data [442]. An algorithm based on edge betweenness successively removes edges with the highest edge betweenness scores, resulting in a "top-down" hierarchical organization of modules in contrast to the "bottom-up" hierarchy resulting from the fast-greedy and walktrap algorithms [309]. The "leading eigenvector" method identifies the eigenvector of the largest positive eigenvalue for the modularity matrix for a network and then separates nodes into communities based on their sign in the eigenvector [310]. The spinglass algorithm [354] originates from the field of statistical mechanics and is based on "spin models" for clustering multivariate data.

Importantly, the community detection methods listed here are designed for unweighted, undirected, connected graphs. Given the realities of measurement error and sampling described in previous chapters, and in particular the directionality of edges that is implicit in bait-prey technologies, straightforward application of these algorithms to an undirected graph representation of protein-protein interaction data should be interpreted with care. Results from these community detection algorithms may be helpful starting points for describing experimental data, but formal inference on quantities of interest may require more sophisticated modeling.

3.3.2   Protein complex estimation

The algorithms just described were designed to identify network communities for which nodes within the community are more densely connected to each other than nodes in different communities. In the specific context of protein interaction data, communities or modules may be interpreted as protein complexes that act as a coordinated unit to perform certain cellular functions. Recognizing the unique nature of TAP or CoIP bait-prey technologies that assay protein complex co-membership as described in the previous chapter, specific statistical procedures have also been developed to estimate protein complex membership, i.e. identify functional modules, using TAP data.

A foundational concept underlying protein complex membership estimation is the bipartite graph. A bipartite graph consists of two sets of nodes and edges identify relationships among the nodes within each set. In the context of protein complex membership, one set of nodes represents proteins and the other protein complexes. Edges connect proteins to the complexes for which they are members. In the previous chapter, Figure 2.4 illustrates a bipartite graph $\mathcal{A}$ for six proteins that are members of two complexes. It also illustrates the underlying affiliation matrix $\mathbf{A}$ and co-affiliation matrix $\mathbf{Y}$ that would result from TAP assays if all proteins were used as baits and if the technology is perfectly sensitive and specific. Note that all proteins that are members of the same complex share edges with all other proteins in that complex per $\mathbf{Y}$; in graph theoretic terminology these completely connected groups of nodes are called "cliques."

Consider a graph with n proteins comprising m total complexes. Also suppose $b \leq n$ proteins are used as baits and all proteins are detectable as prey. Let A be a $n \times m$ matrix where $A_{ik} = 1$ if the protein i is a member of the protein complex k and 0 otherwise. Let Z be a $n \times n$ matrix where $Z_{ij} = 1$ if proteins i and j are co-members of at least one protein complex. Note that $Z = [A \times A^T]$ where $[x] = 1$ if $x > 0$. Let Y be a $b \times n$ matrix where $Y_{ij} = 1$ if protein i is used as a bait and detects protein j as prey in a TAP experiment.

The local modeling algorithm published by Scholtens et al. (2005) [384] starts by using a clique identification algorithm for the underlying undirected graph given the observed data as a preliminary estimate of the protein complexes. Pairwise combinations of the cliques are then proposed and their merit is gauged according to the product of two terms, $P = L \times C$. L is the likelihood and uses a logistic regression model for the probability p_{ij} of detecting protein j as a prey using protein i as bait. Specifically,

$$\log\left(\frac{p_{ij}}{1 - p_{ij}}\right) = \mu + \alpha Y_{ij},$$

where $e^{\mu+\alpha}/(1 + e^{\mu+\alpha})$ is the sensitivity and $1/(1 + e^{\mu})$ is the specificity of the TAP technology. L is the product of $e^{(\mu+\alpha Y_{ij})Z_{ij}}/(1 + e^{\mu+\alpha Y_{ij}})$ over all tested edges from bait to prey. C is a product of $\Gamma(c_k)\phi(c_k)$ over all proposed protein complex estimates c_k, $k = 1, ..., K$, where

$$\Gamma(c_k) = \binom{t_k}{x_k} \frac{e^{(\mu+\alpha)x_k}}{(1 + e^{\mu+\alpha})^{t_k}},$$

t_k represents the number of tested edges in a given proposed complex estimate and x_k represents the number of observed edges. $\phi(c_k)$ is the two-sided p-value from a Fisher's exact test for the distribution of incoming edges for each protein. Taken together, $\Gamma(c_k)\phi(c_k)$ measures the consistency of a given protein complex estimate with the sensitivity of the TAP technology and the extent to which false negative edges are randomly distributed throughout the complex and hence likely represent stochastic error. Using a greedy approach, the algorithm by Scholtens et al. (2005) combines cliques until P is maximized.

Zhang et al. (2008) [517] define A slightly differently, with rows corresponding to proteins and columns corresponding to pull down experiments without regard to the identity of the bait. $A_{ip} = 1$ if protein i is detected in pull down experiment p, and 0 otherwise. For two proteins i and j, let q represent the number of column elements that are 1 for both i and j, let r represent the number of column elements that are 1 for i and 0 for j, and let s represent the number of column elements that are 1 for j and 0 for i. A Dice coefficient D_{ij} is then defined as

$$D_{ij} = \frac{2q}{2q + r + s}.$$

An undirected network of proteins with high values of D_{ij} is then constructed by calibrating against an external database of protein interactions, using an F_1-measure to select the optimal threshold. Once edges are identified in the network, Zhang et al. (2008) identify maximal cliques, and then adopt a clique merging algorithm [136] to allow for potential false negatives in the observed data. The final output can be modeled as a bipartite graph of affiliations of proteins with protein complexes.

Another approach named CODEC, proposed by Geva et al. (2011) [130], adopts a similar definition of A as Zhang et al. (2008), but the identity of the bait is linked to the pull down results being catalogued. CODEC begins with an initial estimate of protein complexes according to *bicliques*, that is, complete bipartite subgraphs in which all edges exist among a subset of nodes representing pull downs and a subset of nodes representing proteins. Similar to Scholtens et al. (2005) and Zhang et al. (2008), CODEC also adopts a greedy heuristic approach to combine bicliques into protein complex estimates, noting the reality of false negative observations in TAP data. A proposed complex c_k is evaluated according to a likelihood ratio score LRS, which can be defined as

$$LRS(c_k) = \sum_{i,j \in c_k} \log \frac{0.9}{d_i d_j / e_k} + \sum_{i,j \notin c_k} \log \frac{0.1}{1 - d_i d_j / e_k}$$

where d_i and d_j are node degrees for proteins i and j and e_k is the number of edges in c_k. While this approach does not specifically assess the distribution of edges among all candidate protein complex members as in Scholtens et al. (2005), CODEC is more computationally efficient and, when compared with published databases of known protein complexes, desmonstrates comparable and often higher sensitivity and specificity [130].

3.4   Software

Similar to analyses of a wide array of high-dimensional omics data, R statistical software (https://cran.r-project.org/) is a foundational component of conducting network analyses of protein-protein interaction data. R packages including *igraph* [62] provide efficient and informative data structures for storing and representing protein-protein interaction data. *igraph* and other packages including *RBGL* [39] contain extensive functionality for calculating the statistical measures, characterizing network topologies, and detecting local communities as described in this chapter. Cytoscape (www.cytoscape.org) [396] is another frequently used computational tool and is designed for both analysis and visualization of complex network data. Other more specialized web-based platforms also exist, some of which are designed specifically for protein-protein interactions, for example PINA: Protein Interaction Network Analysis (omics.bjcancer.org/pina/) [61].

3.5 Integration of protein interactions with other data types

Protein-protein interaction networks have been enormously informative about cellular function when analyzed on their own, however, integration of these data with other "omics" data have yielded additional insight. Integrative analyses of gene expression and protein-protein interaction data have revealed high coexpression of interacting pairs relative to random pairs [27]. Dittrich et al. (2008) [85] use a network optimization algorithm to identify subcomponents of protein networks that are mostly highly associated with cancer subtype based on transcriptomic data. Several publications have integrated experimental protein interaction data with existing pathway databases and, in some cases, gene expression data to predict protein function [69, 323, 443, 492] and to identify functional modules [56, 439]. Systematic approaches and computational tools to integrate multiple sources of high-dimensional data types are urgently needed for a more sophisticated understanding of cellular function.

4

Detection of Imprinting and Maternal Effects

CONTENTS

4.1 Imprinting and maternal genotype effects – Two epigenetic factors

Genomic imprinting and maternal genotype effects are both important epigenetic factors that have profound impacts on mammalian growth and development. An imprinted gene or genetic variant refers to the unequal expression (gene expression or disease expression) of a genotype depending on whether a particular allele (i.e. a particular variation of the gene or genetic variant) is inherited from the mother or from the father. For simplicity, we will not distinguish a gene or a genetic variant hereafter; they will all be referred to generally as genetic variants. If a particular allele inherited from the father leads to the (partially) silencing of the variant, that is, reduced expression of the genotype (compared to the expression when the allele is inherited from the mother), then the genetic variant is said to be paternally imprinted. Conversely, maternal imprinting refers to the reduced expression of the variant when the allele of interest is inherited from the mother. Genomic imprinting is a consequence of the epigenetic process that involves methylation and histone modifications without altering the underlying DNA sequences. It is well recognized that genomic imprinting plays a key role in normal mammalian growth and development [109, 340], such as x-chromosomal inactivation [289]. Its role in the development of complex diseases has also been implicated as well [234, 246]. Maternal genotype effect refers to the situation wherein the phenotype of an individual is influenced by the genotype of the mother either through the prenatal environment or through the mother's genetic materials deposited in the egg cell (not necessarily in the child's DNAs). It has been documented that maternal genotypes may be closely involved in birth defects and some childhood diseases [147, 148, 200].

4.1.1   Imprinting effects on complex diseases

In a genetic association study of a particular trait, the goal is to identify genetic variants that may affect the trait through sequence variation, parental origin of alleles, and even through combinations with maternal genotypes. The roles of sequence variation in complex diseases have been the focus of most association studies, with tens of thousands of unique SNP-trait associations identified at a genomewide significance level (p-value $\leq 5.0 \times 10^{-8}$) (www.ebi.ac.uk/gwas). On the other hand, the roles of epigentic factors, such as imprinting and maternal genotypes, are less explored. Nevertheless, it has been well documented that the etiology of a number of diseases is closely linked to these two epigenetic factors, which may account for a portion of the genetic heritability remaining unexplained [97, 246].

The first imprinted gene for mice was discovered in 1991 [21, 74], while in humans, the first was found two years later in 1993 [132]. In addition to its function in normal mammalian development (e.g. brain development [139]), imprinting has also been found, or suspected, to play an important role in a number of human diseases. The most well-known diseases, for which some of the cases are caused by imprinting, or more generally, parent-of-origin effects, are the Prader-Willi Syndrome and the Angelman Syndrome [246]. These two examples are particularly interesting as both are caused by the same genetic locus on chromosome 15, but one due to maternal imprinting while the other paternal imprinting. Parent-of-origin effects have also been demonstrated to be a cause of a number of other syndromes, including the Beckwith-Wiedeman Syndrome and the Silver–Russell Syndrome [138, 264]. Further, although not yet elucidated conclusively, imprinting has been suspected to play a role in some other complex diseases, such as autism, schizophrenia, obesity, low birth weight, and rheumatoid arthritis [1, 87, 283, 374, 542]. The involvement of imprinting in a number of complex disease should not come as a surprise, as many diseases are due to abnormal epigenetic regulations of genes [26], and 1% of all mammalian genes are estimated to be imprinted [300]. Nevertheless, only a limited number of imprinted genes have been identified thus far. Although scientists are able to carry out mouse experiments to detect imprinting in a genome-wide scale using the Next Generation Sequencing (NGS) technology [139, 471], imprinting detection in humans, especially those that are involved in disease etiology, remains reliant on statistical methods through inferences with observational data. As such, it is highly important to consider powerful statistical methods since imprinting is deemed crucial in understanding the interlocking relationship between the genome and the epigenome [109].

4.1.2 Maternal genotype effects on complex diseases

During pregnancy, some of the mother's RNAs or proteins may be transfered to her fetus, which constitutes the additional genetic materials, apart from the DNAs in the fetus' cell nuclei, that are passed from the mother to the child. This may lead to a child showing a particular phenotype, due to the presence of the mother's genotypes, even if the child does not posess the necessary genotypes in his/her own DNAs. With large-scale data now becoming increasingly available, the presence and importance of the effects of maternal genotype on complex diseases have been increasingly revealed, especially in birth defects, certain psychiatric illnesses, congenital malformations, pregnancy complications, and childhood cancer [147, 148, 200, 317]. Both the genome of the child and that of the mother play important roles. Well-known examples include spina bifida [147, 148, 200], schizophrenia [325] and childhood acute lymphoblastic leukemia [317].

In addition to the main effects of the mother's genotypes, there are two other types of effects that are the results of interactions between mother's genotypes and child's genotypes. The first is typically known as maternal-fetal genotype incompatibility due to the gene products of the mother and the fetus [55, 318, 402]. A particular example is the hemolytic disease induced by Rh incompatibility [415]. This occurs when the mother is Rh-negative (i.e. the mother does not carry a copy of the antigen coding allele to produce the Rh factor, a protein on red blood cells) while the child is Rd-positive. This "mismatch" results in the mother mounting an immune response to combat "foreign" proteins produced by a fetus carrying a copy of the allele. The second type represents a mechanism known as NIMA (non-inherited maternal antigen) owing to the child lacking the protein products derived from polymorphic genes while the mother has a normal expression of the genes. In some cases, rheumatoid arthritis is the consequence of NIMA [160], in which the mother has a copy of the allele coding for an antigen at the HLA-DRB1 locus while the child lacks such an allele (i.e. the allele from the mother was not passed to the child). This could affect the developing immune system of the fetus due to the perinatal environment in which a small amount of the antigen produced by the mother may penetrate the placenta.

4.2   Confounding between imprinting and maternal effects

As discussed above, genomic imprinting has a distinct biological process compared to that of the effect of maternal genotype, yet these two effects are confounded and can mask one another [146, 489, 505]. To see this, let us consider the following main-effect-only generalized linear model, which describes the probability that a child is affected ($D = 1$) by a particular trait given his/her genotype as well as those of the parents.

$$
\begin{aligned}
g(P(D = 1 \mid C, F, M)) \;=\;\; & \beta_0 + \beta_1 I(C = 1) + \beta_2 I(C = 2) \\
+\;\; & \beta_{imp} I(C = 1, F < M) \\
+\;\; & \beta_{m1} I(M = 1) + \beta_{m2} I(M = 2), \quad (4.1)
\end{aligned}
$$

where g is the link function (typical link functions used in this context are log and logit); $I(\cdot)$ is an indicator function taking the value of 0 or 1. Further, C, F and M are the random variables denoting the genotypes of the individual, father, and mother, respectively, and collectively, they are referred to as the trio genotypes. Each of them can take values of 0, 1 or 2, representing the number of copies of the variant allele (which is interchangeably referred to as the minor allele hereafter) in a SNP. In the rest of this chapter, we will use c, f and m to denote the observed genotypic values of C, F and M, respectively; however, for notational convenience (and with slight abuse), we may simply use C, F and M without specifying their realized values, such as in (4.1). The parameter β_0 is the g inverse of the phenocopy rate, which is the probability of being affected given that the trio genotypes are all 0. Disease association with the individual's SNP genotype are described in β_1 and β_2, depending on whether the individual carries 1 or 2 copies of the variant allele. Potential imprinting effect is represented by β_{imp}, which signifies maternal imprinting if it is negative or paternal imprinting if it is positive. Finally, the last two parameters, β_{m1} and β_{m2}, denote the maternal effects when the mother carries 1 or 2 copies of the variant alleles, respectively.

Typically, in a study of association for a Mendelian trait, the probability of being affected (i.e. the penetrance probability) is only dependent on one's

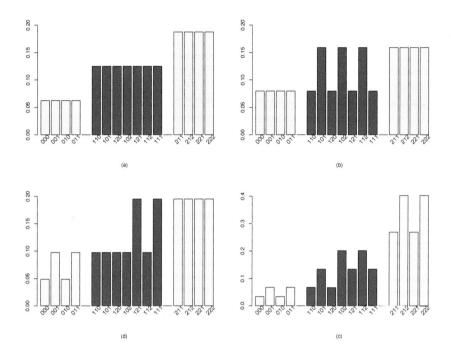

FIGURE 4.1
Barplots showing confounding between imprinting and maternal effects. Each bar represents one of the 15 possible combinations of the trio genotypes in the order of (C, F, M) (indicated below each bar), while the height of each bar gives the probability that an individual is affected given his/her own genotype and those of the parents, $P(D = 1 \mid C = c, F = m, M = m)$, based on a log-linear model. (a) Mendelian trait; (b) trait with paternal imprinting only; (c) trait with maternal effect only; (d) trait with both maternal imprinting and maternal effect.

own genotype. Using the log-linear model as an example (i.e., the link function $g = log$ in equation (4.1)), this property is clearly seen in Figure 4.1(a), which portrays a Mendelian trait by setting the imprinting and maternal effect parameters all to zero. There are 15 possible combinations of the trio genotypes (C, F, M) (marked on the x-axis), with the first four (group 1) depicting the child's genotype being 0 $(C = 0)$, the next 7 (group 2) representing the child having genotype 1 $(C = 1)$ and the last 4 (group 3) with the child's genotype being 2 $(C = 2)$. From the figure, we can see that the penetrance probabilities are all identical for a given individual's genotype regardless of the parental genotypes.

When there is paternal imprinting (in addition to association) but no maternal effect, the penetrance probabilities now depend on the parental genotypes as well (Figure 4.1(b)). Focusing on the middle block (the trio genotype combinations for which the child carries one copy of the variant allele), one can see that when the variant allele is inherited from the father (with trio genotypes $(1, 1, 0), (1, 2, 0)$ and $(1, 2, 1)$), the risk of being affected is smaller than that when the variant allele is inherited from the mother (with correspondong trio genotypes $(1, 0, 1), (1, 0, 2)$ and $(1, 1, 2)$), respectively. Note that when the trio genotype configuration is $(1, 1, 1)$, it is uncertain as to who passed the variant allele to the child.

On the other hand, when there is a maternal effect but no imprinting, the effect can mimic that of paternal imprinting (Figure 4.1(c)). Again, focusing on the middle block, one can see that the penetrance probability is lower for a trio genotype combination where the variant allele is inherited from the father than the correponding probability when the allele is inherited from the mother $((1, 1, 0)$ vs. $(1, 0, 1)$, $(1, 2, 0)$ vs. $(1, 0, 2)$ and $(1, 2, 1)$ vs. $(1, 1, 2))$.

Finally, when both maternal imprinting and maternal effects are present (Figure 4.1(d)), they may cancel each other, leading to "apparent" no imprinting effect (same penetrance probability for $(1, 1, 0)$ and $(1, 0, 1)$, for example), as the "depleted" expression due to imprinting is "replenished" by maternal effect. On the other hand, if there are both paternal imprinting and maternal genotype effects, then the maternal effect will be "magnified" if imprinting effect is assumed absent, or vise versa (scenario not presented in Figure 4.1).

Taken together, if a study aims to detect imprinting effect only (by assuming there is no maternal effect), it may lead to a false positive test (if there is in fact maternal effect but no imprinting effect) or false negative (if there is maternal imprinting and also maternal effect). Similarly, if a study aims to detect maternal effect only (by assuming there is no imprinting effect), then a false positive may result if there is in fact paternal imprinting but no maternal effect. On the other hand, a true maternal effect may be cancelled out by a maternal imprinting effect, leading to a false negative conclusion of no maternal effect. In other words, if the assumption of no maternal effect or no imprinting effect is violated, there can be severe consequences. As such, it is critical that these two confounding factors be studied jointly to evaluate their effects, unless it is known a priori that one of the two types of effects is indeed absent.

4.3   Evolving study designs

As shown in equation (4.1), in order to study imprinting and maternal effects, the basic data unit is a trio (individual and parents). Therefore, such studies are all family based. Nevertheless, there are various and evolving study

TABLE 4.1

Study designs and family types for studying imprinting and maternal effects

Design	Family Type	Detectable Effects[d]	Ref.
Prospective	General peidgrees	A, I, M	[504]
	Nuclear families	A, I, M	[503]
Retrospective	1. CaPT	A, I, M	[478]
	2. CaPT + Co	A, I, M	[99]
	3. CaPT + PoCo	A, I, M	[479]
	4. CaPT + CoPT	A, I, M	[505]
	5. CaPT + CoMP	A, I, M	[457]
	6. CaMP + CoMP	M	[399]
	7. CaPT/CoPT+CaMP/CoMP	A, I, M	[505]
	8. Nuclear families[a]	A, I, M	[156]
	9. Nuclear families[b]	I	[538]
	10. Nuclear families (DSP)[c]	A, I, M	[518]
	11. General pedigrees	A, I, M	[55, 520, 542]

CaPT – Case-Parent Triads; Co – Controls; PoCo – Parents of Controls; CoPT – Control-Parent Triads; CoMP – Control-Mother Pairs; CaMP – Case-Mother Pairs; A – Association; I – Imprinting effect; M – Maternal genotype effect; [a]Father may be missing; [b]One of the parents, either mother or father, may be missing; [c]Discordant sibpair design where additional sibling's information can be included; [d]For Design 6, with additional SNPs from nearby markers, the other effects may also be detectable [521, 522].

designs. We will compare and contrast prospective vs. retrospective designs. Among the retrospective designs, we will discuss the relative merits of affected-family-only designs vs. designs that recruit both affected families as well as control families. We will bring to bear the assumptions needed under different study designs for detecting different effects.

In prospective family-based studies, families are recruited and followed onward over time. This strategy typically results in the inclusion of extended families in addition to nuclear families (Table 4.1), and individuals are not recruited for any particular disease. At a given time, each family member may or may not be affected by a particular disease. As such, data collected from prospective family studies can be used for understanding the genetic contributions to various complex disorders [504], and can be used for studying pleiotropy. There are a number of well-known family-based genetic epidemiology studies that are prospective in nature, including the Framingham Heart Study (FHS) [410], the Hutterite genetic isolates [32, 431] and the Amish populations [96].

In contrast, retrospective family-based association studies recruit families with children affected by a particular disease. A child whose affection status leads to the recruitment of a family into the study is called a proband.

This strategy usually leads to nuclear, instead of extended, families. The earliest retrospective design recruits probands and their parents (one proband per family), leading to the case-parent triad (CaPT) design [477, 478, 483], which continues to be used in recent studies [516] (Design 1 under the "Retrospective" category in Table 4.1). As discussed earlier and as shown in Figure 4.1, there are 15 configurations of CaPT genotypes. Further, according to the main-effect generalized linear model given in equation (4.1), there are six parameters. If interactions between the mother and child genotypes are also considered, then there are six additional parameters [266] (see equation (4.4)). Moreover, implicit in equation (4.1) are the eight mating-type probabilities (probabilities of the parental genotype combinations – there are nine of them but the probabilities sum to 1 hence 8 parameters). Note that if Hardy-Weinberg equilibrium (HWE) is assumed, then there is only a single parameter, but such an assumption is rarely justifiable [505]. A frequently made assumption is mating symmetry, which aserts that the parental genotypes are interchangeable in terms of their mating-type probabilities. Under this assumption, the number of parameters reduces to five, although such an assumption is also questionable [505].

To estimate the parameters without making completely inadequate assumptions and/or reducing the model to a point where important genetic effects are compromised, other designs are also proposed by bringing along additional data. Examples include the addition of unrelated controls [99], parents of controls [479], controls and their parents [504], and controls and their mothers [457], resulting in Designs 2–5 in Table 4.1, respectively. Due to the concerns that fathers are more difficult to recruit, designs that do not recruit fathers or allow fathers to be missing are also considered. Examples include Design 6 in Table 4.1, in which only child-mother pairs are collected with the child being either affected or not [399, 451], and Design 7 where triads are recruited but the fathers' genotypes are allowed to be missing, leading to a combination of complete triads and complete child-mother pairs. A further extension is the inclusion of additional siblings, leading to arbitrary nuclear families with the father being present or missing [156] (Design 8). If either mother or father is allowed to be missing, then the data will be useful for detecting imprinting effect only under the assumption of no maternal genotype effect [538] (Design 9). To address the difficulty of recruiting control families, a recent method makes use of discordant sibpairs and may also include additional siblings [518] (Design 10). Finally, a design utilizing extended families but allowing for missing data has also been entertained for detecting imprinting effect only [520, 542] (Design 11).

As described above, some of the study designs are not amenable for detecting all effects of interest either due to the limitation of data or lack of available statistical methods; the "detectable effects" column provides the details. For example, for Design 6 where child-mother pairs data are collected, it can only be used for detecting maternal ("M") effect [399]. On the other hand, although it is conceivable that a design can be used for detecting other effects (e.g. Design 9), existing methods were only for detecting a certain effect (imprinting "I" in this case). Although multiple designs may be used for detecting the same kinds of effects, some may be more powerful than the others [519], which should be beared in mind when designing a future study.

In the following, we describe several statistical methods for assessing the effects of maternal genotypes and/or imprinting on complex diseases. We first consider data that are obtained prospectively, so that non-uniform family structure will be taken into consideration (Section 4.4). We then take on data from various retrospective designs (Section 4.5)

4.4 Methods for detecting imprinting and maternal effects using data from prospective studies

As discussed above, for prospective studies, families are sampled without conditional on the existence of a specific disease. As such, for a particular disease of interest, the configuration of genotypic and phenotypic values (disease status or measurements) of each family in the study should be modeled jointly. In the following, we describe a Likelihood approach for joint detection of Imprinting and Maternal Effects using pedigree data (LIME-ped) [504]. Consider a binary trait and let $\boldsymbol{Y}_i = (Y_{i1}, Y_{i2}, \cdots, Y_{ic_i})$ denote the binary disease status of all c_i non-founders in the ith family, $i = 1, 2, \cdots, n$. The genotypic data of the ith family is denoted by $\boldsymbol{X}_i = (X_{i1}, X_{i2}, \cdots, X_{in_i}) = (\boldsymbol{X}_{iO}, \boldsymbol{X}_{iM})$, where n_i is the number of individuals in the ith family, and the genotype configuration has been re-organizes in such a way that $\boldsymbol{X}_{iO}$ is the collection of observed genotypes while $\boldsymbol{X}_{iM}$ represents the individuals with missing genotypes. Denote the parameters of the model (say those from equation (4.1)) and those of the mating types collectively as $\boldsymbol{\theta}$. Then the likelihood of $\boldsymbol{\theta}$ given the observed data over all n families in the study is:

$$L(\boldsymbol{\theta} \mid \boldsymbol{X}_{iO}, \boldsymbol{Y}_i, i = 1, \cdots, n) = \prod_{i=1}^{n} P(\boldsymbol{X}_{iO}, \boldsymbol{Y}_i \mid \boldsymbol{\theta})$$

$$= \prod_{i=1}^{n} \left\{ \sum_{\boldsymbol{X}_{iM}} \left[P(\boldsymbol{X}_{iO}, \boldsymbol{X}_{iM} \mid \boldsymbol{\theta}) \prod_{j=1}^{c_i} P(Y_{ij} \mid X_{ij}, X_{iF_{ij}}, X_{iM_{ij}}, \boldsymbol{\theta}) \right] \right\},$$

$$(4.2)$$

where $X_{iF_{ij}}$ and $X_{iM_{ij}}$ are, respectively, the genotypes of the father and the mother of non-founder j in the ith family. Further, the sum in (4.2) is over all missing genotypes within a family, and the products within the square brackets are the consequence of assuming that the disease status for the nonfounders are independent given the familial genotypes. Each penetrance probability within the product, $P(Y_{ij} \mid X_{ij}, X_{F_{ij}}, X_{M_{ij}}, \boldsymbol{\theta})$, can be expressed in terms of the model parameters according to a disease model (e.g. log-linear model, a specific case of (4.1)). Finally,

$$
\begin{aligned}
P(\boldsymbol{X}_{iO}, \boldsymbol{X}_{iM} \mid \boldsymbol{\theta}) &= P(X_{i1}, X_{i2}, \cdots, X_{in_i} \mid \boldsymbol{\theta}) \\
&= \prod_{j=1}^{c_i} P(X_{ij} \mid X_{iF_{ij}}, X_{iM_{ij}}) P(\boldsymbol{X}_{\bar{c}_i} \mid \boldsymbol{\theta}), \quad (4.3)
\end{aligned}
$$

where $\boldsymbol{X}_{\bar{c}_i}$ is the collection of genotypes for all founders. The transmission probabilities in (4.3) are calculated according to Mendelian law of segregation. If HWE is assumed, then the factor representing the probability of the founders is simply the product of the probabilities over individual founders. However, if such an assumption is suspected to be violated, then one may express the probability in terms of the mating type probability parameters specified in $\boldsymbol{\theta}$ for founding couples [504]. For marry-in founders (i.e., those whose spouses are non-founders), the conditional probability of the founder given the genotype of the non-founding spouse can also be expressed in terms of the mating type probabilities [504].

In equation (4.2), the likelihood contribution from each family is calculated by summing over all possible missing genotype configurations. For families with 3–4 generations and a moderate number of missing genotypes, enumerating all the possible configuration is an effective strategy [504]. However, for families with a large number of missing genotypes, it may become cumbersome and too computationally intensive to enumerate. In such a scenario, better computational algorithms (e.g. analogous to reverse peeling [343]) may be pursued.

4.5   Methods for detecting imprinting and maternal effects using data from retrospective studies

For retrospective studies, there are various strategies for detecting imprinting and maternal effects. Recall that these two epigenetic factors are in fact confounded: therefore, joint detection of these two effects is necessary to guard against inflation of type I error rates and/or loss of power, unless there is a priori knowledge that one of the factors does not have an effect on the trait of interest [266]. In the following, we first describe joint detection of imprinting and maternal genotype effects. Then assuming there is no maternal effect, we present methods that can be used to detect imprinting effects. Finally, we also briefly discuss a method that can be used to detect maternal-child genotype interacting effects assuming there is no imprinting.

4.5.1 Joint detection of imprinting and maternal genotype effects

The log-linear model, a specific case under the general model given in (4.1) when g is set to be the *log* function, is frequently used to model the joint effect of association, maternal, and imprinting based on triad genotypes [478, 505]. An equivalent model is that of logistic, where g is the *logit* function, unless there are additional constraints imposed on the log-linear model regarding mating type probabilities [399]. One can view the model specified in (4.1) as the main effect model, since it does not entertain the possibility of any interacting effect between the maternal and child genotypes. As such, one may expand the model to include two more terms, $\beta_{10}I(C = 1, M = 0) + \beta_{01}I(C = 1, M = 0)$, where the first term measures the effect of "genotype incompatibility" (e.g. the Rh-type mismatch as discussed earlier), and the second term makes it possible to study NIMA. However, due to the concern about the dimensionality of the parameter vector (which can lead to over-parameterization), such interaction effects are typically not included in models for joint detection. A special model for detecting interaction effects will be discussed later (4.5.3).

With the model specified in (4.1), there are five parameters of interest: β_1, β_2 for association (both equal to 0 indicating no association), β_{imp} for imprinting (0 indicating no imprinting effect), and β_{m1}, β_{m2} for maternal (both equal to 0 indicating no maternal effect). There is also a nuisance parameter: β_0, the phenocopy rate (related to the probability of getting the disease when neither the child nor the parents carry any copy of the disease allele). For a retrospective study design (can be any of the 10 designs in Table 4.1), the genotypes are collected conditional on the proband's affection status. Therefore, for each trio, the correct probability for its contribution to the statistical analysis is a conditional one [519]:

$$P(C, M, F \mid D) = \frac{P(D \mid C, F, M)P(C \mid F, M)P(F, M)}{P(D)}.$$

As such, one also has to consider mating type probabilities, $\mu_{fm} = P(F = f, M = m)$. Since each can take three possible values, there are eight independent parameters (all nine sum to 1). To reduce the number of parameters, various assumptions have been made. It has been recognized that HWE may not hold for a SNP that is associated with a trait; therefore, the predominant assumption has been a less stringent one, namely mating symmetry, i.e., $P(F = f, M = m) = P(F = m, M = f) = \mu_{fm}$. This reduces the number of independent mating type probabilities to five. Other assumptions have also been made, such as allelic exchangeability [399].

With the above specifications and assuming a log-linear disease model, one can write down the distribution of observing a triad genotype given the disease status:

$$P(C, F, M \mid D = 1)$$
$$= \frac{\gamma_0 \gamma_1^{I(C=1)} \gamma_2^{I(C=1)} \gamma_{imp}^{I(C=1,F<M)} \gamma_{m1}^{I(M=1)} \gamma_{m2}^{I(M=2)} \mu_{FM} P(C \mid F, M)}{P(D = 1)},$$

where each γ parameter is the exponential of the corresponding β, and $P(D = 1)$ is the population disease prevalence, which is typically readily available for most diseases [505].

For the CaPT design (Design 1 in Table 4.1), the observed data can be sorted into 15 categories according to the combinations of the trio genotype vector $(C = c, F = f, M = m)$ (Figure 4.1). (Although each person can take three possible genotypes, 12 of the $3^3 = 27$ combinations are not possible according to Mendelian inheritance.) Assuming all triads are independent, the 15-category observed data vector $\{A_{cfm}\}$ follows a multinomial distribution with the cell probabilities as specified above:

$$\{A_{cfm}\} \sim \text{Multinomial}(A, \{P(C = c, F = f, M = m|D = 1)\}),$$

where $A = \sum_{cfm} A_{cfm}$ is the total number of CaPT families, and $\{P(C = c, F = f, M = m|D = 1)\}$ is the vector of the multinomial probabilities. One can then write down the multinomial likelihood function under the unconstrained model (alternative hypothesis) or under a constrained model (a null hypothesis; e.g. $\beta_{m1} = \beta_{m2} = 0$). Based on the log-linear model, a likelihood ratio test (LL-LRT) can be carried out by comparing the likelihoods of these two nested models [478]. Due to the large number of parameters and limited diversity in the data, the likelihood ratio test is not powerful unless the sample size is very large.

Another class of solutions, LIME-mix, based on partial likelihood and able to accommodate a mixture of complete and incomplete triad families (Design 7 in Table 4.1), seeks to completely eliminate the issue of nuisance parameters from mating type probabilities. With fewer parameters, this class of methods can achieve a greater statistical power for hypothesis testing. However, it comes at a price as it imposes a further requirement in the study design, namely the availability of control families. Data collected from designs 4, 6, and 7, which all include control families, are amenable with LIME-mix. To be specific, let us consider LIME-mix for Design 4, where there are both CaPT and CoPT data.

Analogous to CaPT data, the observed data from CoPT families can also be sorted into 15 categories according to the combinations of the family genotypes $(C = c, F = f, M = m)$. Assuming all triads in CoPT are independent, the observed 15-category data vector $\{U_{cfm}\}$ follows a multinomial distribution as follows:

$$\{U_{cfm}\} \sim \text{Multinomial}(U, P(C = c, F = f, M = m|D = 0)),$$

where $U = \sum_{cfm} U_{cfm}$ is the total number of CoPT families, and $\{P(C = c, F = f, M = m | D = 0)\}$ is the vector of the multinomial probabilities:

$$P(C, F, M \mid D = 0)$$

$$= \frac{\left(1 - \gamma_0 \gamma_1^{I(C=1)} \gamma_2^{I(C=1)} \gamma_{imp}^{I(C=1,F<M)} \gamma_{m1}^{I(M=1)} \gamma_{m2}^{I(M=2)}\right) \mu_{FM} P(C \mid F, M)}{P(D = 0)},$$

where $P(D = 0) = 1 - P(D = 1)$.

The idea of LIME-mix is to consider each of the 15 traid genotype combinations, $(C, F, M) = (c, f, m)$, and compare the count of the case families, A_{cfm}, with that of the control families, U_{cfm}. That is, among the triads that have the (c, f, m) genotypes, $n_{cfm} = A_{cfm} + U_{cfm}$, and

$$A_{cfm} \sim \text{Binomial}(n_{cfm}, p_{cfm}),$$

where

$$
\begin{aligned}
p_{cfm} &= \frac{E(A_{cfm})}{E(A_{cfm} + U_{cfm})} \\
&= \frac{AP(D = 0)P(D = 1 \mid c, f, m)}{AP(D = 0)P(D = 1 \mid c, f, m) + UP(D = 1)[1 - P(D = 1 \mid c, f, m)]}.
\end{aligned}
$$

Note that the above binomial probability is free of any nuisance (mating type) parameters, as are the penetrance probabilities given in (4.1). Therefore, there is no need to assume HWE, or even the less stringent condition of mating symmetry. If we multiply the conditional probabilities over all the 15 (c, f, m) combinations, we will have the following function of the model parameters, $\boldsymbol{\theta} = (\beta_0, \beta_1, \beta_2, \beta_{imp}, \beta_{m1}, \beta_{m2})$:

$$h(\boldsymbol{\theta}) \propto \prod_{(c,f,m)} (p_{cfm})^{A_{cfm}} (1 - p_{cfm})^{n_{cfm} - A_{cfm}}.$$

It turns out that $h(\boldsymbol{\theta})$ is in fact the partial likelihood of the multinomial models [505]. This shows that, by conditioning on each combination of the triad genotypes, the retrospective data are turned into "prospective", as the partial likelihood component of the full likelihood is simply the products of the binomial kernels. Inference of the model parameters can then be made by maximizing $h(\boldsymbol{\theta})$, and likelihood ratio tests can be carried out as before.

The idea of turning a retrospective design (involving the nuisance parameter in the full likelihood) into a prospective one (free of nuisance parameters) by stratifying according to each triad genotype is also applicable to the design without fathers (Design 6), the design including both types (Design 7) [505], the design that can include additional siblings in the case families and control families (Design 8) [156], or the design that recruits only discordant sibpairs without the need to recruit control families (Design 10) [518].

4.5.2   Detection of imprinting assuming no maternal effect

As discussed in Section 4.2, if both imprinting and maternal effects are present but not accounted for, then there could be inflated type I error or loss of power. However, if maternal effect is known, a priori, to be absent, then a test focusing on detecting imprinting effect will be more powerful without incurring inflated type I error rate. In this subsection, we discuss a class of simple, yet powerful, non-parametric methods for detecting imprinting effect assuming no maternal effect. This class of methods was originated from the Parent Asymmetry Test (PAT) proposed for case-parent triad data [477] (Design 1). This class includes the extension of PAT to nuclear families with multiple affected siblings and with potential missing parental genotype data (CPAT) [538]. A further extension to pedigree data (PPAT) involving a Monte Carlo procedure for handling missing data (MCPPAT) widens the applicability of the PAT type approach [542]. Finally, inclusion of unaffected offsprings in the test statistic (leading to PPATu and MCPPATu) provides additional power gains [520].

In the simplest form of PAT, for each CaPT family with genotype vector (C, F, M), its contribution to the score statistic is

$$T_{Ca} = I(F > M, C = 1) - I(F < M, C = 1),$$

where $I(\cdot)$ is the usual indicator function taking the value of 1 or 0. Therefore, T_{Ca} is equal to 1 (-1) if the affected child has a heterozygous genotype and his/her risk allele -allele of interest- comes from the father (mother); in all the other scenarios of the triad genotypes, the contribution is a 0. Under the null hypothesis H_0 (i.e. there is no imprinting effect; note that the alternative hypothesis H_1 is that there is an imprinting effect), the expected contribution for each CaPT family is 0, as the parental origin of the risk allele of a heterozygous child (if it can be determined) is as likely to be from the mother as it is from the father. However, if there is maternal imprinting, then T_{Ca} would likely be positive. Conversely, T_{Ca} would be more likely to be negative if there is paternal imprinting.

For a CoPT family, evidence of imprinting (in the same direction - maternal or paternal - as in the CaPTs) will be captured in a score statistic

$$T_{Co} = I(F < M, C = 1) - I(F > M, C = 1).$$

Under H_0, the expected value of T_{Co} is also 0. Specifically, if there is maternal (paternal) imprinting, then T_{Ca} is expected to be positive (negative).

For an extended pedigree (assuming there is no missing data), each child-parents triad will either contribute a T_{Ca} or a T_{Co}, depending on whether the child is affected or not. Suppose there are A_i CaPTs and U_i CoPTs in the ith family, and the jth CaPT (CoPT) contributes $T_{Ca,j}$ ($T_{Co,j}$). Then the overall contribution of the ith family to the score statistic is

$$T_i = w \sum_{j=1}^{A_i} T_{Ca,j} + (1 - w) \sum_{j=1}^{U_i} T_{Co,j},$$

where w is a weight parameter between 0 and 1, inclusive. When $w = 1$, this reduces to the PPAT statistic [542]. Although any specific weight value may be used for the procedure to remain valid, different weights may lead to different outcomes, some resulting in a statistic more powerful than the others. It appears that setting $w = 1 - \rho$, where ρ is the population prevalence of the disease, works well in a number of simulation studies [520, 540].

The contribution from all data in a study is then $T = \sum_{i=1}^{n} T_i$, where n is the number of families. Under the null hypothesis of no imprinting (H_0), the expected value of T is 0, while its variance can be estimated as $\sum_{i=1}^{n} T_i^2$ [542]. Therefore, to test for H_0, an overall score statistic is

$$S = \frac{\sum_{i=1}^{N} n \left[w \sum_{j=1}^{A_i} T_{Ca,j} + (1-w) \sum_{j=1}^{U_i} T_{Co,j} \right]}{\sqrt{\sum_{i=1}^{n} \left[w \sum_{j=1}^{A_i} T_{Ca,j} + (1-w) \sum_{j=1}^{U_i} T_{Co,j} \right]^2}},$$

which is approximately normally distributed if there is a sufficiently large number of pedigrees.

For extended pedigrees, missing genotype data is the norm rather than an exception. To make use of the data to the maximum extent possible, one may devise a new statistic that is the average of all possible configurations of complete familial genotypes consistent with the observed ones. As in Section 4.4, let $\boldsymbol{X}_i = (X_{i1}, X_{i2}, \cdots, X_{in_i}) = (\boldsymbol{X}_{iO}, \boldsymbol{X}_{iM})$, where n_i is the total number of individuals in the ith family, and the genotype configuration has been further re-organized so that $\boldsymbol{X}_{iO}$ is the collection of observed genotypes while $\boldsymbol{X}_{iM}$ represents the individuals with missing genotypes. Then the new statistic for the ith family can be formed as follows:

$$ET_i \quad = \quad E[T_i(\boldsymbol{X}_{iO}, \boldsymbol{X}_{iM}) \mid \boldsymbol{X}_{iO}],$$

where the original (complete data) statistic, $T_i(\boldsymbol{X}_{iO}, \boldsymbol{X}_{iM})$, is written more fully to reflect its dependence on the genotypes. Since the possible number of configurations of complete familial genotypes is typically very large, a Monte Carlo sampling method can be used to draw samples to provide an approximate statistic:

$$ET_{i,MC} \quad = \quad \frac{1}{M} \sum_{k=1}^{M} T\left(\boldsymbol{X}_{iO}, \boldsymbol{X}_{iM}^{(k)} \right),$$

where $\boldsymbol{X}_{iM}^{(k)}$ is the kth Monre Carlo realization of the missing genotypes given the observed ones. It can be shown that the expected value of $ET_{i,MC}$ is equal to 0 under some regularity conditions [520]. Therefore, one may replace the T_i in the complete data statistic by $ET_{i,MC}$ and proceed with the testing of the null hypothesis of no imprinting effect against the alternative hypothesis of the existence of imprinting effect.

4.5.3   Detection of maternal-child genotype interacting effect assuming no imprinting

For retrospective study designs that recruit child-mother pairs only (Design 6 in Table 4.1), there is very limited information on imprinting, and therefore, the focus is on detecting the maternal genotype main effect assuming no imprinting [399]. This is a special case of what is discussed in Section 4.5.1, and thus, methods therein can accomplish this goal by setting appropriate parameters to 0 in both the null and alternative hypotheses. Due to limited diversity (mother-child genotype combination) in the data, interacting effects are typically not included when studying the main effects.

For some other designs where a greater diversity of data are collected (e.g. Design 11), it is possible to detect maternal interacting effects [55]. Under such a scenario, one can expand the model specified in (4.1) as follows:

$$
\begin{aligned}
log[P(D = 1 \mid C, F, M)] &= ME(C, F, M) + IE(C, M, S_c) \\
&= ME(C, F, M) \\
&+ \sum_{(c,m) \in \mathcal{A}} [\beta_{cm1} I(C = c, M = m, S_c = 1) + \beta_{cm2} I(C = c, M = m, S_c = 2)],
\end{aligned}
$$

$$(4.4)$$

where $ME(C, F, M)$ is the Main Effect (ME) component of the model as given in (4.1) (including all main effects on association, imprinting and maternal); $IE(C, M, S_c)$ represents the Interaction Effects (IE) which may be dependenent on the sex of the child, S_c (1 : male; 2 : female). Specifically, β_{cm1} (β_{cm2}) represents the interacting effect of a male (female) child carrying c copies and a mother carrying m copies, of the variant allele, and $\mathcal{A} = \{(1,0), (0,1), (1,1), (2,1), (1,2), (2,2)\}$ is the set that contains all possible pairs of child-mother genotypes. We note that when $(c, m) - (1, 0)$, the corresponding effect represents Rh-type incompatibility; whereas when $(c, m) = (0, 1)$, this is the NIMA effect. Since the same genotype pairs may have a different effect on the child's susceptibility to a disease as it may depend on some other factors such as sex and the father's genotype, the model specified in (4.4) is completely general that can accommodate such a scenario. Furthermore, for studying certain interaction effect, such as NIMA, there may also exist the main effects of a child's genotype [55]; this scenario can also be investigated through the model specified in (4.4). As already discussed above, there is a concern in power for investigating the full model as specified in (4.1) for full-likelihood-based methods; this concern is further exacerbated with the model in (4.4) since it contains many more parameters. As such, although in theory, all the designs displayed in Table 4.1 can be used for the purpose of detecting interaction effects, it would be more appropriate to consider data from extended families (Design 11) for this purpose. Using the same notation as in Section 4.4, we can write down the retrospective likelihood for the collection of parameter vector $\boldsymbol{\theta}$ as:

$$L(\boldsymbol{\theta} \mid \boldsymbol{X}_{iO}, \boldsymbol{Y}_i, i = 1, \cdots, n) = \prod_{i=1}^{n} P(\boldsymbol{X}_{iO} \mid \boldsymbol{Y}_i, \boldsymbol{\theta})$$

$$= \prod_{i=1}^{n} \left\{ \sum_{\boldsymbol{X}_{iM}} \left[P(\boldsymbol{X}_{iO}, \boldsymbol{X}_{iM} \mid \boldsymbol{\theta}) \prod_{j=1}^{c_i} P(Y_{ij} \mid X_{ij}, X_{iM_{ij}}, \boldsymbol{\theta}) / P(\boldsymbol{Y}_i) \right] \right\}$$

$$(4.5)$$

where $X_{iM_{ij}}$ is the genotype of the mother of non-founder j in the ith family. Further, the sum in (4.5) is over all missing genotypes and the product within the square brackets is the consequence of assuming that the disease status for the nonfounders are independent given the familial genotypes. The denominator $P(\boldsymbol{Y}_i)$ can be expressed as the summation over all possible familial genotype configurations, as in the numerator, except that the summation will be over all individuals rather than just the missing ones. Each penetrance probability within the product, $P(Y_{ij} \mid X_{ij}, X_{M_{ij}}, \boldsymbol{\theta})$ is dependent on the mother's and child's genotypes (but not the father's as the focus is on maternal-offspring interactions), and can be expressed in terms of the model parameters according to a sub-model of disease model (4.4), with some parameters appropriately set to zero according to the effects of interest. For example, except phenocopy rate β_0 and the parameters for studying Rh-type incompatibility, β_{101} and γ_{102}, all the other parameters in the full model may be set to 0 [55]. Finally, as before, the probabilities for the genotypes of a family can be factored into transmission probabilities and mating type probabilities for families whose founders are all couples; how to deal with marry-in individuals without assuming HWE is as discussed in Section 4.4.

4.6 Case studies

We briefly demonstrate the methods described in this chapter in two case studies. In the first study, we analyzed the Framingham Heart Study (FHS) pedigree data in several ways, and showed that utilizing the extended pedigrees led to a more powerful test compared to using only nuclear families or case-control nuclear families extracted from the extended pedigrees. In the second case study, we illustrated the methods by analyzing an United Kingdom rheumatoid arthritis dataset. We once again demonstrated that maximal utilization of available data and "imputing" missing data can lead to greater power than focusing on genotyped individuals only.

TABLE 4.2

Detection of imprinting and maternal genotype effects Using framingham heart study data

SNP	Chr.	Position	Fam. Type	P-value[a]		
				Imp.[b]	Mat1[c]	Mat2[c]
rs6688233	1	9275686	Nuclear	0.01	0.39	0.34
			Extended	0.00	0.00	0.03
			Case-control	0.05	0.29	0.21
rs1459543	4	117729190	Nuclear	0.01	0.29	0.87
			Extended	0.00	0.10	0.26
			Case-control	0.03	0.63	0.28
rs4386830	6	141466657	Nuclear	0.00	0.00	0.00
			Case-control	0.06	0.01	0.02

[a]A p-value of 0.00 is the consequence of rouding, with the actual p-value being less than 0.005. [b]Imp. provides the p-value for either paternal or maternal imprinting effect. [c]Mat1 (Mat2) gives the p-value for the effect of a mother's genotype with a single copy (two copies) of the allele of interest.

4.6.1   Case study 1 – Framingham Heart Study

The Framingham Heart Study (https://www.framinghamheartstudy.org/about-fhs/), beginning in 1948, is a longitudinal prospective study to identify risk factors for cardiovascular disorders. This study is prospective in nature because the subjects were recruited without basing on their affection status for a particular disease. The study subjects were recruited through several cohorts. The original cohort consisted of 5209 individuals from Framingham, Massachusetts. The offspring of the original cohort, the spouses of the offspring, and third generation offspring were recruited in the additional cohorts. Overall, the participants come from around 1000 pedigrees (some are nuclear and some are extended families). Blood pressure measurements and genotypes of common SNPs were made available through dbGaP (https://www.ncbi.nlm.nih.gov/gap). Based on the blood pressure measurements, a binary hypertensive trait can be defined: a person is labelled as hypertensive if his/her highest recorded systolic blood pressure is greater than 140, or if the diastolic blood pressure is greater than 90 mmHg during the study period.

A nuclear family-based study using the logistic model (i.e. when the link function g in (4.1) was set to be logit) was conducted [503]. A number of SNPs were identified to be imprinted. Among them were SNPs rs6688233 on chromosome 1 and rs1459543 on chromosome 4 (Table 4.2). To ascertain whether the use of general pedigree data through the application of LIME-ped can lead to an increase in power for detecting imprinting and maternal effects, a further study was conducted using these two identified SNPs [504]. The reason that these two SNPs were selected was because each of them only exhibited imprinting, but not maternal genotype effect. Therefore it was of interest to see whether significant maternal effects might be discovered with extended pedigrees, which contain more information than reducing to nuclear families. The following describes the results from these two analyses.

For SNP rs668233, a maternal imprinting effect was found (p-value < 0.005) using LIME-ped, confirming the result from the nuclear family analysis. More interestingly, a maternal genotype effect was also detected, with the p-value for mother carrying one copy (two copies) of the allele being less than 0.005 (= 0.03). This result indicates an increase in power with general pedigree data; we also note that the result for the maternal imprinting effect is stronger (with a smaller p-value) compared that from the nuclear family analysis. It is also interesting to note the importance of analyzing imprinting and maternal effects simultaneously in this particular case. As we discussed earlier, the effect of maternal imprinting may be cancelled by maternal genotype efect. As such, an analysis hypothesizing the null effect of maternal genotype might fail to detect the maternal imprinting effect. For SNP rs1459543, a significant maternal imprinting effect was also found (p-value < 0.005) using LIME-ped, which confirms the result when using only nuclear families as with rs6688233, and furthermore, the result is more significant (with a smaller p-value). For this SNP, no significant maternal genotype effect was found, though.

From the above analyses, we can see that LIME-ped is more powerful than an analysis based on nuclear families. It is then of interest to investigate whether further trimming down the nuclear family data to only triads and pairs would lose essential information for testing the effects. To do this, in addition to the two SNPs investigated for general pedigree data, a third SNP, rs4386830 on chromosome 6, was also considered, as it exhibited both imprinting as well as maternal genotype effect. As such, its inclusion here is to see whether both of these two types of effects can still be recovered using the trimmed down CaPT/CoPT+CaMP/CoMP data only. First note that due to the need for having genotype data for both parents or at least for the mother, the number of triad/pair families that can be utilized for the analysis is much smaller; only about half of the number of families (less than 500 triad/pair families) were used for this analysis [503].

Using LIME-mix, for SNP rs688233, the imprinting effect is now only marginally significant, while maternal genotype effects remain undetectable (Table 4.2). On the other hand, for rs149543, the imprinting effect remains significant (although with a larger p-value), and there was also no significant maternal genotype effects detected. Finally, for SNP rs4386830, the maternal genotype effects (both when mother carries one copy or two copies) continued to be significant, however, the imprinting effect became only marginally significant.

Taken together, we see that trimming down the pedigrees to triad/pair families with complete genotype information may lose a significant amount of information, as the number of usable triad/pair families may be much smaller;

in this case study, it is only about $1/2$ of the nuclear families as some contain missing data. This also speaks for the value of "imputing" the missing data based on the observed ones. On the other hand, expanding from nuclear families to extended pedigrees naturally lead to maximal utilization of available data. In this case study, the advantage of using LIME-ped was clearly showcased as the significant results become more significant and additional significant results were detected.

The lessons learned from this case study are not unexpected. Clearly, for a prospective study in which extended pedigree data are available, they should be analyzed accordingly, that is, with respect to its study design, and with maximum utilization of the data if possible. That being said, LIME-mix would be recommendable if the study design is retrospective and/or if any assumption about mating type probabilities is of a concern.

4.6.2   Case study 2 – UK rheumatoid arthritis data

This case study demonstrates a class of simple yet powerful non-parametric methods by applying them to a United Kingdom (UK) Rheumatoid Arthritis (RA) dataset [207]. This class of methods, which are derivatives of PAT [477], are for detecting imprinting effects of markers under the premises that the markers are associated with the disease being studied and that there are no maternal genotype effects on the disease [542].

The UK RA dataset contains information on 807 individuals from 157 pedigrees. Over 10,000 SNPs on chromosomes 1–22 were genotyped, but there are a great deal of missing data – only about 10% of the families have complete genotypes. Given that PAT should only be applied to SNPs whose association with a disease has already been established, we prescreened all the SNPs using a pedigree association testing software that allows for missing data (MCPDT [84]), to identify those that were associated with RA. This initial screening step resulted in 150 SNPs with p-values less than 1%. Four PAT type tests were then executed: PPAT, PPATu, MCPPAT, and MCPPATu. One of the goals of this case study was to gauge the information gain by including contributions from unaffected individuals to the test statistic (PPATu vs. PPAT and MCPPATu vs. MCPPAT) and by "imputing" the missing genotypes (MCPPAT vs. PPAT and MCPPATu vs. PPATu). Using MCPPATu (which makes maximal use of available data), 10 of the 150 SNPs were identified to have significant imprinting effect with a p-value <0.05 (Table 4.3). For MCPPAT that excluded unaffected individuals from contributing to the test statistic, only five of the 150 SNPs led to a p-value <0.05, all of which are included in the 10 SNPs identified by MCPPATu. Finally, if trios with incomplete genotype data were not utilized, then PPAT or PPATu had a p-value of <0.05 only for two of the 150 SNPs, which were among those five detected by MCPPAT, and thus among the 10 detected by MCPPATu. In this dataset, as it turned out, the results for PPAT and PPATu were identical since there

TABLE 4.3

Detection of imprinting effects on Rheumatoid arthritis using a UK dataset.

SNP label	Chr.	Pos. (cM)	MCPPATu	MCPPAT	PPAT(u)
snp615817	13	21.98237	0.0003	0.0002	0.0143
np615818	13	21.98231	0.0050	0.0027	0.0343
np544817	5	95.9142	0.0243	0.0401	–
np76924	7	42.75995	0.0292	0.0150	–
np526432	13	63.37597	0.0340	–	–
np516502	17	60.73835	0.0346	–	–
np58321	3	178.7586	0.0398	0.0318	–
np544814	5	95.91216	0.0404	–	–
np52913	9	27.36454	0.0434	–	–
np544815	5	95.91218	0.0490	–	–

Those with p-values > 0.05 are indicated by –.

was no complete trios for which the child was unaffected [520]; therefore no additional trios were included in the PPATu analysis.

For all the imprinting test methods used in this case study, the type I errors are all well-controlled-for based on extensive simulation studies [520]. As such, the results presented in this case study clearly point to the increase in imprinting detection power when unaffected offspring also contribute to the test statistic and when families with incomplete genotyping data are also utilized. The lessons learned from this case study echo those from the previous case study. It is demonstrated once again that any genotype information is valuable and should not be discarded lightly. Family data should be analyzed in their entirety; missing genotypes imputed conditional on observed data do not lead to bias or inflated type I error rate; rather, they can lead to an increase in statistical power.

4.7   Software

Some of the methods discussed in this chapter have been implemented into software packages, which are listed in Table 4.4. This is by no means an exhaustive list of software packages for detecting imprinting and maternal effects. Although this chapter focuses on binary traits, we also include some software packages for analyzing quantitative traits.

TABLE 4.4
Software for detecting imprinting and maternal effects using SNP and binary trait data

Software	Data Type	Trait	Ref
Mendel	General pedigrees	Binary	[55]
MC-PDT	General pedigrees	Binary	[520, 542]
LIME	CaPT/CoPT + CaMP/CoMP	Binary	[505]
MCPDTI	General pedigrees	Quantitative	[539]

4.8   Concluding remarks

In most of the methods discussed in this chapter, HWE is not assumed, although some may assume mating symmetry, a condition less stringent than HWE. Such assumptions are typically in place to reduce the number of parameters that need to be estimated. Hidden in some of the methods are rare disease assumptions, which have driven other assumptions about population frequency relationships, hidden factors that may contribute to estimation bias and inflated type I errors [502].

This chapter discusses methods for detecting imprinting and/or maternal genotype effect of a single SNP marker. To increase detection power, multiple SNPs can be analyzed jointly through consideration of haplotype effects [538]. On the other hand, haplotypes may be used to help infer missing genotypes or resolve ambiguity of parental origin to increase detection power with single SNP testing methods [49, 234].

We focus on binary traits in this chapter, but quantitative traits are also of great interest and limited methods are available [165, 398, 452]. Environmental factors and other covariates may also play important roles in causal mechanism of a trait. Log-linear and logistic model can accommodate for such effects [522]. For non-parametric approaches, covariates may be accounted for by adjusting quantitative trait values [165]. How to best incorporate covariates with simple non-parametric tests for binary traits remains an open problem. Information from genetic markers tightly linked to the test locus may be utilized to facilitate detectability of effects or to increase detection power [521, 522].

5

Modeling and Analysis of Next-Generation Sequencing Data

CONTENTS

DOI: 10.1201/9781315153728-5

For more than two decades, high-throughput genomic technologies, starting with the microarrays, have made it possible to simultaneously analyze tens of thousands of genes in an organism's genome. In the last few years, the advent of next-generation sequencing (or deep sequencing) has opened a whole new avenue in high-throughput genomics by increasing the coverage, the resolution and the statistical power of such analyses. The RNA-seq technology offers unprecedented information about the transcriptome, but harnessing this information and extracting knowledge from it using bioinformatics tools remain fraught with challenges and present a bottleneck. A great deal of statistical research is now being devoted to this new, interdisciplinary area, resulting in novel methods to extract signals from noisy data and compare signals across multiple experimental conditions. This chapter is intended to provide a brief yet informative overview of the next-generation sequencing technology, the nature of the data it produces, the quantitative issues involved and the statistical challenges/methodologies (including designs and inferential techniques).

5.1   Isolation, quality control and library preparation

For mRNA molecules, which are the primary target of RNA-seq, stability is a major issue. DNA molecules are inherently stable, but mRNA is not. In general, prokaryotic mRNA molecules are unstable and quickly degraded by enzymes called endoribonucleases and exoribonucleases after transcription. Consequently, most of them are short-lived and the average prokaryotic mRNA half-life is less than 10 minutes. This high turnover rate enables prokaryotic cells to respond to environmental changes promptly by making quick changes to the transcription (i.e. mRNA production) process. Eukaryotic mRNA molecules, however, are in general more stable with a half-life of 7–10 hours. The half-lives of mRNA are subject to regulation, based on the developmental stage of the organism or environmental factors. The regulation of eukaryotic mRNA degradation has been known to involve interactions between some sequence elements on the mRNA molecule and proteins or micro-RNAs. Most eukaryotic mRNA decay starts with de-adenylation at the $3'$ end, which is the removal of the poly-adenine tail by the enzyme deadenylase. Then the decay continues either through de-capping of mRNA at the $5'$ end and subsequent degradation by $5'-3'$ exoribonuclease or through direct $3'-5'$ decay from the tail end by a multi-protein complex called exosome.

RNA is usually isolated from freshly dissected or frozen tissue-samples using commercially available and user-friendly kits such as TRIZOL (Life Technologies, Carlsbad, California), RiboPure (Ambion, Austin, Texas) or

RNAEasy (Qiagen, Hilden, Germany). On the other hand, high-throughput RNA isolation systems are primarily based on RNA attached to magnetic particles that facilitate their washing and isolation. Isolation of RNA from formaldehyde-fixed, paraffin-embedded tissues is sometimes carried out, but not recommended due to the degradation problem discussed earlier. Degradation can be prevented by immersing the tissue-sample in RNA storage reagents such as RNAlater (Ambion) or by processing partially and storing as a phenolic emulsion (Trizol). In any case, often RNA samples are separated into size-specific classes (such as mRNA, miRNA, etc.) at this isolation stage using the miRVana column-system by Ambion or something similar. Some researchers prefer isolating RNA initially as total RNA and then separating it into various sizes by means of polyacrylamide gel electrophoresis (PAGE). However, in this latter case, the total RNA sample is typically contaminated with genomic DNA and has to be treated with DNase to digest the contaminating DNA before library preparation, followed by inactivating the excess DNase using reagents.

After isolation, it is important to check the quality of the isolated RNA in terms of purity, quantity and the extent of degradation. Some researchers use devices such as Nanodrop that are user-friendly, require nano-liter amounts of starting material, produce quick readings and have some parallel-processing capability. However, such devices can neither tell the difference between DNA and RNA (thereby failing to provide any information regarding DNA contamination) nor distinguish between degraded and intact RNA. An alternative is to use a system such as QubitFluorometer (by Life Technologies) that employs a more direct method for measuring the RNA and the DNA in the sample, resulting in more specific, accurate measurements of RNA with a wider dynamic range. These systems provide information regarding DNA contamination, but are still unable to measure RNA degradation. To overcome this stumbling block, Agilent Bioanalyzer uses a method different from the fluorescence-based techniques of Nanodrop and QubitFluorometer. Agilent Bioanalyzer is a micro-fluidics capillary electrophoresis-based system for measuring nucleic acids. In addition to being highly sensitive and requiring tiny amounts of starting material, it contains a microchip that is programmed for size control and has space for up to 12 tissue-samples at a time. Samples are mixed with polymers and a fluorescent dye, which are then loaded and measured through capillary electrophoretic movement.

The final step before sequencing is the conversion of the RNA into a cDNA library representing all of the RNA molecules in the tissue-sample. For RNA-seq library preparation, standard library protocols require 0.1–10 microgram of total RNA, although high-sensitivity protocols can work with as little as 10 picogram of RNA. This conversion of RNA to cDNA has two advantages – the chemical stability of DNA as well as the fact that DNA is more amenable to the sequencing chemistry and protocols of various sequencing platforms. Each commercial RNA-sequencing platform has its own library preparation protocol that is made available by the manufacturer. The major steps in library preparation are as follows. First, starting with the pure, intact and quality checked total RNA obtained from the tissue-sample, mRNA has to be purified out of it (the commercial platform ILLUMINA does it in two steps, first by exposing total RNA to magnetic beads and then by washing away the unwanted, non-specifically bound rRNA and other kinds of RNA from the magnetic beads). Secondly, the purified mRNA strands have to be fragmented into smaller pieces by incubation with a fragmentation reagent and the pieces have to be primed with random hexamer primers. Next, the primed mRNA fragments are reverse-transcribed into cDNA fragments using the enzyme reverse transcriptase. For each cDNA fragment thus obtained, the second (i.e. the opposite polarity) strand is then synthesized and the original mRNA fragment is removed, leaving behind only double-stranded cDNA. This double-stranded cDNA is further purified by attaching them to paramagnetic beads and washing those beads to get rid of unwanted stuff such as enzymes, buffers, free nucleotides and residual RNA. Subsequently, the double-stranded cDNA is eluted from the beads and subjected to a process called end-repairing. The 3' ends of the end-repaired double-stranded cDNA fragments are then adenylated (i.e. the nucleotide adenine or A is added to those ends). They are now ready for the next step, which is adaptor ligation. Adaptors (which are short sequences of nucleotides themselves) are ligated to both ends of the end-repaired and adenylated cDNA fragments in such a way that there is a 6-nucleotide difference between the adaptor-sequences of cDNA fragments to be used for two different library reactions. The idea behind using a different adaptor-index for each library reaction is that it allows for pooling libraries later for sequencing and still keeps a way of tracing a sequence-read back to its original library based on its adaptor-index. A new alternative technique called tagmentation is more efficient as it fragments cDNA and incorporates adaptor sequence-tags simultaneously with the help of transposase enzymes. Nextera (under the Illumina platform) is the only commercially available system employing this technique. Thereafter the adaptor-ligated, end-repaired double-stranded cDNA fragments are purified once more (using paramagnetic beads again) and the library is enriched via several (typically 12–16) cycles of polymerase chain reaction (PCR) amplification, with the nucleotide sequences of the adaptors serving as primer-binding sites. Certain types of nucleotides containing triphosphates (deoxynucleoside triphosphate or dNTP) play the role of substrates for this chain reaction. The output of this process, after undergoing another round of purification using paramagnetic beads, is the final library representing the original mRNA in the tissue-sample under study.

5.2   Validation, pooling and normalization

Before being sent to the sequencing machine, the libraries still have to pass through a few quality-control steps. First, the quality of a library has to be validated by either quantifying the yield of double-stranded cDNA in it or visualizing the abundance and size-distribution of the library via PAGE (capillary electrophoresis in the case of Agilent Bioanalyzer). Once validated, several (half a dozen to two dozens) libraries are normalized and pooled together. This is because modern sequencing platforms can handle many libraries simultaneously and normalization evens out the amounts of double-stranded cDNA in different libraries (by, for example, diluting them differently), ensuring that all libraries are equally represented. At last, the normalized libraries are pooled together and passed on to the sequencing machine.

5.3 Sequencing

5.3.1 Single-end vs. paired-end

The sequencing machine reads the base pairs from the ends of the cDNA fragments. Sequencing both ends of each fragment is now routinely done in many platforms and is known as paired-end sequencing. Some platforms carry out single-end sequencing. At a later stage, when the end-sequences are mapped to locations in the genome with the help of mapping algorithms, the sequences obtained by single-end reading often fail to be mapped uniquely. This ambiguity results in decreased efficiency and increased loss of information, as these ambiguous sequences are usually discarded. Paired-end sequencing is a way around this problem, because if one of the two ends has an ambiguous sequence, the other end can first be mapped to a single location in the genome and that will most likely determine the location of the other end uniquely.

5.3.2 Generations of sequencing technology

In the next subsection, we briefly describe the various commercially available second-generation (or next-generation) sequencing platforms. To put things in perspective, first-generation high-throughput sequencing basically referred to Sanger dideoxy sequencing that used capillary electrophoresis for resolving nucleic acid fragment lengths. A standard run of this system would typically involve 96 capillaries and produce a sequence of length 600–1000 bases per capillary, which amounts to about 60,000 – 100,000 bases of sequence. Second-generation sequencing platforms mostly use a similar sequencing method by synthesis chemistry of individual nucleotides, but do so in a massively parallel way involving numbers of sequencing reactions in a single run that are several orders of magnitude more than first-generation systems (tens or hundreds of billions of bases). This incredibly high throughput of these more advanced systems gives them unprecedented sensitivity and the ability of discovering novel transcripts, small non-coding RNAs and transcription factor binding sites. However, these systems tend to produce short reads (read-lengths in the hundreds of nucleotides). More advanced sequencing technologies (third generation) have recently begun to be introduced that use individual molecules of DNA or RNA as the starting template and produce longer reads per sequencing reaction (upwards of 1500 nucleotides), albeit carrying out fewer sequencing reactions per run.

5.3.3   Various next-generation sequencing platforms

In the last decade, a number of manufacturers have introduced next-generation sequencing platforms some of which are still in use and others have been discontinued. The following is a brief description of each major platform. See, for example [236] and [470] in this context.

5.3.3.1   Illumina

First introduced by Solexa about a decade ago and later marketed under Illumina's brand name, it is one of the most widely used systems today. It includes several versions of the Genome Analyzer as well as the more recent MiSeq and HiSeq. It uses fluorescently labeled nucleotides passing through a flow cell with many micro-fluidic channels or lanes. That is where the sequencing reaction takes place and detection signals are obtained via optical scanning. For simultaneous detection of nucleotide incorporation in millions of sequencing reactions, the four nucleotides are labeled with four different fluorescent labels. It is based on the Sanger "sequencing by synthesis" principle (mentioned earlier) but differs from the first-generation Sanger dideoxy sequencing in a number of significant ways. The top and bottom surfaces of each micro-fluidic channel are covered with oligonucleotide sequences that are complementary to the anchor sequences in the adaptors ligated to the cDNA fragments (see library preparation). When the libraries are passed through these lanes, the ligated adapters bind to these oligonucleotide sequences, thereby fixing or immobilizing the associated cDNA fragments onto the lane surfaces. Subsequently, each of these cDNA fragments serves as a template and is clonally amplified through a process called bridge amplification, whereby up to a thousand identical copies of each template are generated in close proximity, forming a cluster. These clusters (and not the individual cDNA fragments in the library) are used as the basic detection units, because otherwise the fluorescent signal intensity would be too low for the optical scanner. After cluster generation and the removal of one strand from the double-stranded cDNA fragments in the cluster, reagents are passed through the flow cell to carry out sequencing by synthesis. In each synthesis round, the addition of a single nucleotide (A, C, G or T) takes place and the corresponding fluorescent signal is imaged. The specific nucleotide is identified by this stored signal – a process known as *base calling*. A reconstruction of the sequence of additions at a particular location on the flow-cell surface (that corresponds to a particular bridge-amplified cluster) produces the sequencing read for an original double-stranded cDNA fragment in the library. This reconstruction process can be performed at one end or both ends of the cDNA fragment, producing single-end or paired-end reads. Depending on the instrument model and the sequencing length, a single run of the Genome Analyzer can take anywhere between 70–280 hours, that of MiSeq takes 5–55 hours (producing up to 30 million paired-end reads) and that of HiSeq 2500 takes 7–265 hours (producing up to 6 billion paired-end reads).

Ideally, the simultaneous addition of nucleotides to the many identical copies (or bridge-clones) of a cDNA template in a cluster should be in perfect synchronization from one step to the next (i.e. should be in phase). In practice, a non-zero percentage of templates lose sync with the majority of templates in a cluster, that is, they either fall behind by a few bases or are a few bases ahead (known respectively as phasing and prephasing). This causes more background noise in the resulting dataset and decreasing accuracy (i.e. decreasing quality of base-calling) as more and more sequencing cycles are completed. Despite this, the Illumina systems have one of the lowest error rates (smaller than 1%) among all the commercially available systems, the most common type of errors being single nucleotide substitution.

The majority of the run time of an Illumina sequencing machine goes to imaging the clusters on the flow-cell surface tiles. For imaging, the fluorescent labels on the nucleotides are illuminated with red and green lasers and scanned through four different filters, producing four images on each tile of the flow-cell surface after every cycle. The raw images captured after each cycle are analyzed (using the proprietary software that comes with the instrument) to record the location coordinates of each cluster, its signal intensity and noise level. Next, this information is used in the base-calling phase by the Real Time Analysis software that makes the base calls and computes a quality score for each call, filtering out low-quality reads. Finally, the base-call files (bcl files) generated in the previous step are converted to FASTQ files by the proprietary software CASAVA.

5.3.3.2   SOLiD

It stands for Sequencing by Oligonucleotide Ligation and Detection and is commercially made available by Applied Biosystems of Carlsbad, California. Instead of the bridge amplification used by Illumina, this system uses a special process (emulsion polymerase chain reaction or emPCR) to amplify the numbers of copies of the DNA templates. Also, the sequencing chemistry is based on ligation instead of synthesis. Another important feature that distinguishes this platform from its competitors is a unique double-interrogation strategy that results in potentially greater sequencing accuracy and makes it the instrument of choice for single nucleotide polymorphism (SNP) detection. In this platform, a library of cDNA fragments is attached to magnetic beads (one molecule per bead). The DNA on each bead is then amplified by PCR in an emulsion, with the amplified products still remaining attached to the bead. The amplified products are then covalently bound to a glass slide. A sequencing round begins with the addition of a universal primer to all cDNA fragments attached to the same magnetic bead. This renders the starting sequence of all those fragments known and identical. Using several primers that hybridize to that universal primer, di-base probes with fluorescent labels are competitively ligated to the primer. That is the start of a circle. If the bases at the first and second positions of the di-base probe (starting from the 3′ end) are complementary to the DNA that is being sequenced, then the ligation reaction takes place and the fluorescent label generates an optical signal. The remaining unbound probes are washed out. Subsequently, the primer and the probes are all reset for the next round. Now the 5′ end of the new universal primer will match to the base just preceding the earlier base. In this way, primers are reset by a single nucleotide five times so that at the end of the cycle, at least four nucleotides will have been interrogated twice (due to the di-base or dinucleotide probes) and the fifth nucleotide at least once. Ligation of subsequent di-base probes ensures a second interrogation of even that fifth nucleotide. The entire sequencing step consists of five rounds and each round consists of five cycles. This double interrogation method causes a single nucleotide polymorphism (SNP) to produce a two-color change, whereas a measurement error results in a single color change. This is a reliable way of distinguishing between a true SNP and an error in sequencing. However, there is a price to pay for this. Decoding the raw data from SOLiD has an added layer of difficulty as it encodes by two bases instead of one, which necessitates an alternative representation of the nucleotide sequence (called the "color space"). Instead of each single nucleotide being represented by one specific color, as was the case with Sanger-type sequencing chemistry, here each color stands for four potential two-base combinations. This makes any attempt to directly translate color-reads to base reads more error-prone. The best solution to this problem is converting the base reference sequence itself into the color space and then converting it back to the nucleotide space once the sequence has been aligned to a reference genome encoded in color space. All this is, however, easier said than done and results in a substantially higher error rate for SOLiD than Sanger-type sequencing platforms, despite the potential for increased accuracy coming from the double interrogation strategy. The latest instruments (e.g. the 5500 W) have gotten rid of magnetic bead amplification and used flow-chips in the place of amplifying templates. A machine using two flow-chips can produce up to 320 gigabytes of data in a single run.

5.3.3.3   Ion Torrent semiconductor sequencing

It is the first next-generation sequencing platform that does not involve chemically modified nucleotides, fluorescence labeling and image scanning. This increases the speed, decreases the cost and leaves a smaller equipment footprint on the final output. This platform uses the adaptor-ligated library preparation step (involving clonal amplification by emulsion PCR) followed by the synthesis-based sequencing chemistry of other platforms. But instead of detecting optical signals or photons from fluorescently labeled nucleotides, it detects changes in the pH (a measure of proton or hydrogen ion concentration) in a well when a nucleotide is added and protons are released. When a nucleotide is newly incorporated into a DNA strand, the chemical reaction catalyzed by DNA polymerase releases a pyrophosphate group and a proton, with the latter causing a pH change in the vicinity of the reaction. However, the extent of change is not specific to any particular nucleotide, so to determine the DNA sequence, each of the four substrate nucleotides is added to the reaction in sequential order. If a pH change is detected after the introduction of a nucleotide, it strongly indicates that the template DNA strand contains its complementary base at the latest position. In order to detect these very small pH changes, this platform resorts to semiconductor technology.

The sequencing takes place on a set of ion semiconductor chips, each of which contains an array of micro-wells. Each micro-well has in it one single-stranded template DNA and one molecule of DNA polymerase. Underneath each well is an ion sensitive layer. The micro-wells are flooded with a particular type of deoxyribonucleotide triphosphate or dNTP (N = A or T or C or G) sequentially, one after another in order. In each round, one of these dNTPs gets affixed to the template strand, depending on whose complementary base there is at the latest position on the template. The associated chemical reaction releases a proton that triggers the ion-sensitive material underneath. The electric pulses from those sensors are directly transmitted to a computer which immediately translates them into a DNA sequence. The final steps of signal processing and DNA sequence assembly are accomplished via embedded software.

For all the advantages of the above-mentioned technique, there are drawbacks too. The overall error rate of this platform is higher than that of the Illumina platform. The primary reason behind that is indels caused by homo-polymers. When the DNA template contains a homo-polymeric region (i.e. a stretch of identical nucleotides), the pH change is stronger and proportional to the number of repeats of that nucleotide. Consequently, as the number of repeats (n) increases, there is a gradual decrease in the signal-strength ratio between n repeats and n-1 (or n+1) repeats. This limits the ability of the system to detect the total number of repeats correctly. In the current version of the system, the error rate for detecting a 5-base homo-polymer (i.e. when n = 5) is about 3.5%. Another significant drawback is the short size of the reads (35–400 base pairs per run)

5.3.3.4   Pacific biosciences single molecule real-time sequencing

The Pacific Biosciences single molecule real-time (SMRT) sequencing platform is considered as a third-generation technology, due to its higher sensitivity than the second-generation platforms described above and the resulting capability to sequence single DNA molecules. Not only does it render any form of amplification unnecessary, but also it generates much longer reads (median length 8–10 kilobases, longest around 30 kilobases) than most other platforms. Like the previously mentioned platforms, it is also based on sequencing by synthesis, but a crucial difference is that it uses nucleotides carrying fluorescent labels linked to their end phosphate group but no terminator group (as would have been the case with Illumina). When a new nucleotide gets added to an elongating DNA strand, with the cleavage of the end pyrophosphate group, the fluorescent label is released at the same time. This makes real-time signal detection possible. The sequence-detecting signal is continuously recorded as a movie at a speed of 75 frames per second instead of using separate scanner images.

SMRT uses what are called *zero-mode waveguides* (ZMW). These are space-restricted chambers that allow light energy and reagents to be guided into extremely small volumes (of the order of zeptoliters or 10^{-21} liters). To be more precise, a ZMW is a hole only tens of nanometers in diameter that is microfabricated on an ultra-thin metal film (100 nanometers in thickness). The metal film is deposited onto a glass substrate. This provides a single chamber that contains a single molecule of DNA polymerase and a single DNA molecule that is to be sequenced in real time. The diameter of a ZMW being smaller than the wavelength of light, only the bottom 30 nanometers of the ZMW is illuminated by the light coming through the glass substrate. The resulting ultra-miniscule detection volume leads to a substantial reduction in background noise and makes it possible to detect nucleotide incorporation into a single DNA molecule. Using specific flurescent nucleotide triphosphates, the addition of an A, C, G or T to an elongating nucleotide chain can be detected while it is being synthesized. Because of this speed advantage, the runtime can be very short (i.e. of the order of 1–2 hours). The current version of the instrument (PacBio RS II) with its current movie-length of 3 hours can produce up to 375 megabases of sequence in a single run.

Since it does direct sequencing of single DNA molecules, it has another advantage over some of the earlier platforms. It has the capability of detecting nucleic acid modifications (such as DNA methylation, that is, 5-methyl cytosine formation). While the presence of such modifications causes consistent delays in the kinetics of the DNA polymerase used in sequencing, this has been exploited by this platform for the detection of DNA modifications (currently claiming to detect up to 25 base modifications in a single run). Two notable disadvantages of this platform are its high error rate (10–15%, with the most common error-type being indels) and high run cost. Paired-end sequencing is not possible on this platform. Also, the amount of DNA sample required at the beginning is on the high side (1000 nanograms). The library preparation steps for SMRT are similar to some other platforms and involves shotgun fragmentation of DNA into appropriate sizes, fragment end repair and adapter ligation, annealing to sequencing primers and binding of DNA polymerases with it.

5.3.3.5 Nanopore technologies

Nanopore sequencing is a third-generation, single-molecule technology that uses a single enzyme to separate a DNA strand and guides it through a protein pore embedded in a membrane. The simultaneous passage of ions through the pore generates an electric current which is measured. The specific nucleotides passing through the pore (that is, A, G, T or C) impede this current flow differently. This makes the current sensitive to the specific nucleotides and produces a signal that is detected in the pore. This, although conceptually simple, is technologically quite challenging as it involves measuring very small changes in electric current at the single-molecule scale. However, once this technical barrier is overcome and this platform is successfully commercialized, one of the greatest advantages of it will be a small device size (like a cellular phone or even a USB stick). At least that is the claim by the companies that are currently developing it. Overall, this approach is quite promising and will likely have a greater impact on genomics in the future.

5.3.3.6 Choosing a platform

The key factors that go into the choice of a platform are accuracy, the number and lengths of reads, the preference for paired-end over single-end, the amount of sample material needed, cost and run time. If the goal is to detect single-nucleotide polymorphisms (SNP), then choosing a platform with a very low error rate (and hence very high accuracy, such as SOLiD) is essential because of the very low frequency of SNP occurrence and the need for distinguishing the true SNPs from sequencing errors. However, it is important to remember in this context that one can compensate for low accuracy by generating more reads, that is, by repeatedly sequencing the same piece of RNA. On the other hand, if one is trying to quantify gene expression to detect differential expression, identify known protein-coding genes or discover new genes, the need for accuracy is much less stringent. In that case, most of the platforms described above (Illumina, SOLiD, Ion Torrent) can be used. One can see, for example, [236] for more details on this issue.

Regarding lengths of reads, longer reads are needed to reduce the percentage of reads mapping to multiple locations of the reference genome. A relatively long read (e.g. 50 nucleotides) will bring down that percentage below 0.01%, which is good enough for detecting differential expression. Reads longer than that will be essential for purposes such as annotating novel genes in a species for which neither a reference genome nor any other sequence data-source is available. For this, the newer generation models produced by Pacific Biosciences (PacBio RS II) is suitable. The question of how many reads are needed should be put in perspective by considering the overall genome size of the organism under study and the proportion of its genome that protein-coding genes occupy. For humans, those two numbers are about 3 billion nucleotides and 3.3%. This means that 1 million paired-end reads of length 50 nucleotides each (or single-end reads of length 100 nucleotides each), which amount to 100 million nucleotides of sequence data, will cover all the protein-coding genes once. So, if a particular platform has an output of 20 million reads, it will provide 20 times the coverage which means a high to decent amount of coverage for a vast majority of genes and possible omission of a few rarely expressed or low-expression genes. These days the typical output from the common technology platforms is about 30 million, which is good enough to capture a vast majority of (but probably not all of) the genes expressed in a sample. When better coverage is crucial for some reason, one should choose the platforms that yield a large number of reads more easily. Two things to keep in mind are that no platform may ever be able to provide enough coverage to obtain every single transcript from each locus, and there is no consensus on the question of how many reads are necessary to confirm the existence of a transcript.

The issue of single-end vs. paired-end reads has been discussed earlier, so next we comment on the amount of sample material needed. Sequencing platforms that use amplified, double-stranded cDNA basically have no lower limit on the amount of material needed because of the amplification. Also, platforms are available that can sequence the total mount of RNA from a single cell. So any discussion on the amount of sample material may seem irrelevant, except for one reason – the fact that providing more than the minimum required amount of tissue sample to a platform increases the representation of the RNA species in that sample. So in general it is not recommended to supply only the bare minimum needed.

The cost aspect, similarly, has lost some of its relevance over the years as the cost of sequencing has gone down significantly during the past decade. Still, it remains an important issue as not all research projects are equally well-funded and the standards for data quality have also gone up over the years. Purchasing a laboratory sequencer for oneś personal use is now more feasible than ever (e.g. the Personal Genome Machine by Ion Torrent, MiSeq by Illumina, etc.), but outsourcing the sequencing job by sending tissue samples or RNA-seq libraries to commercial NGS facilities still remains an effective way of cost mitigation. Regarding the downward movement of sequencing cost over time, it is widely believed that the bottom has not been reached yet. This belief is reinforced by the increasing competition among commercial and non-profit core NGS facilities.

In a fast-moving field such as genomics, it is neither desirable nor ideal that a sequencing run gets delayed. The unfortunate reality, however, is that a number of platforms experience delays. It is even more frustrating to know that the delay is not caused by the sequencing machine running slowly, but by the insufficiency of libraries to fill a flow cell for a single run and the resulting wait for more libraries to be submitted. If this can somehow be avoided by careful planning, such delays will not occur. Also, further downstream, the huge amount of data generated from an NGS platform often take a really long time to be preprocessed and analyzed. Compared to it, any delay at the sequencing stage may seem insignificant.

5.4 Factors affecting NGS data accuracy

In addition to any errors in base calling (i.e. the final identification of an A, T, C or G done by an NGS platform), there may be biases creeping into the earlier steps of the whole process. Biases can affect both the library preparation step and the sequencing step. It may not be possible to completely eliminate these biases and other factors contributing to inaccurate signals, but we can still be mindful of them while choosing the experimental design, the statistical model, the analysis method and algorithm so as to minimize their adverse effects on the final outcome. See, for example, [470] in this context.

5.4.1 At the library preparation stage

At the very beginning of the library preparation stage, DNA fragmentation by sonication and nebulization tends to break the DNA strands more often than expected after a cytosine (C), compared to the other three. This violates the complete randomness assumption that we make about the fragmentation process. Subsequently, the size selection process of the DNA fragments also introduces bias in its own way. For example, the use of a high gel-melting temperature in the gel extraction method is biased towards recovering DNA fragments with higher GC content.

Next, the ligation step introduces some bias that affects the sequencing of both long-stranded RNA species and short-stranded ones, though in different ways. Size-selected DNA fragments (double stranded) are first subjected to repair and adenylation (i.e. creation of A tails) at both ends, followed by ligation of adapters carrying $5'$ T overhangs. This adapter ligation process has been found to be biased against DNA fragments starting with a thymine. This bias affects the cDNA molecules obtained via reverse transcription from messenger RNA or long non-coding RNA, but it does not affect short-stranded RNA molecules (e.g. siRNA) because adapter ligation for them precedes reverse transcription to cDNA. However, the small RNA adapter ligation process brings in another type of sequence-specific bias for some small RNA species, depending on their secondary and tertiary structures (which, in turn, are affected by the temperature and chemical composition of the ligation reaction mixture.

Afterwards, the adapter-ligated DNA fragments are amplified using polymerase chain reaction (PCR) that involves DNA polymerases. This process in known to be biased against highly GC-rich or AT-rich DNA fragments, resulting in an under-representation of the genomic regions that are of this type. One way this problem can be partly dealt with is by optimizing PCR conditions for GC-rich or AT-rich regions, but the only way to completely get rid of it is to use a library preparation process that does not involve PCR.

5.4.2   At the sequencing stage

As was described earlier, many NGS platforms are based on the sequencing by synthesis principle, which uses DNA polymerases. Consequently, the coverage bias against highly GC-rich or AT-rich genomic regions that was mentioned above is present at the sequencing stage too, and it is difficult to completely get rid of. In addition, other activities involved in the sequencing process often introduce their own biases and artifacts. For example, misalignment during scanning or unintended light reflections can lead to inaccuracies in imaging. The presence of lint, dust particles, crystals and air bubbles in the buffers could generate artificial signals. Fortunately, these problems can be avoided to a large extent by being sufficiently careful.

At the signal processing and base-calling steps, some platforms such as the Illumina Genome Analyzer suffer from problems that their proprietary software packages are effective in dealing with, but other commercial or open-source software often used by researchers have different algorithms for these tasks than the proprietary software and they produce varying results (due to making different assumptions on the signal distribution and other reasons). For example, some of them make the assumption that the signals from the four detection channels for A, C, G and T are independent. Some assume that signals from different cycles are independent. However, in reality, the A and C signal channels have some dependence due to the overlap between the emission spectra of their fluorescent labels. So do the G and T signal channels. Also, due to phasing and prephasing, signals from a cycle are dependent on those from cycles preceding and succeeding it.

5.5   Applications of RNA-Seq

The primary objectives of RNA-seq are the determination of the nucleotide sequence (i.e. the particular order of the A, C, G and U residues), learning of the gene structure (i.e. locations of promoters and enhancers, 5′ and 3′ untranslated regions, exon-intron junctions, poly-adenylation sites, etc.) and quantification of the abundance of RNA molecules (i.e. the absolute and normalized numerical amount of each specific sequence) in a tissue sample. Each of these, in turn, opens up other avenues of important applications. For example, knowledge of the sequence enables the identification of known protein-coding genes as well as the discovery of new genes or long non-coding RNA species. It also provides valuable information about the secondary structure (e.g. hairpin bends, bulges, etc) and tertiary structure (i.e. the three-dimensional shape of the molecule) which are crucial in determining the class and function of it (e.g. whether it is a transfer RNA or micro-RNA, etc.). Quantification of abundance enables us to detect differential gene expression between two tissue samples or two experimental conditions or two organisms. Below we elaborate a little more on some of these.

The sequence reads from an RNA-seq platform are usually mapped at first to known protein-coding genes archived in existing databases. This not only helps us confirm known intron-exon boundaries but also discover completely new exons. In addition, this enables precise identification of a gene's important structural features such as the 3′ untranslated region (UTR), the 5′ transcription start site (TSS) or polyadenylation sites. Based on RNA-seq read counts, it is possible to compare the usage of one axon to that of another (i.e. which one was busier in the transcription process and how much busier). Due to the massively parallel nature of the RNA-seq platforms, it is possible to do all of these in a genomewide manner.

Prior to RNA-seq, annotations of protein-coding genes used to depend on computational predictions based on genomic sequences. A high through-put technology such as RNA-seq not only allows us to verify many of those previous predictions but also to discover novel protein-coding genes with no previous prediction. This is crucial when there is no genome sequence database available for an organism and we are trying to build its transcriptome solely on the basis of its RNA-seq reads.

It has already been mentioned that RNA-seq makes the detection of differential gene expression between two tissue-types (or experimental conditions or organisms) possible. Another important purpose that RNA-seq serves is the study of quantitative traits. Just as genome-wide association studies (GWAS) link single-nucleotide polymorphisms (SNP) to various physical or physiological traits that are quantifiable, eQTL is the study of association between gene expression changes and SNPs. As is now well-known, such association can be a direct or indirect causal relation (direct or local if, for example, the SNP is located in the enhancer region of a gene, thereby changing the gene expression; indirect or distal if, for example, the SNP is located far away from the gene but structurally changes a transcription factor that was needed for the gene's expression and renders it non-functional). As RNA-seq quantifies gene expression levels, it can be used to find these associations. A newer branch of association studies known as sQTL tries to find the association between SNPs and the location and usage of gene-splicing sites.

With the advancement of the RNA-seq technology and its ability to yield more and more reads that are longer, it has now become possible to identify rare transcripts that are potentially important. One example is the transcripts produced by "fusion genes." Such genes come into existence when two previously separate genes contribute parts of their own structures (e.g. protein-coding region, 3′ poly-A region or 5′ UTR) to a fused structure. This is quite common in cancer tissues. However, recently this phenomenon has been detected in normal tissue via RNA-seq studies, implying that it is not necessarily indicative of a disease condition.

The advent of RNA-seq has greatly enhanced our capability of finding long and short non-coding RNAs that are abbreviated as lncRNA and miRNA respectively. Their existence was known before the RNA-seq era, but this technology has opened our eyes to their plentifulness and ubiquity. An lncRNA is a transcript that is longer than 200 nucleotides, produced by a region that is not a part of (or overlapping with) a protein-coding exon, and does not belong to other known non-coding RNA species such as transfer RNAs and ribosomal RNAs. They are now known to play some role in epigenomics by controlling transcription as enhancers and by binding with histone proteins to change their functions. Short non-coding RNAs can actually be of several different types, the most well-known of which is micro-RNAs or miRNA. The approach developed for sequencing miRNAs (called miRNA-seq) can be implemented on most common sequencing platforms, once the miRNAs are converted to double-stranded cDNAs. However, the steps needed to convert the starting material (either size-selected and fragmented small RNAs, or the total RNA which is all RNA species combined) to double-stranded cDNAs differ from the pre-sequencing protocols of the RNA-seq platforms described earlier in this chapter. The miRNA-seq method can also be used to sequence other species of short non-coding RNA species. Micro-RNAs are known to play a role in regulating gene expression by binding with and degrading messenger RNAs and preventing their translation.

Recently, RNA-seq has been used for "exome sequencing" or "exome capture" which is not RNA-seq in the strictest sense of the word. The purpose is to identify variations in the protein-coding gene sequences from genomic DNA samples. The idea is to sequence fragmented genomic DNA enriched for exons by means of hybridization to exonic sequences. Since the primary motivation behind this has been studying diseases in humans, and SNPs (as well as other variations) need to be identified from large human cohorts, sequencing only the exonic sequences of an individual is a convenient way to keep the cost down. Another advantage is that, focusing on just the exons automatically means focusing mostly on protein-coding genes, so the variations (SNPs, etc.) found in this way are directly relevant to protein structure modification.

Interacting with a variety of proteins (e.g. transcription factors) is a crucial aspect of the day-to-day normal functioning of a genome. Many of these interactions take place in a region-specific manner. Figuring out which specific regions of the genome the interacting proteins bind to (known as *transcription factor binding sites* (TFBS)) starts with the capturing of the protein-bound regions via a mechanism called *Chromatin ImmunoPrecipitation* (ChIP) and then sequenced using an NGS platform. This is one of the most important uses of the NGS technology.

We conclude this section with a few more applications of next-generation sequencing that are proving to be increasingly important. As was mentioned earlier in the context of Pacific Biosciences Single Molecule Real-Time sequencing, some NGS platforms are useful in epigenomics because of their ability to detect DNA methylation (i.e. conversion of the cytosine residue to methylcytosine). Lately, RNA-seq has started replacing Sanger sequencing (once the gold standard) as the preferred technology platform for *de novo* genome assembly of an organism – even for those with large and complex genomes. And last but not least, recently RNA-seq has started being used in metagenomics (the study of all genomes present in a community of organisms) as it is able to quickly sequence everything that there is in a metagenome and give us an overall profile of the composition and functional state of a microbial community.

5.6   RNA-Seq data preprocessing and analysis

Generally speaking, RNA-seq data analysis involves three major tasks. The first one is base calling, which is based on a deconvolution of the optical or physicochemical signals produced by the sequencing mechanism. Almost all sequencing platforms store these base-call results in the FASTQ file format, each FASTQ file containing a huge number of reads (i.e. the A-T-G-C orderings of DNA fragments sampled from a library). Next comes the data quality check or quality control step where the reads in a FASTQ file are checked for their quality and preprocessed before being mapped to a reference genome. Depending on the results of a number of quality metrics that are examined, the FASTQ files are preprocessed to weed out low-quality reads, eliminate portions of reads containing low-quality base calls and get rid of any unwanted items (such as PCR primers and adapter sequences). This is followed by the mapping or alignment of the preprocessed reads to a reference genome in an attempt to find the genome locations where the reads most likely came from. Once this is done, application-specific downstream analysis can begin.

5.6.1   Base calling

Base calling is carried out using algorithms in the proprietary software packages that are platform-specific and come with the instruments (e.g. Illumina's *Bustard* algorithm).The output of such an algorithm is a base call (i.e. identification of a nucleotide) for each sequencing cycle along with a confidence score for that call. These are stored in a file format such as FASTA, CFASTA, QUAL or FASTQ, with the last one being the most widely used. Conversion tools are available for the other file formats to be converted to FASTQ which is a text-based format. The confidence score or quality score (Q-score) that comes with a base call is obtained as $Q = (-10)log_{10}P_E$ where P_E is the probability of an erroneous base call. Typically a Q-score of 20 (which corresponds to $P_E = 0.01$) is the minimum requirement for a base call to be deemed reliable, although the value of Q can go up to 60. The reporting of this Q-score in a FASTQ file, however, is not done by numbers. They are usually encoded with ASCII characters, the most commonly used encoding scheme being the one introduced by Sanger sequencing.

Here are some more details about Illumina's base calling mechanism and the quality scores that come with it from [293]. During each cycle of Illumina's "sequencing by synthesis" process, images produced by a charged coupled device (CCD) record fluorescence intensities in each of the four nucleotide channels. These are stored in an intensity matrix whose columns correspond to the cycles and rows correspond to the channels. Bustard converts these observed intensities into concentrations by multiplying them with the inverse of an estimated "crosstalk" matrix to adjust for the correlation among the four channels. As more and more cycles go on, loss of fragment copies in the library results in reduced intensities which, in turn, leads to reduced concentrations. As a way arond this, Bustard rescales the concentrations in each cycle by a factor proportional to the reciprocal of the average concentration for the cycle. So all cycles end up having the same average concentration. Next, Bustard uses a Markov chain model and its transition probabilities to estimate the probability of one base being correctly synthesized during a cycle, that of no new base being synthesized during a cycle (known as *lagging* or *phasing*), and that of two bases being synthesized during a cycle (known as *leading* or *pre-phasing*). The algorithm adjusts the rescaled concentrations using these estimated probabilities. Subsequently, these adjusted concentrations are used to make base-calls and assign quality scores for the called bases. By assigning such quality scores, it is possible to assess the performance of a base-calling algorithm and compare one algorithm with another (some other algorithms are briefly mentioned below). The conventional quality-scoring algorithm that most people use is *Phred*. It involves a four-phase procedure to estimate a number of parameters regarding peak shape and peak resolution. Then it takes these parameter estimates and searches for a quality score that corresponds to those estimates in some known table of quality scores. The *Phred* quality scores have been found to be quite accurate for several different types of sequencing platforms

Among the recent alternatives to Bustard is *Alta-Cyclic* [100] that takes a support vector machine (SVM) approach to build a classifier which classifies each newly synthesized nucleotide as one of the four possible bases. Like every supervised machine-learning algorithm, this SVM must be trained first, which is done using a known reference genome in one of the flow cell lines. Another alternative is *Rolexa* [370] that is used in the Solexa platform. Its probabilistic algorithm corrects for positional bias, phasing, rephasing and crosstalk. Next, it estimates the conditional probability of each base given a quadruple of intensities (the quadruple being modeled as a mixture of four multivariate normal densities). The one with the highest conditional probability is called. This algorithm also involves a procedure for the identification and removal of ambiguous base-calls based on entropy calculations. Other alternatives are *BayesCall* [220] that uses a full Bayesian model for the four bases, concentrations of active templates and the observed fluorescence intensities with cycle-dependent parameters and estimation using variations of the EM algorithm (Monte Carlo Expectation Maximization and Expectation

Conditional Maximization), *BING* [237] that uses pixel-based base-calling as opposed to cluster-based base-calling, [33] who probabilistically model the log intensities in such a way that it includes a read effect and a base-cycle effect as well as latent indicator variables for each possible base in each read and cycle, *Ibis* which uses a multi-class SVM classifier and makes phasing in a given cycle dependent on the intensity values from the two adjacent cycles (before it and after it), *AYB* or All Your Base [285] that fits a cluster-specific multivariate regression model for the intensity matrix via the iteratively reweighted least squares (IRLS) algorithm, *freeIbis* [357] which is claimed to be a more efficient version of *Ibis* with calibrated quality scores for the Illumina platform.

Now, back to the discussion on Q-scores. Because each sequencing platform has its own calibration of the Q-score, if the scores obtained from different platforms are to be compared or combined, it is necessary to have an idea about each of their calibration methods and re-calibrate all of them so they become comparable. The platforms use either a control lane or a precomputed calibration table to come up with their own P_E. To make the resulting Q-scores comparable, they are re-calibrated in the following way. A subcollection of the reads is used that map to regions of the reference genome containing no SNPs, so any mismatch between the reference sequence and the reads can be attributed to a sequencing error. Depending on the rate of such mismatches at each base position of the reads, a new calibration table is constructed, which is then used for recalibration.

5.6.2 Quality control and preprocessing of reads

The first step is a general quality control analysis which examines the overall quality of the millions of reads. The overall quality assessment includes scanning the reads for low-confidence bases, biased nucleotide composition, duplicates, adapters, etc. The output of this step are the number of reads and some quality metrics, which are also the input for the second step (i.e. preprocessing). The primary goal of preprocessing is not only the removal of low-quality bases but also that of various artifacts from the individual reads (e.g. adapters, poly-adenine tails, microbiome, etc.). It may also involve trimming and filtering. Once all this is done, the data will be ready for the subsequent step of read alignment to a reference genome.

5.6.2.1 Quality control

The three main data quality metrics are Q-scores, the read length distribution and the percentage of each base across base positions. Examining Q-scores can be done across all base positions of all reads from the first to the last sequenced base. Or it can be done by plotting the average Q-score of each read and looking at their distribution pattern. Either way, the goal is to have a vast majority of reads with an average Q-score > 30 and minimize the percentage of reads with an average score < 20. For platforms based on sequencing by synthesis, due to reasons mentioned earlier, usually the base positions that are in the early phases of a sequencing run tend to have higher Q-scores than those coming later in the process. It is necessary to have a median Q-score of at least 20 even for these late-phase base positions. If the median Q-score drops significantly below that threshold, the affected base positions must be scrutinized and low-quality bases have to be trimmed from the reads concerned. Also, an increasing number of the ambiguous call "N" (which happens when none of the four bases can be called with enough confidence) is indicative of diminishing base-call quality.

Regarding the read length distribution, it is less of a worry for platforms that produce reads with a high degree of homogeneity in their lengths. For some platforms such as the Pacific Biosciences, however, there is considerable heterogeneity in the read lengths, and so the read length distribution is something to watch carefully. In addition to indicating the total volume of useful data generated in a sequencing run, it also shows the relative amounts of shorter and longer reads. If the cumulative distribution function (CDF) of the read lengths from a platform is stochastically larger than that from another platform (both having comparable base-quality and equal data volume output), then the first platform is preferable because longer reads have an advantage in the upcoming steps of this process. A CDF being stochastically larger than another means that the graph of it is always below that of the competing CDF.

If DNA fragmentation is truly random during the library preparation process, the probability of observing each of the four bases at each base position should be the same. As a result, if one plots the percentage of each base across all base positions, the plots for adenine, guanine, thymine and cytosine should be approximately parallel to each other. Also, the overall percentage visible in each plot should be proportional to the overall frequency of each base in the initial library. Any significant departure from a parallel configuration points to anomalies in the library preparation process, such as nonrandom DNA fragmentation or undesirable over-representation of some RNA species in the library.

The quality control steps are usually carried out using software packages such as *NGS QC Toolkit* [335], *FASTX-Toolkit* or *FastQC*.

5.6.2.2   Preprocessing

If the quality control step detects low-quality reads and/or unwanted artifacts, the preprocessing step gets rid of them. Low-quality base calls at the 3′ end of the sequence are trimmed away, along with artifacts such as adapters and poly-A tails as well as duplicated portions. Some platforms routinely do sequence filtering as a part of their FASTQ file generation process, while others don't. In either case, if the quality scores are found to be below the acceptable threshold 20 in the quality control step, further filtering and/or trimming will be required using software tools such as *Trimmomatic* [30], *ngsShoRT* [52] or

Sickle which is a sliding-window, adaptive, quality-based trimming tool for FASTQ files. In addition to these specialized tools, preprocessing can also be performed using options available in the quality control software packages mentioned earlier.

5.6.3 Read alignment

This step is aimed at finding the point of origin for each individual read by mapping them to a reference genome. If a reference genome is not yet available, reads can be mapped to a transcriptome (created from the reads themselves via a *de novo* assembly method). Mapping a read to a reference genome involves a sequence alignment. Mapping is computationally intensive due to several reasons – the huge number of reads, the large size of a reference genome, the need to map spliced reads non-contiguously, etc. Because of this, often the genome sequence is transformed and compressed into an index to expedite the mapping process. For example, the commonly used Burrows-Wheeler transform – an idea borrowed from string matching theory. Nevertheless, simultaneous mapping of millions of reads (often short in length) to a big reference genome remains a challenging task, unlike mapping a single or a few sequence(s) of moderate length(s) using BLAST or other similar tools. A major source of difficulty is figuring out whether the deviation of a read from the reference genome is due to sequencing errors or it is a true sequence deviation (i.e. meaning that the genome sequence of the tissue sample under study truly differs from the reference genome as a result of indel mutations and/or polymorphisms). Another major challenge is the identification of novel splice junctions in RNA-seq data.

Several alignment algorithms have been put forward in the last one-and-a-half decades. Older algorithms such as BLAST used hash tables and seed-and-extend methods. Examples of newer algorithms based on the optimization of and improvement upon BLAST's original approach are Efficient Larg-scale Alignment of Nucleotide Databases (ELAND), Novoalign and Short Oligonucleotide Alignment Program (SOAP). As explained in [257], SOAP is based on reference genome indexing. So is Novoalign. However, ELAND and another one called Mapping and Assembly with Qualities or MAQ [256] are based on indexing the reads from the sample. It uses an ELAND-like hashing method for alignment, followed by a Bayesian statistical model to produce Phred-scaled quality scores for the resulting alignments (which is actually $10log_{10}(P_E)$ where P_E is the posterior probability of incorrect assignment). MAQ is quite efficient in combining the mapping quality information with the Q-scores and utilizing mate-pair information for paired-end read alignment in diploid samples. To understand the basic difference between the approach used by algorithms such as SOAP and Novoalign and that used by algorithms such as ELAND and MAQ, one needs to understand the original seed-and-extend approach of BLAST. In it, if a match happens between short nucleotide sequences (or "words") in the query sequence and the reference genome, that matched region is used as a "seed" to extend the alignment to adjacent regions. BLAST only uses seeds that are consecutive sequences of nucleotides and exactly match portions of the reference genome. Since this seed selection criterion is a bit too stringent, resulting in lower alignment sensitivity (as it is not sensitive to sequence variations), some newer algorithms started using non-consecutive or spaced seeds, thereby enhancing the probability of finding a match. All four algorithms mentioned above do this, but some of them implement it differently than the others. SOAP and Novoalign start out by chopping the reference genome into small, equal-sized pieces and saving them in a giant hash table which is subsequently used to search for near-matches with similarly chopped pieces of the reads from the tissue sample. This is called reference genome indexing. ELAND and MAQ, on the other hand, construct the hash table from the reads and extract short subsequences from the reference genome to look for near-matches with the ones in the hash table.

In order to achieve more efficient indexing and faster searching, the algorithms that use the Burrows-Wheeler transform with an efficient backward search are *Bowtie* [242] and *BWA* [255]. Although *Bowtie* is a sub-optimal greedy algorithm, it does produce high-quality alignments with double indexing to prevent excessive backtracking. It also has built-in options for the users to choose their own balance between efficiency and accuracy. *BWA* allows for inexact matching and gapped alignment.

The software called *Tophat* [440] was designed for the discovery of novel splice junctions ab initio. This two-step algorithm starts by mapping all reads to the reference genome using Bowtie. The reads that do not map to the genome are designated "initially unmapped" or IUM. The next step is to assemble the mapped reads using the MAQ algorithm mentioned above and construct an initial consensus. Once this is done, the sequences that flank potential donor/acceptor splice sites within neighboring regions are joined to construct potential splice junctions. Finally, the IUM reads are indexed and aligned to these splice junction sequences. A more recent version of this algorithm that can handle variable-length reads and variable-length indels with respect to the reference genome is called *TopHat2*. Another relatively new and fast-running spliced alignment algorithm is *STAR* or Spliced Transcripts Alignment to a Reference. While it runs faster than TopHat, its memory-space requirement is considerably larger than that of TopHat. In addition to speed, the other advantages of STAR include its ability to perform an unbiased search for splice junctions as it does not need any *a priori* information regarding their locations, sequence signals or introl lengths. It can align a read having any number of splice junctions, mismatches and indels as well as those having poor-quality ends. It uses the so-called "maximum mappable length" approach which chops a read into 50-base long pieces and finds the best portion that can be mapped for each piece. Next, it maps the remaining portion of a piece (which may end up being mapped far away in the case of a splice junction). This sequential "maximum mappable seed" search looks for exact matches and uses the genome in the form of uncompressed suffix arrays. Subsequently, it stitches the seeds together within a given genomic window, allowing for indels, mismatches and splice junctions. While doing so, it deals with seeds from paired reads concurrently to increase sensitivity.

The combined effect of all these new developments is that the run time for aligning tens of millions of reads to a big and complex genome is now minutes instead of hours. Clearly, the two central issues in sequence alignment are sensitivity and run time. With the advent of newer sequencing platforms capable of producing longer reads, another important issue is efficient alignment of longer reads. One more thing to keep in mind is the reference bias that creeps in if we stick to the same reference genome for all our alignment jobs. Some algorithms such as SOAP2 (a newer version of SOAP) and Bowtie are preferable if lowering run-time is the primary concern. On the other hand, a greater emphasis on sensitivity will dictate that we choose algorithms such as *SHRiMP2* [67] or *Stampy* [277]. The algorithm *GenomeMapper* [380] is capable of reducing reference bias by using multiple reference genomes simultaneously. Algorithms such as BWA-MEM, which is a modified version of BWA [254], LAST [228] and Basic Local Alignment with Successive Refinement or BLASR [45] are designed for efficiently aligning longer reads.

The output from this step is an alignment file listing the mapped reads and their mapping positions in the reference sequence. These are usually stored in a SAM (Sequence Alignment/ Map) or BAM (Binary Alignment/Map) file. The former is in a tab-delimited text format and the latter is a compressed binary version of it. A SAM/BAM file has a header section followed by an alignment section containing the fields "Query sequence read name (or Query template name)," "Bitwise flag," "Reference sequence name," "Leftmost mapping position (on the reference sequence)," "Mapping quality," "CIGAR string," "Reference name of the next read (for paired-end reads)," "Position of the next read (for paired-end reads)," "Observed template length," "Segment sequence" and "ASCII of Phred-scaled base quality."

The contents of the SAM/BAM files are carefully checked to see the percentage of aligned (in particular, uniquely aligned) reads, detecting and filtering out multi-reads (i.e. reads that map to multiple locations in the reference genome) and doing the same to duplicate reads. Currently, even the top-of-the-line alignment algorithms are able to match only 70–80% of the reads to unique positions in the reference genome due to a variety of reasons – biological, technological and computational. Increasingly longer reads produced by the latest platforms and increasingly efficient algorithm development may eventually do away with the technological and computational reasons, but the biological ones (e.g. DNA polymorphism, mutation, presence of repetitive sequences in genomes, etc.) will still remain. Regarding multi-reads, which cause problems in the downstream analysis, there are two possible avenues. They can either be filtered out, leading to the removal of a substantial percentage of the reads and information loss, or recycled by algorithms such as *BM-MAP* [515] that probabilistically allocate a multi-read to one of the candidate positions in the reference genome that match it. Duplicate reads, whose existence is usually a low-probability event unless there is PCR over-amplification, should also be detected after the mapping step and removed for the sake of performance enhancement in the later steps. However, their removal carries a risk. It is impossible to distinguish between biological duplicates (that exist naturally in a genome) and technical duplicates (created by PCR over-amplification). We actually want to remove the latter, but in the process, end up removing the former too and thereby lose true biological information. All three of the tasks mentioned above can be performed using software packages such as *Picard* or *SAMtools*, along with other necessary tasks (e.g. SAM-to-BAM conversion, indexing of those two file-types, merging of multiple BAM files to one, alignment visualization, etc.). Alignment visualization can be either using a text-based alignment viewer or a direct graphical visualization by the superimposition of mapped reads on top of the reference genome.

5.6.4   Genome-guided transcriptome assembly and isoform finding

Once the alignment is done, one important purpose for which it can be used is the discovery of novel genes and splice variants. Most sequencing platforms produce short reads which are much smaller than the length of an mRNA produced by a gene (with the exception of a few platforms such as PAC BIO II that can produce reads as long as a mature mRNA). So a single read does not shed much light on the detailed structure of a gene, such as its intron-exon organization, poly-adenylation sites and transcription start sites. Most exons are quite short (shorter than a couple of hundred base-pairs), so the order of alternative exons in a gene and their use must be reconstructed via mapping to the reference genome and then linking alignments from one region to another. This is called a mapping-based assembly of the transcriptome, as opposed to *de novo* assembly. Both assembly approaches involve constructing a graph for each gene locus based on RNA-seq reads. The graph serves as a starting point for resolving isoforms. However, construction of these graphs is tricky because it involves splitting the data in such a way that a single graph represents only a single locus.

Here we explain the graphical approach in the context of mapping-based assembly. First, any mapping algorithm that allows split (i.e. spaced or non-consecutive) reads can be used to align the RNA-seq reads to the reference genome. If gene models (i.e. structural details of genes) were available for the reference genome, this alignment itself would provide information about which exons belong to which genes. If no gene model is available for the reference genome, mapped reads must first be segmented in order to represent gene loci. Subsequently, a *splicing graph* (also often called an *exon graph*) is constructed for each locus. A *graph* is nothing but a collection of nodes (or vertices) and a collection of edges connecting some pairs of nodes. If you imagine trying to travel from one node to another in a graph, it will only be possible if those two nodes are connected via a sequence of consecutive edges (called a path). In a splicing graph, each node is an exon, each edge represents an exon junction and a path represents an isoform. By applying a path-finding algorithm on such a graph, one can find out some or all of the paths that exist in it. The number of such paths depends on the edge-set of the graph. If the edge-set is so rich that the graph is fully connected (i.e. every pair of nodes is linked via a path), then all conceivable isoforms are possible. On the other hand, a sparse edge-set will make only some isoforms possible.

So, what kind of a topology (i.e. edge-set) should a particular splicing graph have? That is where the RNA-seq reads provide the crucial data-based guidance. The task is to choose a topology which best corresponds to the data. Those possible splice junctions (or edges) for which there is no support from the RNA-seq reads are removed from the graph, and only the edges with significant support are kept. "Support" here means split-reads and paired-end information. In the case of a split-read, if the beginning of the read is mapped to one exon and the end of the read to another exon, this provides evidence for these two exons to be adjacent in an mRNA sequence. In a paired-end case, the same applies to the two ends of the read-pair (i.e. one end mapped to one exon and the other end to another). This latter case is considered somewhat weaker evidence for the existence of an exon junction than the split-read case.

As [288] points out, accurate estimation of isoform abundance is very challenging if not all isoforms are known, since the read-pairs from unknown isoforms can affect the accuracy of the abundance estimation of the known ones. So, isoform finding is a crucial step before what we describe in the next subsection – abundance estimation and gene expression quantification.

The output of this transcriptome assembly step is gene models and transcript models. Assembled transcripts from different samples are merged and combined with reference annotation in order to produce more complete gene models. These can subsequently be used for gene expression quantification.

5.6.5 Quantification and comparison of expression levels

In this step, each single read is associated with a gene based on its mapping location. Expression of novel or familiar genes and their transcript production can be quantified using the gene models and transcript models from the previous step. When studying an organism whose genome is already well-annotated, obtaining gene models and transcript models via the method described in the previous section is not necessary. Instead, the reference annotation for that well-studied organism can be directly used. However, this restricts the quantification of expression levels to only known genes and transcripts. Expressions of novel genes for such organisms will still have to be quantified with the help of the previous section's output.

Abundance estimates can be reported in the form of raw read-counts or in normalized units such as FPKM (fragments per thousand nucleotides per million mapped reads) or RPKM (reads per thousand nucleotides per million reads). At this stage, the sequencing data simply take the form of a table of genes and their read counts or FPKM or RPKM values. One of the important purposes that it serves is the comparison of abundance between two different (or among several different) tissue samples or organisms or experimental conditions. This is known as *differential expression analysis* and it involves various types of statistical methods. Some kind of *normalization* of the raw counts is necessary in order to create a "level playing field" for comparison before such statistical methods are applied, because of possible differences in read numbers between two libraries and/or between two sequencing runs on different days using different batches of reagents and/or between two organisms due to the inherent differences in their transcriptome compositions. That is why the concept of RPKM [301] was originally introduced. Based on transcript lengths and the sequencing depth, this RPKM can compare the expression levels across different genes and samples. The main challenge in using RPKM to estimate transcript abundance is the fact that mapped reads are frequently shared by multiple isoforms. The abundance models that used raw counts before the introduction of the RPKM concept had to assume that each transcript had a single isoform and reads were uniquely mappable to transcripts. As it became clear that this naiive assumption was far from reality, some *ad hoc* approaches were put forward, such as the "rescue" method (i.e. allocating fractions of the reads in proportion to the coverage of uniquely mapped reads) or that of allocating fractions of multi-reads (i.e. reads that map to several places in the reference genome) equally to the target transcript isoforms.

Some of the first departures from such simplistic, *ad hoc* approaches are seen in [497], [204], etc. [497] proposed an expectation-maximization algorithm for probabilistic reconstructions of full-length isoforms from splice graphs. [204] took a model-based approach assuming that the number of reads coming from an exon of a certain length followed a Poisson distribution with its mean being a normalized function of the exon length. Maximum likelihood estimates of the relative abundances of different transcripts were found using a concave optimization algorithm. The next type of statistical models was necessitated by the advent of paired-end sequencing and are generally known as *insert length* models. The lengths of the fragments used in the sequencing process as well as transcript lengths play an important role in them. In paired-end sequencing, reads correspond to both ends of the sequenced fragments. In this case, information about insert-size must be utilized to ensure that the two exons truly form a junction, as opposed to the possibility that the two exons are merely in the same transcript but something else is between them. In other words, when the reads in a pair map upstream and downstream of an alternatively spliced exon, isoforms will be indicative of different intervening insert lengths and this information should be utilized to maker the abundance estimates more accurate. The insert-size distribution depends on the RNA-seq library. Typically, the average insert-size is used for each read-pair and if the fragment-size variability is large in the library, the estimates of the insert-size for any particular read-pair will also have high variability (i.e. cannot be very accurate). [440] introduced the earliest insert-length model by extending [204] approach to paired-end reads and made their algorithm available through the widely used software package *Cufflinks*. This was followed by other such models. For example, [225] came up with a Bayesian algorithm (*MISO*) for RNA-seq data and in it, they used an insert-length distribution that was based on the implied lengths of read-pairs that mapped to large intron-less regions (e.g. the 3′ untranslated regions). [373] is an excellent review of this approach and these models. An insert-length model tacitly assumes that the filtering of reads is based on their lengths and independent of their nucleotide sequences. Conditional on insert lengths, transcripts are assumed to be sampled from a uniform distribution. Using the relative probability of observing the given insert-length as a weight, the read-pairs are assigned (using a weighted probabilistic assignment scheme) to isoforms consistent with both individual reads. We conclude this subsection by pointing the reader to the algorithm developed by [108], called *IsoInfer/IsoLasso*, to deal with the problems of alignment, isoform finding and abundance estimation. First it computes a large set of possible isoforms and then weeds out many of them using the Lasso method to select a best subset.

5.6.6 Normalization methods

As was mentioned in the previous subsection, we often observe many artifacts and biases that adversely affect the quantification of expression levels via abundance estimation. So normalization is a very important step in RNA-seq data analysis. The purpose of normalization is to identify and remove systematic technical differences (i.e. technical biases) among samples that are often there. Among the plethora of normalization algorithms developed for RNA-seq data are total read count normalization (*RC*), upper quartile normalization *UQ*), relative log expression (*RLE*), trimmed mean of M-value normalization (*TMM*), median normalization (*med*), quantile normalization (*Q*), *DESeq*, *RPKM* and *FPKM*, *RSEM* and *Sailfish*. Some of these are global normalization procedures while others are not. In a global normalization method, only a single factor is used to scale the read-counts for all the genes from each sample.

Let n_{gj} be the observed read-count for gene g in the jth sample, N be the number of samples and G be the number of genes. Let $D_j = \sum_{g=1}^{G} n_{gj}$ denote the total number of read-counts for the jth sample. Also, let C_j be the normalization factor used for the jth sample. In *total read-count normalization*, the assumption is that the read-counts are proportional to the expression levels of genes and to the sequencing depth. Since we want to ensure that the scaled total number of reads in each sample is the same (call that number K), the C_j's must be chosen in such a way that $C_j D_j = K$ for $j = 1, \ldots, N$. So it is clear that C_j must be K/D_j. As a result, the normalized read-count for gene g in sample j is $C_j n_{gj} = K n_{gj}/D_j$. Typically, $K = 10^6$ is used.

Clearly, the above method does not take into account the distribution of the read-counts for each sample. Bullard et al. (2010) [37] proposed a normalization method that tries to match the read-count distributions across samples. Let $Q_{j(p)}$ be the upper $100p^{th}$ percentile of the read-counts in the jth sample. The *upper quantile normalization* method seeks to ensure that $C_j D_j Q_{j(p)} = (\Pi_{i=1}^N D_i Q_{i(p)})^{1/N}$ for $j = 1, \ldots, N$. From this, it is easy to find the normalization factor C_j for the jth sample. A commonly used value of p is 0.75.

In a differential expression study, a common experience is to find that a vast majority of genes are not differentially expressed. If a gene g is not differentially expressed between the jth and the j'^{th} samples, then the ratio $n_{gj}/n_{gj'}$ of the counts of that particular gene g in those two samples would be expected to be equal to $C_j/C_{j'}$. Often it is observed that a few highly differentially expressed genes dominate in their influence on the total read-counts. When that is the case, using the total read-count D_j of the jth sample to compute that sample's normalization factor C_j (as was done in the two methods described above) runs the risk of inducing bias into the estimation of read-counts for other genes in that sample. It is better to artificially create a *pseudo-reference sample* and assume that the expression level for each gene g in this artificial sample is equal to the geometric mean $(\Pi_{i=1}^N n_{gi})^{1/N}$. We will then have, for a particular gene g,

$$\frac{n_{gj}}{(\Pi_{i=1}^N n_{gi})^{1/N}} = \frac{C_j}{C_{Pse}}$$

where C_{Pse} denotes the normalization factor for the pseudo-reference sample. If we take C_{Pse} to be equal to 1, then C_j ends up being equal to the left-hand side of the above equation. Now, this was for a particular gene g. If we take the median of the left-hand side of the above equation over all genes (call it M_j), and define C_j^* as $C_j^* = (\Pi_{i=1}^N M_i)^{1/N}/M_j$, we can use C_j^* as the normalization factor for the jth sample (clearly the C_j^*'s multiply to 1). This is called *relative log expression normalization*, because $(\Pi_{i=1}^N M_i)^{1/N} = \frac{1}{N}\sum_{i=1}^N log(M_i)$ and the geometric mean appearing in the expression level for the pseudo-reference sample can also similarly be expressed as an average of log-values.

It has been mentioned above that the total read-count of a sample is often dominated by a few highly expressed genes. In the RLE method, we dealt with this reality by creating a pseudo-reference sample. Another way out would be to remove genes from the upper and the lower ends of the expression levels. This idea leads to the *trimmed mean of M-value* normalization method. First, we define two quantities – the log fold-change and the absolute intensity. The log fold-change for gene g in the jth sample compared to a reference sample *ref* is defined as

$$M_{g,ref}(j) = log_2 \left\{ \left(\frac{n_{gj}}{D_j}\right) \Big/ \left(\frac{n_{g,ref}}{D_{ref}}\right) \right\}$$

where $n_{g,ref}$ and D_{ref} denote the read-count for gene g in the reference sample and the total read-count for the reference sample respectively. The absolute intensity of gene g in the jth sample is defined as

$$A_{g,ref}(j) = \frac{1}{2}\left\{log_2\left(\frac{n_{gj}n_{g,ref}}{D_jD_{ref}}\right)\right\}$$

Suppose we trim off a certain percentage (say, $100q\%$) of the ordered $M_{g,ref}$ values from the top end and the bottom end, and also trim off a certain percentage (say, $100q^*\%$) of the ordered $A_{g,ref}$ values from the two ends. Let G^* be the set of genes that still remain after the trimming. Using those, the normalization statistic will be defined as

$$TMM_{j,ref} = \frac{\sum_{g \in G^*} W_{g,ref}(j)M_{g,ref}(j)}{\sum_{g \in G^*} W_{g,ref}(j)}$$

where the weight $W_{g,ref}(j)$ is the inverse of the variance of the $M_{g,ref}(j)$ values for $g \in G^*$. Since the variance of the $M_{g,ref}(j)$ values can be shown to be approximately equal to $\{(D_j - n_{gj})/D_jn_{gj} + (D_{ref} - n_{g,ref})/D_{ref}n_{g,ref}\}$, the reciprocal of this expression is the weight used in the formula for $TMM_{j,ref}$. Finally, letting $B_j = 2^{TMM_{j,ref}}$, we define the normalization factor for the jth sample as

$$C_j = \frac{exp\left(\frac{1}{N}\sum_{i=1}^{N}log_e(B_j)\right)}{B_j}$$

and clearly the C_j's multiply to 1.

We conclude this subsection with a discussion of RPKM. As its name suggests, the RPKM approach normalizes for the total transcript length and the number of reads. Let N be the total number of mappable reads in an experiment, L be the sum of the numbers of base-pairs in the exons of the genes and C_{ex} be the number of mappable reads that were mapped to those exons. Then RPKM is defined as

$$RPKM = C_{ex}/(\frac{N}{10^6})(\frac{L}{10^3}).$$

5.6.7 Differential expression analysis

An enormous volume of research has been conducted in the last decade to devise suitable statistical methods for detecting differential expression of genes between tissues, organisms or experimental conditions. Here we provide a glimpse of some of it. Lorenz et al. (2014) is an excellent reference for further details.

5.6.7.1 Binomial and Poisson-based approaches

Suppose that we are quantifying the expressions of G genes in L populations, and X_{ijkg} denotes the number of reads that mapped to gene g in replicate

j of population i along lane k of the sequencing platform ($i = 1, \ldots, L$, $j = 1, \ldots, J_i$, $k = 1, \ldots, K$, $g = 1, \ldots, G$). [37] proposed a Poisson generalized linear model:

$$log_e(E(X_{ijkg}|\delta_{ijk})) = log_e(\delta_{ijk}) + \lambda_{ijg} + \theta_{ijkg} \qquad (5.1)$$

where λ_{ijg} is the gene-specific rate in the jth replicate of the ith population, δ_{ijk} is an offset term adjusting for variation in lane depths (and also depending on the population and the replicate) and θ_{ijkg} is an unexplained/unspecified technical effect. Two testing approaches for differential expression was used – t test for the maximum likelihood estimates of the λ_{ijg} and a likelihood ratio test. The t test-statistic had its denominator computed in two different ways – one based on the variance of the maximum-likelihood estimate of λ_{ijg} and the other on an approximate variance obtained via the delta method. Both, however, showed inferior detection rates (i.e. true positive rates) than the likelihood ratio test as well as the Fisher's exact test. The reason behind this is partly the dilution effect of keeping (i.e. not filtering out) many, many genes with low counts. Screening out low-count genes before running those tests should improve the true positive rates of all of them, possibly reducing the difference in performance between the t tests and the likelihood ratio test. although the choice of the screening threshold will surely play an important role. [465] took a different approach in their software package *DEGseq* (available via CRAN). They came up with two tests based on the thresholding of plots of log(fold change) as a function of the mean log(expression), known as Bland-Altman plots or MA plots, for two scenarios – one without technical replicates (i.e. a single sample for each population) and one with technical replicates. They assumed that the read counts had a binomial distribution and used a normal approximation to the conditional distribution of the log count ratio between populations (M) and the average of log counts (A) between populations.

When the total number of sequenced reads becomes large, the binomial distribution can be approximated by a Poisson distribution. With this in mind, let X_{ijg} have a $Poisson(a_{ij}\lambda_{ig})$ distribution where a_{ij} is a replicate-specific scalar coefficient (needed to account for variation in read intensity among biological replicates) and λ_{ig} is the relative rate parameter for gene g in the ith population. Following usual notational convention, let $X_{i.g} = \sum_{j=1}^{J_i} X_{ijg}$ be the within-population read count for the ith population and $X_{..g} = \sum_{i=1}^{L} \sum_{j=1}^{J_i} X_{ijg}$ be the overall read count for gene g. Then $X_{i.g} \sim Poisson(\sum_{j=1}^{I_j} a_{ij}\lambda_{ig})$ and $X_{..g} \sim Poisson(\sum_{i=1}^{L} \sum_{j=1}^{I_j} a_{ij}\lambda_{ig})$. Now the question of differential expression between two populations boils down to testing, for every gene g, the null hypothesis $H_{0g} : \lambda_{1g} = \lambda_{2g}$, which ultimately is a conditional binomial test. In other words, under the null hypothesis, the conditional distribution of $X_{i.g}$ given $X_{..g}$ is binomial with the number of "tosses" $= X_{..g}$ and success probability $p_0 = \sum_{j=1}^{I_j} a_{ij} / \sum_{i=1}^{L} \sum_{j=1}^{I_j} a_{ij}$ which no longer involves the subscript g and is therefore common to all the genes.

This test can be carried out using the exact binomial probability mass function or using a normal approximation, the test statistic being $X_{i.g}$. And of course, since it is a per-gene test, some sort of multiplicity adjustment for the p-values is needed for false discovery rate control. The methodology described above is simple and easy to implement but, unfortunately, the underlying distributional assumptions (binomial, Poisson) have turned out to be unrealistic for real-life read count data. The variability in counts from one replicate sample to another is usually much greater than that allowed by an equidispersed distribution such as Poisson, even after the inclusion of the constants a_{ij}'s for read intensity adjustment. This necessitates a departure from the Poisson assumption.

The first such departure was the use of a generalized Poisson model by [411]:

$$P(X_{ijg} = x) = \frac{\lambda_{ig}(\lambda_{ig} + x\eta_{ig})^{x-1}e^{-\lambda_{ig}-x\eta_{ig}}}{x!}$$

where the newly introduced parameter η_{ig} is the average bias caused by the sample preparation and sequencing process, which also serves as a shrinkage parameter for the moments of X_{ijg} as the mean of X_{ijg} is $\lambda_{ig}/(1-\eta_{ig})$ and the variance is $\lambda_{ig}/(1-\eta_{ig})^3$. Under this model, the test for differential expression is a likelihood ratio test where the denominator of the test statistic has the data likelihood maximized freely (i.e. in an unrestricted way) over λ_{ig} and θ_{ig}, and the numerator has the data likelihood maximized under the constraint $\lambda_{2g} = w\lambda_{ig}$. The coefficient w here is a normalization constant that is intended to account for the variation in sequencing depths between populations. It is estimated as the ratio of the total amount of sequenced RNA in the two populations. For each population, the total amount of sequenced RNA is estimated as a weighted sum (over all genes) of the unrestricted maximum-likelihood estimate of λ_{ig}, with the weights coming from the gene lengths. The likelihood ratio test statistic mentioned above follows approximately a chi-squared distribution with 1 degree of freedom under the null hypothesis. There have been studies indicating that this test has superior detection power than the Poisson test described earlier. Unfortunately, however, the generalized Poisson distribution allows negative values of the λ_{ig}'s which is a problem because, first of all, the likelihood ratio test cannot be carried out if any λ_{ig} estimate is negative as the likelihood becomes zero and, secondly, a negative intensity estimate has no practical interpretation. [15] took a different approach that does not suffer from such problems. In their two-stage procedure, the first stage consists of screening the gene counts to find out which are overdispersed. Then in the second stage, depending on whether a gene is overdispersed or not, the test for its differential expression is carried out using different test statistics. To be more specific, the first stage starts by filtering out genes with low cumulative counts (over populations and replicates) using a threshold that can be chosen differently for different experiments depending on the number of replicates and overall read intensity values. For the genes that survive the filtering, a random effects Poisson model is fitted to the gene counts. The model involves an overdispersion parameter ϕ_g for gene g so that the null hypothesis of no overdispersion for that gene boils down to $H_{0g} : \phi_g = 1$. This test is carried out via an adjusted score statistic whose quantiles are compared with those of a chi-squared distribution with d.f. = 1 in the following way. A Working-Hotelling (WH) simultaneous confidence band is constructed for the theoretical χ_1^2 quantiles and genes for which the adjusted score statistic exceeds the upper boundary of the WH simultaneous confidence band are considered as overdispersed. For all other genes, in the second stage, a standard likelihood ratio test based on a Poisson model is carried out. For the genes identified as overdispersed, the test performed is a likelihood ratio test derived from fitting overdispersed quasi-likelihood models under the null hypothesis of no differential expression and the alternative hypothesis of differential expression. Finally, multiplicity adjustment for false discovery rate control is carried out separately for overdispersed genes and non-overdispersed genes using a procedure such as Benjamini-Hochberg.

5.6.7.2   Empirical Bayes approaches

In the Bayesian realm, an empirical Bayes approach was proposed by [347] which uses two different tests – a likelihood ratio test based on a Poisson model and another test based on a quasi-likelihood model that adjusts for overdispersion. The *assumption adequacy averaging* method estimates the empirical Bayesian probability of no differential expression for each gene in the following way. The estimate is based on a weighted average of the empirical Bayesian probabilities of no differential expression for that gene under the Poisson model and the quasi-likelihood model, with the weights coming from the empirical Bayesian probability of no overdispersion. In other words, it uses the law of total probability. On the other hand, the *empirical best test* method seeks to select the best test based on the empirical Bayesian probabilities of no overdispersion for each gene. This method then applies the adaptive histogram estimator to obtain the empirical Bayes probabilities of no differential expression based on the set of p-values for the tests of differential expression, using the best test for each individual gene.

5.6.7.3   Negative binomial-based approaches

An alternative to modifying/generalizing the Poisson model or screening genes first for overdispersion before testing for differential expression is to use an overdispersed probability model in the first place. The *negative binomial* distribution offers such a model. It can handle overdispersion due to technical variation as well as biological variation. Let us assume that the same biological sample is being repeatedly sequenced multiple times, creating technical replicates. Suppose that p_{gi} is the fraction of all cDNA fragments in the ith sample that are from the gene g. This p_{gi} does not vary between technical replicates (because they are the same biological sample) but it does vary among different biological samples (biological replicates). As p_{gi} varies from one biological replicate to another, let us assume that the ratio between its variance and mean-squared remains unchanged for each gene g. Let θ_g be this constant ratio between its variance and mean-squared (i.e. $\sqrt{\theta_g}$ be its biological coefficient of variation). Suppose that the total number of mapped reads in the ith sample is n_i and among them, the number of reads mapped to the gene g is X_{gi}. Then, using the formulas $E(U_1) = E(E(U_1|U_2))$ and $Var(U_1) = E(Var(U_1|U_2)) + Var(E(U_1|U_2))$ for any two random variables U_1 and U_2, it can be shown that $E(X_{gi}) = n_i p_{gi}$ (call it μ_{gi}) and $Var(X_{gi}) = \mu_{gi} + \theta_g \mu_{gi}^2$ which clearly shows overdispersion (i.e. variance greater than mean).Now, dividing both sides of $Var(X_{gi})$ by μ_{gi}^2, we get the squared coefficient of variation of X_{gi} on the left-hand side and $1/\mu_{gi} + \theta_g$ on the right-hand side. Since for a fixed biological sample (i.e. given a fixed p_{gi}), the variation of X_{gi} among the technical replicates is conditionally modeled by a Poisson with mean $n_i p_{gi}$ (i.e. μ_{gi}), the coefficient of variation of this conditional Poisson distribution is $\sqrt{1/\mu_{gi}}$. So ultimately we get the following equation:

$$CV^2 \text{ of } X_{gi} = \text{Technical } CV^2 + \text{Biological } CV^2.$$

If we assume that $X_{gi} \sim Poisson(\lambda)$ but the mean parameter λ itself is a random variable (varying from one technical replicate to another) that follows a Gamma density with shape parameter θ_g^{-1} and rate parameter $(\theta_g \mu_{gi})^{-1}$, then it can be shown that

$$P(X_{gi} = x_{gi}) = \frac{\Gamma(x_{gi} + \theta_g^{-1})}{\Gamma(\theta_g^{-1}) x_{gi}!} \left(\frac{1}{1 + \theta_g \mu_{gi}} \right)^{x_{gi}} \left(\frac{(\theta_g \mu_{gi})^{-1}}{1 + (\theta_g \mu_{gi})^{-1}} \right)^{\theta_g^{-1}}$$

which is a Negative Binomial probability mass function with mean μ_{gi} and variance $\mu_{gi}(1 + \theta_g \mu_{gi})$.

[286] proposes a log-linear model for the mean parameter of this Negative Binomial distribution for X_{gi}:

$$log(\mu_{gi}) = \boldsymbol{y_i}^t \psi_g + log(n_i)$$

where $\boldsymbol{y_i}$ is a vector of covariates capturing the characteristics of the ith sample or the treatment condition applied to the ith sample, ψ_g is a p-dimensional vector of coefficients associated with the gene g and the superscript "t" stands for transpose. The log-likelihood $l(\psi_g, \theta_g)$ is proportional to

$$\left(\sum_{i=1}^{N} x_{gi} \right) log(\theta_g)$$

$$+ \sum_{i=1}^{N} \left\{ x_{gi}(\boldsymbol{y_i}^t \psi_g + log(n_i)) - (x_{gi} + \theta_g^{-1}) log(1 + \theta_g n_i e^{\boldsymbol{y_i}^t \psi_g}) \right\})$$

where N is the total number of samples and the maximum likelihood estimation of the parameters ψ_g is carried out using a Newton-Raphson iteration procedure that proceeds like $\psi_g^{new} = \psi_g^{old} + (YWY^t)^{-1} Y \boldsymbol{z}_g$, with

$Y = [y_1, \ldots, y_N]$, $z_{gi} = \frac{x_{gi} - \mu_{gi}}{1 + \theta_g \mu_{gi}}$ and $\mu_{gi} = n_i e^{y_i^t \psi_g}$ for $i = 1, \ldots, N$, as well as $z_g = [z_{g1}, \ldots, z_{gN}]$. The matrix W is a diagonal matrix with the ith diagonal entry being $\mu_{gi} x_{gi} \theta_g (1 + \theta_g \mu_{gi})^{-2}$ for $i = 1, \ldots, N$. One potential problem with this Newton-Raphson algorithm is the possibility of the likelihood function decreasing after an iteration (instead of increasing). This motivates a slightly modified version of the above, which will ensure that the likelihood function always increases. We define the new iteration procedure as $\psi_g^{new} = \psi_g^{old} + \nu (YWY^t)^{-1} Y z_g$, where ν plays the role of a step-size constant. Taking $(YWY^t)^{-1} Y z_g$ as a search direction and using a linear search, we can determine the step-size ν such that the iteration always increases the likelihood. The computation of the search direction can be simplified if instead of this iteration procedure, we use $\psi_g^{new} = \psi_g^{old} + \nu^* Y z_g$ where ν^* is the constant determining the step-size in each iteration.

All of the above discussion on iterative maximum likelihood estimation of the parameters of a log-linear model for the mean of a Negative Binomial distribution is based on the assumption that the dispersion parameter θ_g is known. In practice, however, it needs to be estimated. This brings us to the *Cox-Reid adjusted profile likelihood* method for θ_g. The adjusted profile likelihood for θ_g is defined as the penalized log-likelihood

$$APL_g(\theta_g, \hat{\psi}_g) = l(\theta_g; x, \hat{\psi}_g) - (1/2)log(||YWY^t||)$$

where $||.||$ stands for the determinant of a square matrix and $l(\theta_g; x, \hat{\psi}_g) =$

$$\sum_{i=1}^{N} \left\{ \sum_{l^*=1}^{x_{gi}-1} log(1 + l^* \theta_g) - x_{gi} log(\theta_g) - log(x_{gi}!) + x_{gi} log(\theta_g \mu_{gi}) \right.$$

$$\left. - (x_{gi} + \theta_g^{-1}) log(1 + \theta_g \mu_{gi}) \right\}.$$

Using it, the parameter θ_g is estimated by the Newton-Raphson iteration procedure

$$\theta_g^{new} = \theta_g^{old} - \frac{\left[\frac{\partial (APL_g(\theta_g, \hat{\psi}_g))}{\partial \theta_g} \right]_{\theta_g^{old}}}{\left[\frac{\partial^2 (APL_g(\theta_g, \hat{\psi}_g))}{\partial \theta_g^2} \right]_{\theta_g^{old}}} = \theta_g^{old} + \zeta, \text{say}$$

Once again, as we saw earlier, it is possible that some of the iteration steps in this Newton-Raphson algorithm leads to a decrease (instead of increase) in the adjusted profile likelihood function. And like before, to ensure that the iterations always lead to an increase, we can slightly modify the above iteration step to

$$\theta_g^{new} = \theta_g^{old} + \nu^{**} \zeta$$

where ν^{**} is the "slack" constant determining the step-size in each iteration. Notice that in our discussion of how to estimate θ_g, we are assuming that

$\hat{\psi}_g$ is available and can be plugged into the adjusted profile likelihood for θ_g. However, as we saw earlier, in order to obtain $\hat{\psi}_g$, we need an estimate of θ_g. So what we need to do is iterate between computing $\hat{\psi}_g$ assuming an initial estimate of θ_g, then using this $\hat{\psi}_g$ to compute $\hat{\theta}_g$, and then using this $\hat{\theta}_g$ to compute a new $\hat{\psi}_g$, etc.

Once this estimation is done, we will be interested in testing whether ψ_g is the zero vector or not (i.e. whether the log-linear model for the Negative Binomial mean using $\boldsymbol{Y}_i$'s as the covariates is significant or not). [184] came up with the following test procedure. If there was no overdispersion, the likelihood ratio test statistic for testing $H_0 : \psi_g = \boldsymbol{0}$ vs. not zero would be

$$2\left\{l(\hat{\psi}_g, \hat{\theta}_g) - l(\psi_g^0, \hat{\theta}_g)\right\}$$

where ψ_g^0 implies plugging in 0 for each component of the vector ψ_g in the log-likelihood, and $l(\hat{\psi}_g, \hat{\theta}_g)$ is equal to

$$\left(\sum_{i=1}^N x_{gi}\right)log(\hat{\theta}_g) + \sum_{i=1}^N \left\{x_{gi}(\boldsymbol{y_i}^t\hat{\psi}_g + log(n_i)) - (x_{gi} + \hat{\theta}_g^{-1})log(1 + \hat{\theta}_g n_i e^{\boldsymbol{y_i}^t\hat{\psi}_g})\right\}.$$

However, when overdispersion is there, the likelihood ratio test statistic for testing $H_0 : \psi_g = \boldsymbol{0}$ will have to be

$$\frac{2\left\{l(\hat{\psi}_g, \hat{\theta}_g) - l(\psi_g^0, \hat{\theta}_g)\right\}}{\hat{\theta}_g}$$

which, under the null hypothesis, will have an $F(p, N - p - 1)$ distribution, whereas the test statistic for the "no overdispersion" scenario will have a central $\chi^2_{(p)}$ distribution.

It is worth pointing out that throughout this subsection, we have assumed a gene-specific dispersion parameter θ_g that can vary from one gene to another. Earlier, [367] thought of a simple way of sharing information among genes, by assuming that all genes share the same dispersion parameter θ. Given estimates $\hat{\psi}_g$ of ψ_g for $g = 1, \ldots, G$ (G being the number of genes), the common θ can be estimated by maximizing the common adjusted profile likelihood $APL_{com}(\theta) = (1/G)\sum_{g=1}^G APL_g(\theta)$. In this estimation process, all genes contribute equally to the estimation of θ. This approach can be implemented via the "*estimateGLMCommonDisp*" function in the R package "*edgeR*" developed by [364]. This common dispersion scenario can be thought of as a weighted likelihood scenario where the likelihoods of the individual genes are all weighted equally in the objective function that is maximized. In practice, however, this assumption of a common dispersion parameter for all genes proves to be unrealistic as it has been observed in many RNAseq datasets that genes with lower expression levels tend to have higher variability (or dispersion). This dependence of dispersion on the expression level was modeled by [8] and in the "*edgeR*" package, the dispersion values obtained from this mean-dispersion trend has been referred to as *trended dispersion*, which can also be estimated via the weighted likelihood approach. See [52] for details.

All three scenarios described above (common dispersion, trended dispersion and gene-specific dispersion) are examples of the weighted likelihood empirical Bayes approach. In general, the empirical bayes approach involves first estimating the prior distribution from the data and then applying the standard Bayesian approach with this estimated prior to obtain posterior estimates. However, [469] has shown that an empirical Bayes estimator is equivalent to an estimator obtained by maximizing a weighted likelihood function on a set of observations.

5.6.8   Classification

A wide range of supervised and unsupervised learning methods have been considered for RNAseq data, be it to infer coordinated patterns of gene expression, to discover molecular signatures of disease subtypes, or to derive various predictions.

5.6.8.1   Linear discriminant analysis

Here we describe classification of tissue samples via linear discriminant analysis, which is a supervised learning method.

We observe the class labels $y_1, \ldots, y_n$ where $y_j = i$ if the jth tissue sample comes from the ith class ($i = 1, \ldots, I$). Our aim is to construct a classifier $C(x)$ for predicting the unknown class labels y of a tissue sample x. For example, when $I = 2$ classes. class 1 may be healthy and class 2 may be metastatic.

A linear classifier looks like $C(x) = \beta_0 + \boldsymbol{\beta}^T x = \beta_0 + \beta_1 x_1 + \ldots + \beta_p x_p$ for the production of the group label y of a future entity with feature vector x. Fisher's linear discriminant function: $y = sgn(C(x))$, where $\boldsymbol{\beta} = S^{-1}(\bar{x}_1 - \bar{x}_2)$ and $\bar{x}_1$, $\bar{x}_2$ and S are the sample means and pooled-sample within-class covariance matrix found from the training data.

In high-dimensional data where the number of genes far exceeds that of tissue samples, the above linear discriminant analysis (LDA) method cannot be directly applied because the pooled-sample within-class covariance matrix becomes singular and its inverse cannot be computed. As a way out, a modification of LDA has been proposed by researchers where the assumption is that the features are independent, that is, an observation in the ith class has a $N(\mu_i, \Sigma)$ distribution where Σ is diagonal. This is referred to as *diagonal LDA* and has been shown to perform well in high-dimensional settings. It is implemented as follows: one assigns an object with feature vector x to the ith class if

$$-\frac{1}{2}\Sigma_{k=1}^p (x_k - \bar{x}_{ik})^2 / \hat{s}_k^2 + log(\hat{\pi}_i)$$

is the largest among $i = 1, \ldots, I$, where $\bar{x}_{ik}$ is the sample average of the kth feature in the ith class (i.e. i^{th} population), $\hat{s}_k^2$ is the pooled-sample within-class variance for the kth feature ($k = 1, \ldots, p$) and $\hat{\pi}_i$ is the proportion of observations from the ith class. However, one drawback of diagonal LDA is its use of a decision boundary that involves all p features of the p-dimensional feature vector. For large p, a sparse classifier that directly involves only a subset of the features would be more desirable for the sake of interpretability, simplicity and reduced variability.

Another key component of the LDA and diagonal LDA is the assumption of normality. In the case of RNSseq data, however, the entries of the feature vectors are counts which are nonnegative integers. For such count data, a Poisson or Negative Binomial model is more appropriate. So [488] has put forward the *Poisson LDA* method. It uses an approach that is not only suitable for nonnegative count data but also involves only a subset of the features (i.e. brings in sparsity). Before describing it, first we provide the details on an extension of diagonal LDA that is still based on the normality assumption but uses a sparse decision boundary. [433] and [434] proposed the method *nearest shrunken centroids* (NSC), which is named this way because it is a modification of the *nearest centroids classifier* (a simplified version of diagonal LDA obtained by assuming that $s_1 = s_2 = \ldots = s_p$ and $\pi_1 = \pi_2 = \ldots = \pi_I$ where π_i is the prior probability that an observation belongs to the ith category). NSC modifies the expression that is maximized in diagonal LDA (see above) so that the $\bar{x}_{ik}$ in it is replaced with a statistic (call it $\hat{\mu}_{ik}$) that satisfies $\hat{\mu}_{1k} = \ldots = \hat{\mu}_{Ik}$ for the kth feature for some k. Doing so ensures that the kth feature is not involved in classification any longer. Let

$$d_{ik} = \frac{\bar{x}_{ik} - \bar{\bar{x}}_{.k}}{m_i(\hat{s}_k + s_0)}$$

where $\bar{\bar{x}}_{.k} = \frac{1}{n} \sum_{l=1}^{n} x_{lk}$ is the overall mean of the kth feature (n being the sample size), $m_i^2 = 1/n_i - 1/n$ with n_i being the sample size for the ith class, and s_0 is some positive constant. This d_{ik} is in some sense a Student's t-like statistic that measures the difference between the mean of the ith class for the kth feature and the overall mean for the kth feature. Rewriting the above definition of d_{ik}, we get $\bar{x}_{ik} = \bar{\bar{x}}_{.k} + m_i d_{ik}(\hat{s}_k + s_0)$. In order to facilitate the occurrence of $\hat{\mu}_{1k} = \ldots = \hat{\mu}_{Ik}$ for some values of the index $k \in \{1, \ldots, p\}$, we apply soft thresholding on d_{ik} and convert it to $d_{ik}^* = sgn(d_{ik})max(|d_{ik}| - \xi, 0)$, with ξ being a tuning parameter to be chosen via cross validation. Finally, we define $\hat{\mu}_{ik}$ as

$$\hat{\mu}_{ik} = \bar{\bar{x}}_{.k} + m_i d_{ik}^*(\hat{s}_k + s_0)$$

and use this $\hat{\mu}_{ik}$ in the decision rule for diagonal LDA described earlier. If ξ is sufficiently large, then $\hat{\mu}_{1k} = \ldots = \hat{\mu}_{Ik}$ will be true for some features k. Those features of the feature vector won't play a role in the classification process.

Now, about Poisson LDA. It mimics NSC except that it is based on the assumption that the entries of the feature vector are independently drawn from Poisson distributions instead of Normal. More specifically the assumption is that, given that the jth individual in the sample belongs to the ith class $(i = 1, \ldots, I)$, the kth component of its feature vector $x_{jk} \sim Poisson(s_j g_k e_{ik})$, where s_j allows for variability in the number of counts per sample, g_k allows for variability in the number of counts per feature and e_{ik} is a measure of differential expression for the kth gene in the ith class. Under this distributional assumption, ultimately the decision rule is as follows: assign a test object or individual with feature vector $\boldsymbol{x}$ to the ith class if the quantity

$$\sum_{k=1}^{p} x_k log(\hat{e}_{ik}) - \hat{s}_i^* \sum_{k=1}^{p} \hat{e}_{ik} \hat{g}_k + log(\hat{\pi}_i)$$

is the largest for the ith class among $i \in \{1, \ldots, I\}$. In order to bring in sparsity, [488] recommends soft thresholding of $\hat{e}_{ik}$ towards 1 using a tuning parameter whose value is to be chosen via cross validation (in a manner similar to the soft thresholding for NSC described above).

5.6.8.2   Support vector machine classifier

The support vector machine classifier was originally introduced by Vladimir Vapnik (see, for example, [455]). RNAseq data can be viewed as a matrix of expression levels. For M tissue samples, for each of which we measure the expression levels of N genes, the results can be represented by an $N \times M$ matrix. Each row of the gene expression matrix can be regarded as a point in an M-dimensional expression space. Conceptually, a straight-forward way to build a binary classifier is to construct a hyperplane separating one group of points from another in this space, with one group lying on one side (call it the positive side) of the hyperplane and the other group lying on the other side (call it the negative side). If this is indeed possible, we say that the points are separable in the M-dimensional space. Unfortunately, most real-world situations involve nonseparable data-points for which no such hyperplane exists that flawlessly separates the positive from the negative data-points. One solution to this inseparability problem is to map the data into a higher dimensional space and construct a separating hyperplane there. This higher dimensional space is typically called the feature space. With an appropriately chosen feature space of adequate dimensionality, any consistent training set (i.e. set of data-points for which the relationship between the classification category and the corresponding feature vector is known) can be rendered separable. However, mapping the training set into a higher dimensional space incurs some cost – both computational and theoretical. In addition, artificially separating the data in this way exposes the learning system to the risk of finding trivial solutions that overfit the training data.

Support vector machines (SVM) provide an elegant remedy for both of these difficulties. First of all, the decision function for classifying points with respect to the hyperplane only involves inner products or dot products between points in the feature space. Since the SVM algorithm for finding a separating hyperplane in the feature space can be stated entirely in terms of vectors in the input space (the space of the data-objects that are being classified) and dot products in the feature space, the SVM method can construct the hyperplane without ever representing the space explicitly. This is accomplished simply by defining a function (the *kernel function*) that plays the role of the dot product in the feature space, thereby avoiding the computational burden of explicitly representing the high-dimensional feature vectors. If for a dataset the SVM method is unable to find such a cleanly separating hyperplane in the feature space, it could either be due to the poor quality of the training data (i.e. the training data containing too many mislabeled examples) or due to the chosen kernel function being inappropriate for the training data. In the former case, one has to use a "soft margin" allowing some training examples to fall on the wrong side of the separating hyperplane. This is done by introducing the so-called *slack variables*. In the latter case, one will just have to change the kernel fuction. To sum it up all, completely specifying an SVM requires the specification of two items – the kernel fuction and the amount of the penalty for violating the soft margin.

Starting with an expression vector $\boldsymbol{X_i}$ for the ith gene ($i = 1, \ldots, n$), the simplest kernel $K(X_i, X_j)$ that we can use to measure the similarity between the ith and the jth genes is the dot product in the input space: $K(X_i, X_j) = \sum_{k=1}^{M} x_{ik} x_{jk}$. For some technical reasons, often the kernel that is used in practice is $K(X_i, X_j) = \sum_{k=1}^{M} x_{ik} x_{jk} + 1$. With this kernel, the feature space is essentially the same as the M-dimensional input space and the SVM method constructs a separating hyperplane in this space. Raising this kernel to a higher power, that is, using $K(X_i, X_j) = (\sum_{k=1}^{M} x_{ik} x_{jk} + 1)^d$ for some $d > 1$ leads to polynomial separating surfaces of higher degrees in the input space. In the feature space corresponding to this kernel, for any gene X_i there are features for all d-fold interactions between gene expression measurements, represented by terms of the form $x_{ik_1} x_{ik_2} \ldots x_{ik_d}$ where $1 \le k_1, \ldots, k_d \le M$. For example, $d = 2$ produces a quadratic separating surface in the input space and the corresponding separating hyperplane in the feature space includes features for all pairwise expression interaction terms $x_{ik_1} x_{ik_2}$ with $k_1, k_2 \in \{1, \ldots, M\}$. Other than polynomial kernels, a different kind of a popular kernel is a radial basis kernel: $K(X_i, X_j) = exp(-\|X_i - X_j\|^2/2\sigma^2)$ where σ is the so-called width of the Gaussian form being used.

So, with the commonly used quadratic kernel, the binary SVM classification procedure in the general nonseparable case is $\boldsymbol{C(x)} = \beta_0 + \boldsymbol{\beta}^T \boldsymbol{x} = \beta_0 + \beta_1 x_1 + \ldots + \beta_M x_M$ where β_0 and $\boldsymbol{\beta}$ are obtained as follows:

$$min_{\beta_0, \boldsymbol{\beta}} \frac{1}{2} \|\boldsymbol{\beta}\|^2 + \gamma \sum_{j=1}^{n} \xi_j$$

subject to $\xi_j \ge 0$ and $y_j \boldsymbol{C(x_j)} \ge 1 - \xi_j$ for $j = 1, \ldots, n$, where the y_j's are either $+1$ or -1. The ξ_j's are the slack variables and the case with $\gamma = \infty$ corresponds to the cleanly separable case.

We conclude this section by referring the reader to [425] which mentions other supervised learning (i.e. classification) methods with RNAseq data such as principal components classification and partial least squares classification.

5.6.9   Further downstream analysis

The final output of the analysis steps described above is typically a list of known genes (that are differentially expressed between two samples, or are found to cluster together for some reason, or something like that). In addition, the list may include some novel genes that have been discovered. Also, the output may contain more details about gene structure (e.g. new exons, alternative transcription start sites, etc.). Along with protein-coding genes, other types of novel transcripts can also be found in this process, such as long and short non-coding RNAs.

When the genome of an organism is being sequenced for the first time, usually an automated pipeline is set up to annotate the genes. Since there is no well-annotated reference genome for such an organism to compare the sequencing output with, annotation is performed on computationally predicted genes constructed from the reads that are aligned to create long transcripts. First the transfer RNAs, ribosomal RNAs and other structural RNA molecules are annotated and gotten rid of. This is followed by the alignment of protein-coding genes to databases of known genes and inference is drawn regarding their functions based on sequence similarity.

If a differential expression analysis has been conducted, the output will have a long list of genes and quantitative differences between their expression levels in different tissue-samples or under different experimental conditions. To shorten the list, first it is subjected to some kind of a stringent filter (e.g. a smaller significance level or a bigger fold-change). With the genes still remaining after such filtering, *gene set enrichment analysis* (GSEA) is performed. It is a way of grouping those genes on the basis of their annotations and testing whether any group is over-represented in comparison with a background (which could be the entire gene set or some suitable portion of it). For instance, if a short-list of 1000 genes has 30 that are known to be members of a certain biological pathway (e.g. vitamin D metabolism or activation of the beta cells in the pancreas that secrete insulin) whereas the entire gene set of 20,000+ contains only 75 belonging to that pathway, the question is whether this is "sufficiently unusual" for the short-list. This question is answered using a statistical hypothesis test (called *Fisher's exact test*) that is based on a hypergeometric probability distribution. Before doing this, however, the genes must be grouped on the basis of their annotations. The most widely used annotation database for this purpose is *Gene Ontology* or *GO* [12]. GO being a hierarchical and controlled vocabulary, it is possible to annotate a gene at different levels of detail. For the same reason, it is possible for a gene to have multiple annotations too. GSEA produces a list of over-represented biochemical processes, molecular functions and cellular pathways/locations in our short-list of genes. This enables us to test hypotheses such as whether genes belonging to those processes, functions or pathways/locations are all responding to a certain intervention (e.g. a drug) or not, or whether they are naturally "fired up" in a species but not in another.

6

Sequencing-Based DNA Methylation Data

CONTENTS

DOI: 10.1201/9781315153728-6

6.1   DNA methylation

An eukaryote genome may have many trillions of cells of different types, each carrying essentially the same genome in its nucleus. The differences among different types of cells are determined by where and when different sets of genes are turned on or off. For example, red blood cells turn on genes that produce proteins to carry oxygen from the lung and deliver it to the body tissues, while liver cells turn on those that break down fats and produce energy in fat metabolism, among other functions. Therefore, the static genome alone is insufficient for performing these functions; it is the dynamic epigenome that instructs the genome to change its functions without changing the underlying DNA sequences.

DNA methylation is the the most well-known and best characterized epigenetic marks in eukaryotes. It was in fact the first discovered epigenetic mark, and remains the most studied. There are several types of DNA methylation, inclduing 5-methylcytosine (5-mC), 5-hydroxymethylcytosine (5-hmC), 5-formylcytosine (5-fC), and 5-carboxylcytosine (5-caC). This chapter focuses on 5-mC DNA methylation (simply referred to as DNA methylation for simplicity hereafter), as it is the most common type of DNA methylation and highly relevant for investigating the role of methylation in disease etiology. DNA methylation is characterized by the addition of a methyl group to the cytosine residue in a CpG dinucleotide pair (we will use CpG and CG interchangeably). This elusive and dynamic phenomenon has been referred to as the fifth letter of the code of the "book of life" [269], which accounts for about 1–6% of the nucleotides in mammalian and plant genomes [296]. It is well understood that DNA methylation plays a pivotal role in regulating, silencing, or partially silencing gene transcription in normal cells, including embryonic development and cell differentiation [185, 251, 355]. Normal DNA methylation is also crucial for genomic imprinting, X-chromosomal inactivation, silencing of transposable elements, regulation of gene transcription, aging, and autoimmunity [289, 324, 361, 407, 546].

Aberrant DNA methylation, especially those acquired later in life and/or occurring in gene promoters, can lead to aging, diseases, and malignancy through transcription repression. A general rule of the relationship between DNA methylation and transcription regulation is that hyper-methylation is associated with gene repression, although there are exceptions to this rule [111]. Cytosine methylation can directly block the binding of transcription factors; methylated bases can also be bound by methyl-binding-domain proteins that recruit chromatin-remodeling factors [29, 168]. DNA methylation plays a crucial role in the development of nearly all types of cancer [105, 280, 435]. It is also hypothesized that aberrant DNA methylation patterns can be used to define leukemia subtypes [114]. Although hyper-methylation commands a great deal of attention, hypo-methylation may also be associated with the DNA of abnormal tissues. As such, the sites and regions of differential methylation (either hypo- or hyper- comparing diseased to normal samples) are believed to play an important role in complex diseases and finding them is the subject of many current studies. Indeed, genome-wide profiling of DNA methylation has provided a great deal of insight on how epigenetic changes are associated with aberrant expression patterns in certain regions of the genome under specific biological context [66].

6.2 Evolving technologies for measuring DNA methylation

There are a variety of platforms for measuring DNA methylation. Prior to the use of massively parallel sequencing technology (also known as next generation sequencing (NGS)) for analyzing DNA methylation, studies on DNA methylation relied primarily on techniques such as low throughput traditional methods, and more recently, microarrays or chips. Broadly, methodologies for detection of methylation can be classified according to the chemistry they rely on — either bisulfite conversion based or capture based. Examples of traditional techniques, microarray platforms, and NGS-based technologies for each of the categories are given in Table 6.1. These three types of methods reflect the various levels of coverage, from a handful of specific CpGs, to primarily CpG-islands covering only a fraction of CpG sites, to all CpG sites in the entire genome.

Great leaps in the NGS technology in the past decade have fueled the development of whole-genome bisulfite sequencing (WGBS), which was regarded as the gold-standard for measuring single-nucleotide resolution cytosine methylation levels. However, WGBS is very expensive and requires a great deal of resources, and thus remains out of research for most clinical studies. Even for studies with the necessary resources for carrying out some GWBS experiments, typically only a handful of samples are processed due to cost consideration. To reduce cost yet retain some of the attractive features of WGBS, Reduced-Representation Bisulfite Sequencing (RRBS) [287] and Double-Enzyme RRBS [464] have been developed. They are still bisulfite sequencing based to retain the advantage of single-base resolution, but they only cover a subset of the CG sites in the genome to reduce the cost so the technology can be adopted for large-scale studies. Collectively, we refer to this set of platforms and techniques as BS-seq.

Throughout the history of DNA methylation analysis, capture-based assays have been developed in parallel to those that are bisulfite conversion based. There is no exception in the NGS era, as methylated DNA enrichment techniques (collectively referred to as Cap-seq) have emerged as viable competitors to BS-seq because they are much more resource friendly. For example,

TABLE 6.1

Evolving DNA methylation detection technologies

Chemistry	Method	Technique
Bisulfite	Traditional	Bisulfite genomic sequencing [124]
		Methylation specific PCR (MSP) [169]
		MethyLight [94]
	Microarray	Infinium HumanMethylation450 BeadChip [28]
		Infinium MethylationEPIC BeadChip [298, 432]
	NGS-based	Whole genome bisulfite sequencing (WGBS) [59, 269]
		Reduced representation bisulfite sequencing (RRBS) [287]
		Single-cell resolution WGBS/RRBS [104, 144, 403]
Capture	Traditional	Southern blot [210]
	Microarray	Differential Methylation Hybridization [187]
		MeDIP-chip [476]
	NSG-based	MeDIP-seq [89]
		MethylCap-seq [34]
		MBD-seq [391]

cap-seq analysis is feasible for large-scale clinical studies as these approaches require less patient material; only a fraction of the sequencing depth is needed in comparison to single-base resolution BS-seq. Despite the cost advantage of cap-seq, there has been some reluctance to adopt these methods owing to the lack of nucleotide-level methylation signal. In fact, because the base unit of information is a DNA fragment captured by a methylcytosine binding protein or antibody and then sequenced [424], it is assumed that at least one of the CpGs within each sequenced fragment is methylated because of the capture/enrichment step, but it is uncertain precisely which CpG dinucleotides within the fragment are methylated [120]. It has been increasingly documented that DNA methylation patterns may very from cell to cell, even among cells of the same type [9, 403]. Therefore, multiple assays for DNA methylation profiling at the single cell level have been developed, including scBS-seq, scWGBS, and scRRBS-seq [104, 144, 403], as well as those for joint profiling of multimodal genomic features, including linking gene expression and chromatin accessibility with DNA methylation, and others [9, 57, 179, 248, 291]

In the following sections, we describe several statistical methods for analyzing sequencing-based data to detect differentially methylated cytosines (DMC) and differentially methylated regions (DMR) between two conditions, typically diseased and normal. Although some of the methods discussed can accommodate multiple conditions, the focus will be on settings with two conditions. The focus on detecting differential methylation for studying diseases is justified because methylation is a flexible genomic parameter that can change genome function under external influence, and is hypothesized to be the primary missing link between genetics, diseases, and the environment. Studying the differences in methylation patterns between healthy and diseased individuals may help identify disease etiology and treatment targets. Although there may be some interest in finding DMCs, DMRs are considered as a more representative feature for prediction such as diagnostics/prognostics [157, 270, 365]. Regardless of whether the data are generated from the BS-seq or the Cap-seq technologies, a common feature is that methylation signals in dense CpG regions (for example, CpG islands) are spatially correlated, say, within 1000 bases [95, 157, 195]. However, such correlations are variable [268] and should be accounted for in a data-adapted manner. DMR detection methods need to take this into account.

6.3 Methods for Detection of DMCs using BS-seq data

6.3.1 BS-seq data

The first step in obtaining BS-seq data is bisulfite treatment of the DNA, which converts an unmethylated nucleotide "C" to an "U" but leaves a methylated "C" unchanged. Then, the converted DNA is sequenced, with the "U" read as a "T". At each CpG site, each nucleotide resolution read is the outcome of a binary variable, and the number of C's from each site can be regarded as a sample drawn from a binomial distribution $Bin(n, p)$, with n being the number of reads that cover this position and p being the probability that this CpG site is methylated. Figure 6.1 provides a schematic diagram for BS-seq data in a DNA segment (depicted by the horizontal line) containing a number of CpG dinucleotides (marked by the short vertical lines). The data displayed exhibit greater similarity between close neighbors than those that are farther apart. There is also unequal coverage, with potentially different n values at each of the different sites. Missing data may also exist; for example, there are no reads covering the 4th CpG site.

The fact that the nucleotide resolution data from BS-seq at each CpG site can be treated as coming from a binomial distribution has led to the use of simple methods for detecting DMCs, including Fisher's exact test and simple logistic regression analysis. However, such methods ignore between-sample variability as well as correlation among neighboring sites. As such, beta-binomial distributions have been used to take care of between-sample variability, while smoothing and data-tiling have been adopted to incorporate correlating signals at neighboring sites. Since detection of DMRs is the ultimate goal for many biological applications, a further step is typically

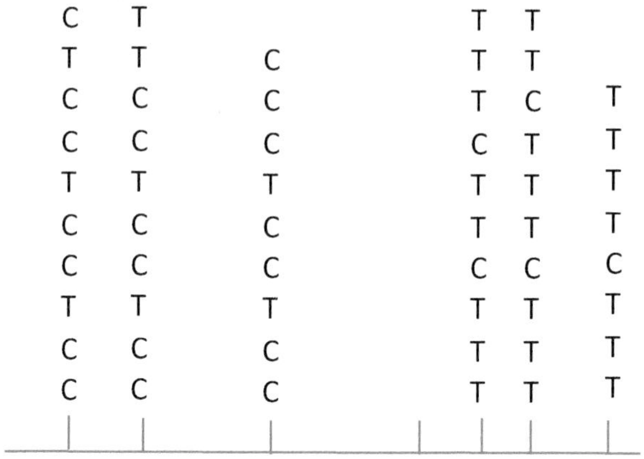

```
C   T                           T   T
T   T           C               T   T
C   C           C               T   C   T
C   C           C               C   T   T
T   T           T               T   T   T
C   C           C               T   T   T
C   C           C               C   C   C
T   T           T               T   T   T
C   C           C               T   T   T
C   C           C               T   T   T
 |   |           |             |   |   |   |
```

FIGURE 6.1
Schematic diagram of BS-seq data on CpG's.

followed after the detection of DMCs to find DMRs by aggregation. Two ways of aggregation have emerged: (1) aggregating the DMC signals in predetermined regions, such as CpG islands, CpG shores, and gene promoters, and (2) aggregating by finding dynamic regions through setting thresholds for DMC significance, minimum number of (consecutive) "significant" CpG sites, and size of a region. Methods have also been developed to take into account the correlation in the aggregation step by a multivariate transformation of the p-values. A single-step method also exists to construct credible bounds and obtain DMRs directly, taking sample variation and correlation into consideration directly and jointly. These methods will be described in the following subsections.

6.3.2   Fisher's exact test

Unless otherwise specified, the description and notation are for a single CG site. Let n_i and C_i be, respectively, the total number of reads and methylated reads covering a particular CG site for samples under condition $i, i = 1, 2$. Then

$$C_i \sim \text{Binomial}(n_i, p_i), \quad i = 1, 2,$$

where p_i, as defined earlier, is the probability that a read is methylated, which is interpreted as the methylation level at the CG site. The objective is therefore to test the following null hypothesis (H_0) against the alternative hypothesis (H_a):

$$H_0 : p_1 = p_2 \quad \text{vs.} \quad H_a : p_1 \neq p_2,$$

that is, the equality of two binomial probabilities against the two sided alternative. DMC will be declared if H_0 is rejected at a predetermined α level. One frequently used method for testing the equality of two binomial proportions is Fisher's exact test, utilized by Lister et al. [268] and also recommended in the software methylKit [3] and other more recent publications [480]. To apply Fisher's exact test, the data are organized into a 2×2 contingency table with two ways of classification: condition and methylation. The margins are assumed to be fixed: on one margin, they are the numbers of reads (n_1 and n_2); on the other margin, they are the methylated reads ($C_1 + C_2$) and the unmethylated reads ($n_1 - C_1 + n_2 - C_2$). Then the number of methylated reads for condition 1, C_1, among all the methylated reads $m = C_1 + C_2$ (assumed fixed) follows a hypergeometric distribution such that

$$P(C_1 = c \mid m, n = n_1 + n_2) = \frac{\binom{m}{c}\binom{n-m}{n_1-c}}{\binom{n}{n_1}},$$

where $\max\{0, m - n_2\} \leq c \leq \min\{m, n_1\}$.

One can compute the p-value by finding the probabilities of all combinations of methylated reads in the two samples, keeping the total number of methylated reads fixed; the p-value is the sum of the probabilities that are at least as large as the one observed:

$$\frac{\binom{C_1+C_2}{C_1}\binom{n_1+n_2-C_1-C_2}{n_1-C_1}}{\binom{n_1+n_2}{n_1}}.$$

Since tens of millions of CG sites are potentially considered in a study of the human genome, a much smaller comparison-wise p-value than 5% is needed to achieve a reasonable False Discovery Rate, say 5% as in Lister's study (p-value < 0.0002) [268].

There are a number of potential issues associated with the use of Fisher's exact test in this context. First, it is known that the test can be conservative when applied to testing the equality of two binomial proportions due to the fixed margin assumption (conditioning argument) [63, 271, 448]. Whether it is appropriate to assume that the total number of methylated reads is fixed is the subject of a recent investigation [278]. Second, if there are multiple samples within each group (e.g. biological replicates), then the between-sample variability is not being accounted for since all observations are combined within each group. This is a non-negligible issue as methylation is known to be extremely variable even for cells of the same type. Third, since each CG site is analyzed separately, spatial correlation among the sites is also ignored. This can lead to difficulty in interpreting the results. It also reduces the strength of information contained in the data, especially when coverage is low [157].

6.3.3   Logistic regression

We now consider a more general setting where covariates may be available; a second index is thus introduced to denote samples within a group. Let n_{ij} and C_{ij} be, respectively, the total number of reads and methylated reads for sample j in group i, $i = 1, 2; j = 1, \cdots, N_i$. Further, let $\boldsymbol{X}_{ij}$ denote the designed matrix (without a column of 1's) coding for a vector of covariates. Then

$$C_{ij} \sim \text{Binomial}(n_{ij}, p_{ij}),$$

where p_{ij} is the probability that a read from sample j in group i is methylated given its covariates. One can then link p_{ij} with the samples by means of a logistic regression model:

$$\text{logit}(p_{ij}) = \beta_0 + \beta_1 I(i = 2) + \boldsymbol{X}_{ij}\boldsymbol{\gamma}. \tag{6.1}$$

In the special case where there are no covariates [3], (6.1) becomes

$$\text{logit}(p_i) = \beta_0 + \beta_1 I(i = 2),$$

where p_i is the probability that a read is methylated and is the same for all samples in the same group, and β_1 is the logarithm of the ratio of the odds of a read being methylated for group 2 vs. group 1. To detect differential methylation between the two groups (i.e. $p_1 \neq p_2$), the equivalent hypotheses in terms of the model parameters are as follows:

$$H_0 : \beta_1 = 0 \quad \text{vs.} \quad H_a : \beta_1 \neq 0.$$

The likelihood for the parameters β_0 (nuisance parameter) and β_1 (parameter of interest) is simply

$$L(\beta_0, \beta_1) \propto \left(\frac{e^{\beta_0}}{1 + e^{\beta_0}}\right)^{C_1} \left(\frac{1}{1 + e^{\beta_0}}\right)^{n_1 - C_1} \left(\frac{e^{\beta_0 + \beta_1}}{1 + e^{\beta_0 + \beta_1}}\right)^{C_2} \left(\frac{1}{1 + e^{\beta_0 + \beta_1}}\right)^{n_2 - C_2},$$

where $C_i = \sum_{j=1}^{n_i} C_{ij}$ and $n_i = \sum_{j=1}^{n_i} n_{ij}$. From the above likelihood, we can see that the information (total number of reads and number of methylated reads) is combined in each group, and thus variability between samples within each group is not being accounted for. The likelihood can be directly maximized to obtain the maximum likelihood estimate (MLE) $\hat{\beta}_1$; standard error of $\hat{\beta}_1$ can also be obtained. Likelihood ratio test and the Wald test can be used to assess evidence against the null hypothesis. The score test can also be constructed without even needing to find the MLE. Confidence intervals may also be constructed for β_1, or equivalently, $p_2 - p_1$. However, all these procedures invoke large sample theory and require the sample size to be large, which may not be satisfied because the sample size is typically small and the coverage is low. Bootstrap sampling can be used, but it is likely not practical for such large-scale analysis as there are tens of millions of CG sites.

Under the logistic regression framework, one can easily extend (6.1) to handle samples from more than two groups, and furthermore, covariates characterizing the samples can be incorporated, which are advantages over Fisher's exact test [3]. When individual level covariates are available for each sample, the methylation probability for each sample will be different, which can account for some of the between-sample variability. However, there may still be variability (e.g. due to technical sources) not due to the measured covariates, which cannot be accounted for with the logistic regression procedure.

6.3.4 Beta-binomial formulations

Variation in methylation from sample to sample within a group could be attributed to a variety of technical and biological sources, such as natural variation and unequal cytosine conversion rates. To account for variability between samples (e.g. biological replicates) within the same group, the beta distribution has been utilized to accommodate this. Using the same notation as in Section 6.3.3, we let

$$C_{ij} \sim \text{Binomial}(n_{ij}, p_{ij}) \quad \text{and} \quad p_{ij} \sim \text{Beta}(\mu_{ij}, \phi_{ij}), \tag{6.2}$$

where μ_{ij} and ϕ_{ij} are unknown parameters representing the mean and variance of p_{ij} : $E[p_{ij}] = \mu_{ij}$ and $var(p_{ij}) = \mu_{ij}(1 - \mu_{ij})\phi_{ij}$. Note that this is a different parameterization from the usual formulation with parameters α_{ij} and β_{ij}. However, $\mu_{ij} = \alpha_{ij}/(\alpha_{ij} + \beta_{ij})$ and $\phi_{ij} = 1/(\alpha_{ij} + \beta_{ij} + 1)$, or conversely, $\alpha_{ij} = \mu_{ij}(1 - \phi_{ij})/\phi_{ij}$ and $\beta_{ij} = (1 - \mu_{ij})(1 - \phi_{ij})/\phi_{ij}$; thus, the two parameterizations are equivalent. One can integrate over p_{ij} to obtain the marginal distribution $P(C_{ij} \mid n_{ij}, \mu_{ij}, \phi_{ij})$, which is known as the beta-binomial distribution:

$$P(C_{ij} \mid n_{ij}, \mu_{ij}, \phi_{ij}) = \binom{n_{ij}}{C_{ij}} \frac{B(C_{ij} + \alpha_{ij}, n_{ij} - C_{ij} + \beta_{ij})}{B(\alpha_{ij}, \beta_{ij})},$$

where B denotes the usual beta function. Further,

$$E(C_{ij}) = n_{ij}\mu_{ij} \quad \text{and} \quad Var(C_{ij}) = n_{ij}\mu_{ij}(1 - \mu_{ij})(1 + (n_{ij} - 1)\phi_{ij}).$$

Assuming $n_{ij} > 1$, then $(1 + (n_{ij} - 1)\phi_{ij}) > 1$ unless $\phi_{ij} = 0$. As such, ϕ_{ij} is referred to as the over-dispersion parameter, as it "inflates" the variance of an ordinary binomial distribution.

Suppose $\boldsymbol{X}_{ij}$ denotes the design vector coding for a set of covariates (including a 1 for the intercept) and assume that

$$g(\mu_{ij}) = \boldsymbol{X}_{ij}\boldsymbol{\gamma},$$

where $\boldsymbol{\gamma} = (\gamma_0, \gamma_1, \cdots)$ is the corresponding vector of regression coefficients and g is the link function. Typical link functions are logit as in RADMeth [86] and probit as in Biseq [166].

Suppose the only covariate for each sample is its group membership and there are only two groups being considered. Then $g(\mu_{ij}) = \gamma_0$ if $i = 1$ and $g(\mu_{ij}) = \gamma_0 + \gamma_1$ if $i = 2$, which also implies that $\mu_1 = \mu_{1j}$ for $j = 1, \cdots, N_1$ and $\mu_2 = \mu_{2j}$ for $j = 1, \cdots, N_2$. As such, for each CG site, testing whether the locus is differentially methylated is basically testing the hypotheses

$$H_0 : \gamma_1 = 0 \text{ vs. } H_1 : \gamma_1 \neq 0 \quad [86, 166].$$

This is equivalent to testing

$$H_0 : \mu_1 = \mu_2 \text{ vs. } H_1 : \mu_1 \neq \mu_2 \quad [107, 332].$$

The advantage of the beta-binomial regression formulation is that, if there are multiple groups, then their effects can be accommodated by setting up indicator variables. Further, other covariates, if they do exist, can be factored into the model as well. Several existing beta-binomial based methods are quite similar, although each has its own unique features. For instance, MOABS [421] and MethylSig [332] can be regarded as special cases of RADMeth [86] in that the latter can handle situations when there are covariates beyond group labels, although the methods for parameter estimations are different (see below). However, each offers additional features. For example, MethylSig uses smoothing and has the option of performing tiling analysis [332].

6.3.4.1 Parameter estimation.

Various methods may be used to estimate the parameters so that statistical inferences can be drawn. In the following, we discuss both likelihood-based and Bayesian-based methods.

Methods of Moment (MOM) estimation. When the only covariate for each sample is its group membership, then one can use the MOM to estimate the parameters μ_i and ϕ_i, as can be seen below in the empirical Bayes esimation procedure below. Although simple, they are not useful in the testing step if the likelihood ratio test or Wald test are used for statistical inference.

Likelihood-type estimation. For statistical inference, the overdispersion parameter is usually assumed to be the same for each group [107]; that is, $\phi_i = \phi_{ij}, j = 1, \cdots, n_i$. A more extreme setting, in which there is only a single common overdispersion parameter for all groups, has also been entertained [86, 332].

A number of methods have used maximum likelihood estimates (MLEs) to carry out tests, but some do not elaborate on how the MLEs were obtained [86, 421]. In the general setting of g groups with additional covariate information potentially available, the likelihood of the model parameters of the beta-binomial regression can be written down:

$$L(\gamma_i, \phi_i, i = 1, \cdots, g \mid C_{ij}, n_{ij}, j = 1, \cdots, n_i, i = 1, \cdots, g)$$

$$= \prod_{i=1}^{g} \prod_{j=1}^{n_i} \binom{n_{ij}}{C_{ij}} f(C_{ij}, n_{ij}, g^{-1}(\boldsymbol{X}_{ij}\boldsymbol{\gamma}), \phi_i, j = 1, \cdots, n_i, i = 1, \cdots, g),$$

where $f(\cdot)$ is a function of the data and parameters.

One maximum likelihood estimation method is given in Griffiths [140]. Another is discussed in Tripathi et al. [441]. It is known that there is no closed-form solution to obtain the MLEs. If n_{ij} are all the same for each group, then the MLEs may be found via direct numerical optimization using methods for fitting multinomial Polya distributions [292]. However, for BS-seq data, the n_{ij}'s are typically not all the same. One possible solution is to assume that ϕ_i is fixed and estimate γ using Newton-Raphson [118]. Since the MLE of γ is a function of ϕ_i, one can then construct a profile likelihood of ϕ_i to obtain its estimate [118].

If group membership is the only covariate (back to the simple beta-binomial setting), there is still no closed-form solution. For this setting, an alternative method has been proposed [332]. The group mean is first used to estimate μ_i: $\hat{\mu}_i = \sum_{j=1}^{n_i} C_{ij} / \sum_{j=1}^{n_i} n_{ij}$. Then an MLE is obtained for ϕ, assuming there is a common dispersion parameter, with the constraint that it is non-negative. Now plug this estimate into the likelihood again to obtain the MLE for μ_i directly. Among the several methods used for obtaining approximate MLEs, their relative performances are unknown.

Empirical Bayes estimation. For empirical Bayes-type estimators, the idea is to cast the problem into the Bayesian setting and use information across the entire genome (or genomic regions) to estimate the hyper-parameters of the prior distribution. Specifically, consider the setting in which there are no other covariates apart from the group label for each sample. Then $\mu_{ij} = \mu_i$ and $\phi_{ij} = \phi_i, j = 1, \cdots, n_i, i = 1, \cdots, g$. The Beta-binomial formulation in Section 6.2 can be interpreted as a Bayesian formulation in which $p_{ij} \sim \text{Beta}(\mu_i, \phi_i)$ is the prior distribution for p_{ij}. In empirical Bayes, one uses information from all CG sites to estimate the hyperparameters. To do this, we need to further expand our notation to include the CG site explicitly, as all CG sites in the genome (or genomic region being analyzed) will be considered jointly: we use μ_i^s and ϕ_i^s to denote the mean and over-dispersion from group i and CG site s. Let $\phi_i^s \sim F(\tau_i^0)$ be the prior distribution of the over-dispersion parameter, where F is some probability measure that is parameterized by τ_i^0, a vector of hyper-parameters that is only dependent on the group label, not specific CG sites.

To find the distribution of F and the hyper-parameter vector τ_0^i, one may use MOM to estimate the mean and the over-dispersion parameter for each CG site s and each group i, noting that the observed binomial data are for the particular site s (even though it is not explicitly written):

$$\hat{\mu}_i^s = \frac{\sum_{j=1}^{n_i} C_{ij}}{\sum_{j=1}^{n_i} n_{ij}},$$

$$\hat{\phi}_i^s = \frac{\sum_{j=1}^{n_i} C_{ij}^2 - (\hat{\mu}_i^s)^2 \sum_{j=1}^{n_i} n_{ij}^2 - \hat{\mu}_i^s(1 - \hat{\mu}_i^s) \sum_{j=1}^{n_i} n_{ij}}{\hat{\mu}_i^s(1 - \hat{\mu}_i^s) \sum_{j=1}^{n_i} n_{ij}(n_{ij} - 1)}.$$

A parametric distribution may be fitted to the collection of overdispersion estimates $\{\hat{\phi}_i^s, s \in \mathcal{S}\}$, where $\mathcal{S}$ is the set of all CG sites in the genome. For example, using publicly available BS-seq data on an early mouse embryo [406], a log-normal model was fitted to the overdispersion data across the CG sites, with estimated hyper-parameter values $\boldsymbol{\tau}_i^0$ (containing mean and variance) [107]. One may similarly fit a parametric distribution to $\{\hat{\mu}_i^s, s \in \mathcal{S}\}$ to specify the prior distribution for μ_i. With the prior for μ_i and ϕ_i completely specified, one can then find the posterior distribution of μ_i and ϕ_i for each CG site.

A simpler procedure was suggested in [107], in which the focus was on the estimation of the overdispersion parameter, and thus only the empirical Bayes prior for ϕ_i was obtained. The mean parameter, μ_i, was pre-estimated based on the sample mean for each CG site: $\hat{\mu}_i = \sum_{j=1}^{n_i} C_{ij} / \sum_{j=1}^{n_i} n_{ij}$. The posterior for ϕ_i was then obtained and maximized to obtain an estimate $(\hat{\phi}_i)$ through Newton-Raphson method and plugging in the estimates $\hat{\mu}_i$ and $\boldsymbol{\tau}_i^0$. This was performed for each of the groups separately.

In summary, in beta-binomial based methods, the goal is to accommodate variability of methylation levels between samples within a group, which can be driven by underlying covariates that may or may not be explicitly modeled. If they are explicitly modeled, this becomes a beta-binomial regression problem. In this case, the mean parameter for the beta-binomial distribution is different for each sample; the coefficient associated with the group label is the main parameter of interest. The covariates are used to account for confounding and their associated parameters may be of interest if their effects are being focused on and interpretable (e.g. whether age modifies the methylation level). If they are not explicitly modeled, then this is simply a beta-binomial problem. The methods are all similar, differing only in the details. One example where the methods differ in details is the assumption on the over-dispersion parameters. Some assume that there is a single over-dispersion parameter across all the groups and samples [86, 166, 332]; others allow for different over-dispersion parameters for different groups [107, 421]. There are also a variety of ways for estimating the model parameters. Some are likelihood-based [86, 166, 332], while some are based on empirical Bayes [107, 421].

6.3.4.2 Statistical inference – hypothesis testing

For methods that are based on beta-binomial without considering additional covariates [107, 166, 332, 421], after the parameters are estimated, various procedures have been proposed to test the hypotheses. In the two-group setting, the null hypothesis of no group difference is tested against the alternative hypothesis of having a difference in the two group means:

$$H_0 : \mu_1 = \mu_2 \text{ vs. } \mu_1 \neq \mu_2.$$

For example, for DSS [107], after the over-dispersion parameter ϕ_i is estimated for each group i based on its posterior distribution, the variance of the mean estimate (obtained assuming a common mean for all samples in a group) can then be estimated. This leads to a Wald test:

$$T = \frac{\hat{\mu}_1 - \hat{\mu}_2}{\sqrt{v\hat{a}r(\hat{\mu}_1) + v\hat{a}r(\hat{\mu}_2)}}.$$

The T statistic is assumed to follow a normal distribution, which is only an approximation based on empirical results [107]. Note that this test is performed site by site.

Under the beta-binomial regression setting when covariates are utilized, the main hypotheses are now

$$H_0 : \beta_1 = 0 \text{ vs. } \beta_1 \neq 0,$$

where β_1 is the coefficient representing the difference between the two groups. The testing procedure under RADMeth [86] falls into this category, with the assumption that the over-dispersion parameter is the same for both groups. The parameters for the full model (i.e. with β_1 as a free parameter to be estimated) and those for the constrained model (i.e. setting $\beta_1 = 0$) are estimated separately, and the estimated likelihood (after plugging in the estimated parameter values) are denoted as L_1 and L_0, respectively. Then

$$T = -2\ln(L_0/L_1)$$

is the likelihood ratio test statistic, and it follows the χ_1^2 distribution asymptotically. The null hypothesis H_0 is rejected in favor of the alternative hypothesis H_1 if the statistic T is large. The eventual p-value for each site is found by combining not just the p-value from its own likelihood ratio test but also from neighboring sites through the Z-transformation [86, 166].

6.3.5 Smoothing

It is known that methylation signals within 1000 bases in dense CpG regions, including those in the promoter regions, are spatially correlated [95, 157, 195]. Therefore, a number of methods have been proposed to take such correlation into account by applying kernel smoothing to borrow information from neighboring loci and help improve the measurement accuracy of the methylation signals. This is of particular relevance in situations where there are missing data or low coverage. Two of the methods that are beta-binomial based apply this technique to boost their measurements [166] or estimation [332]. A t-type statistic is also proposed after kernel smoothing [157].

6.3.5.1 Smoothing as part of the beta-binomial procedures

For each CG site, the methylation level for a sample may be improved by borrowing information from its neighboring sites. The neighboring sites constitute a local region ($\mathcal{R}$) surrounding the locus. The region is typically defined to be within a distance (number of base pairs) from the CG site. For example, in BiSeq [166], kernel smoothing is used to estimate the mean at each CG site (s^*) using its own observed data as well as data from other CG sites within a region less than 80 base pairs from s^* (therefore, 80 base pairs is referred to as the bandwidth of the kernel smoothing method). The smoothed methylation level, $\hat{p}_{ij}^{s^*}$, is the one that maximizes the following weighted log-likelihood function:

$$l(p_{ij}^{s^*} \mid C_{ij}^s, n_{ij}^s, s \in \mathcal{R}) = \sum_{s \in \mathcal{R}} K(s, s^*) \log \left[\binom{n_{ij}^s}{C_{ij}^s} (p_{ij}^{s^*})^{C_{ij}^s} (1 - p_{ij}^{s^*})^{n_{ij}^s - C_{ij}^s} \right],$$

where C_{ij}^s and n_{ij}^s are the observed methylated and total reads at site s, and K is the kernel function, which may be Gaussian or triangular, for example. In Biseq, a triangular kernel is used. The smoothing step is performed for each sample and each CG site s^*, and the estimated $\hat{p}_{ij}^{s^*}$ is now treated as coming from a beta distribution.

On the other hand, in [332], kernel smoothing is used to help improve the estimation of model parameters. Assuming the over-dispersion parameter is the same for both groups, the likelihood of ϕ^{s^*} at site s^* is evaluated not only at the site itself but also at all sites in the local region ($\mathcal{R}$) surrounding the site (set to be within 200–300 base pairs of s^* [332]). The objective function to be estimated is then the sum, with each term weighted by the kernel:

$$A(\phi^{s^*} \mid \hat{\mu}_1^s, \hat{\mu}_2^s, s \in \mathcal{R}) = \sum_{s \in \mathcal{R}} K(s, s^*) l(\phi^{s^*} \mid \hat{\mu}_1^s, \hat{\mu}_2^s),$$

where $K(\cdot)$ is the kernel (triangle kernel is also used in MethylSig [332]), and $\hat{\mu}_1^s$ and $\hat{\mu}_2^s, s \in \mathcal{R}$ are the estimated means at all sites within the local region $\mathcal{R}$. After the dispersion parameter is estimated, the mean parameter will also be estimated using the kernel smoothing technique by maximizing:

$$B(\mu_1^{s^*}, \mu_2^{s^*} \mid \hat{\phi}^s, s \in \mathcal{R}) = \sum_{s \in \mathcal{R}} K(s, s^*) l(\mu_1^{s^*}, \mu_2^{s^*} \mid \hat{\phi}^s).$$

6.3.5.2   BSmooth

As discussed at the beginning of this chapter, WGBS is very expensive, as a coverage of 30 times is typically required for the data to be viewed as informative enough to detect differential methylation. This cost-prohibitive requirement needs more than 350 lanes of sequencing on an Illumina GA II instrument. To use the data more efficiently so a lower coverage is acceptable, BSmooth [157] was developed to make use of neighboring signals to improve estimation accuracy, which led to an observation that the coverage can be cut down to only 4 times. Using the same notation as above, we let $Y_{ij} = C_{ij}/n_{ij}$ denote the observed methylation level for observation j in group $i, i = 1, \cdots, g$. We further include the site notation such that Y_{ij}^s is the methylation level at

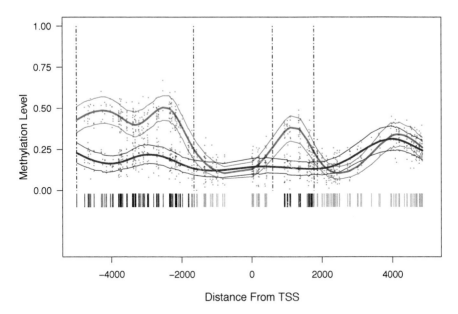

FIGURE 6.2

Example data and schematic diagram of BCurve. The CG sites in a Transcription Start Site (TSS) region, also known as a promoter region, are marked by short vertical lines. Each circle represents a proportion of methylated read for a sample at a CG sites. There are three diseased samples (marked by red circles) and three normal samples (marked by blue circles). The thick orange curve is the estimated overall methylation level in this region for the diseased group, which is bounded by the credible band within the two thin orange curves. Similarly, the thick green curve and the region bounded by the two thin green curves are the estimated overall methylation level and the confidence band, respectively, for the normal group. A region in which the two confidence bands do not overlap is identified as a differentially methylated region and marked by two dotted vertical line at its boundaries.

site s; such values are shown as red and blue circles in Figure 6.2, where each CG site s is marked by a short vertical line at the bottom.

In BSmooth, C_{ij}^s is assumed to follow a binomial distribution; that is, $C_{ij}^s \sim \text{Binomial}(n_{ij}^s, p_{ij}^s)$, with Y_{ij}^s being the estimate of p_{ij}^s whose standard error is $\sqrt{p_{ij}^s(1 - p_{ij}^s)/n_{ij}^s}$. It is assumed that the p_{ij}^s along a chromosomal region falls on a smooth curve f; that is, $p_{ij}^s = f(s)$. To estimate f, a kernel is chosen (specifically, a tricube kernel is used in BSmooth [157]), with the local region ($\mathcal{R}$) required to include at least 70 CG sites and be at least 2 kilo-bases in length. A weighted logistic model is fitted to data in $\mathcal{R}$, with two components in its weights: the first component is inversely proportional to the estimated standard deviation of the methylation signal, and the second component is obtained from a tricube kernel. The "smooth" estimate at each CG site is predicted from the weighted logistic model. In other words, the estimated methylation level $Z_{ij}^{s^*}$ is not only dependent on $Y_{ij}^{s^*}$, but also Y_{ij}^{s} for all $s \in \mathcal{R}$, with s closer to s^* contributing more and those with smaller variability getting more weights. The smoothing is done for each sample in each group. Suppose $g = 2$, then the difference of the estimated averages of the two groups is divided by the estimated standard deviation of the difference to obtain a t-like test statistic. Differentially methylated loci can then be declared by using a threshold corresponding to a certain significance level of a t-distribution.

6.4   Methods for detection of DMRs using BS-seq data

6.4.1   Rule-based procedure – follow-up on DMCs

A rule-based procedure is typically the second step of a method that identifies DMCs in its first step. As motivated earlier, biologically, it is frequently of greater interest to find an entire region that is differentially methylated than individual DMCs. However, because of the nature of the data being binomial, it is natural for many methods to be proposed for detecting DMCs. The detection of such can be thought of as the first of a procedure for detecting DMRs. Indeed, after DMCs are detected, there are rule-based procedures to define DMRs. In BSmooth [157], after all DMCs are found, DMRs will be formed by aggregating consecutive DMCs and requiring that a DMR region has at least three consecutive DMCs and any two consecutive CG sites cannot be separated by more than 300 base pairs. A further requirement is that the average of the numerators of the t-statistics needs to be greater than 0.1. In other words, on average, the methylation levels in these CpGs need to differ by more than 10% of maximum methylation. The application of BSmooth in the comparative study performed in RADMeth [86] used an even more stringent rule. Many other methods also use rule-based procedures to form DMRs [107, 166, 332], although there is little consistency among the methods, as such rules are ad-hoc in nature.

6.4.2   Credible band procedure – a single step approach

As discussed above, most methods for detecting DMRs are two-step procedures. The first step finds the DMCs, which is followed by a rule-based procedure to define DMR regions. BCurve, a single-step method that finds DMRs directly, provides a different approach [329]. BCurve operates in each CpG cluster, which is either obtained from a clustering algorithm or in predefined regions, such as gene promoters. For CpG clusters formed by executing a clustering algorithm, the goal is to find regions in which any two consecutive CG sites whose methylation levels are measured are not far away from each other. In a nutshell, BCurve assumes that the mean methylation levels within each cluster fall on a smooth curve with a confidence band. In the two-group setting, regions with a cluster for which the two confidence bands from the two groups do not overlap are taken to be DMRs. Figure 6.2 provides a schematic diagram for the BCurve method for a promoter region.

The mathematical formulation unfolds as follows. In addition to the notation used above, let us introduce an indicator variable, $Y_{ij(k)}$, which takes the value of 1 if the kth read in sample j within group i is methylated; otherwise it is 0. Therefore, $C_{ij} = \sum_{k=1}^{n_{ij}} Y_{ij(k)}$ and $p_{ij} = P(Y_{ij(k)} = 1)$, where p_{ij}, C_{ij}, and n_{ij} are as defined above. Instead of working with p_{ij} directly, BCurve further introduces a latent Gaussian representing the underlying methylation level, $Z_{ij(k)} \sim N(\mu_{ij}, 1)$, which is appropriately standardized such that $Y_{ij(k)} = I(Z_{ij(k)} > 0)$, where $I(\cdot)$ is the usual indicator function. Then $p_{ij} = \Phi(\mu_{ij})$, where Φ is the standard normal cumulative distribution function. Suppose we use $\mathcal{R}$ to denote a CG cluster, which is essentially a region in the genome. It is at the latent methylation level where a smooth curve is introduced, which has much more flexibility in modeling. Specifically, we model the mean methylation level for sample j in group i at each CG site s in $\mathcal{R}$, μ_{ij}^s, by means of a cubic spline, but also allow for between-sample variability within a group. Mathematically,

$$\mu_{ij}^s = \boldsymbol{b}^s \boldsymbol{\beta} + \alpha_{ij}, \ s \in \mathcal{R},$$

where $\boldsymbol{b}^s$ is a row vector of cubic splines (basis functions), and $\boldsymbol{\beta}$ is a column vector of parameters. Moreover, α_{ij} is a random effect to capture the within-group, between-sample variability, and it is assumed to follow $N(0, \sigma_{i,\alpha}^2)$ independently and identically for all samples within a group. However, note that the variability can be different in different groups. Therefore, the notion that cancer samples are more variable than normal samples can be accommodated, for example. In other words, this modeling scheme allows for flexibility so there is an overall mean methylation curve for each group; each sample within a group, in turn, has its own mean curve to allow for between-sample variability. The overall mean and confidence band at the underlying methylation level can then be translated back to the corresponding ones for the probability of methylation, as depicted in Figure 6.2, which shows one confidence band for the diseased group (orange curves) and one for the normal group (green curves). In this schematic diagram, there are two DMRs in this CG cluster, each bounded by a pair of dashed vertical lines. As one can see from the figure, the DMRs are regions in which the two confidence bands do not overlap.

6.4.3   Summary of methods for BS-seq data – Which methods to choose?

There are a number of methods and software available for analyzing BS-seq data, and therefore, it would be important to be equipped with the knowledge of which method(s) are appropriate for a particular problem. A recent comprehensive comparison of four beta-binomial-based methods was carried, which points to the consistent performance of RADMeth [151]. In general, if multiple samples are present in each group, then Fisher's exact test or MethylKit are not recommended since between-sample variability is not accounted for. If covariates are available and are believed to be potential confounders such that their effects should be accounted for, then beta-binomial regression models such as those implemented in RADMeth and BiSeq, are appropriate choices. An update of DSS, DSS-general [333], also takes covariance into consideration through regression modeling. If the focus is on detecting DMRs between two groups, and it is important to account for spatial correlation among methylation levels at neighboring CG sites, then BCurve would be a good choice, as it is a procedure based on statistical inference rather than based on ad-hoc rules to define DMRs. Finally, it is noted that the methods described in this chapter are by no means the only ones available in the literature. A list of 22 procedures and software packages were reviewed in a study [392], which includes the beta-binomial methods described in this chapter, as well as methods that are based on hidden Markov models [422, 513] or Shannon entropy [272, 417, 529].

6.5   Methods for detection of DMRs using Cap-seq data

6.5.1   Cap-seq data

Cap-seq data are inherently different from BS-seq data. Cap-seq data do not go through the bisulfite conversion process; rather, fragments containing methylated CG sites are "pulled down" and sequenced. There are two types of protocols for obtaining Cap-seq data. One type relies on methyl-binding protein, with MethylCap-seq as an example [120]. Briefly, DNA extracted from a sample is sonicated to fragments of 100–300 bps in length. Those containing at least one methylated CpG site are then captured by biotinylated methyl-CpG binding protein and pulled down, forming the methylated fragment library. Another type relies on antibody immunoprecipitation, with MeDIP-seq as a prime example [424]. Pulldown DNA fragments are then purified and sequenced. Typically, a single-end read is obtained from each pulldown fragment and mapped to the genome. For instance, short reads of 36 base pairs are obtained for MethylCap-seq; a total of approximately 40 million reads can result from a sample.

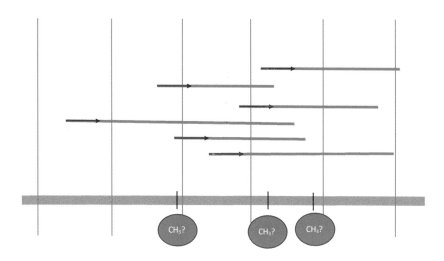

FIGURE 6.3
Schematic diagram of Cap-seq data on CpG's.

Figure 6.3 provides a schematic diagram for Cap-seq data in a DNA segment (displayed as a thick gray horizontal bar) harboring three CpG sites (marked by short black vertical line segments). Each pulldown DNA segment is depicted by a blue horizontal line (differing in length). Recall that a pulldown fragment contains at least one methylated CpG site; as we can see from the figure, each blue line segment is over at last one CpG sites. However, it is unknown, from the experimental protocol, which of the CG sites (if the segment covers more than one) is methylated and thus responsible for the pulldown of the segment. To complicate things further, the read may not cover any CG site because it is only a small portion at one end of the pulldown DNA segment. Furthermore, the pulldown fragment length is also lost after the sequencing step. In other words, what are available from the Cap-seq data are simply the reads – the short red arrowed segment of the same length at one end of every pulldown fragment – we do not observe the blue pulldown fragments in their entirety. However, note that the distribution of the lengths are available from the library of pulldown fragments, which are 100 – 300 base pairs in length (with an average of 150 to 200). Therefore, Cap-seq data are not of nucleotide resolution; they are data providing a 150 to 200 base pairs resolution of the methylome [424]. The distributions superimposed on the 3-CpG-sites region in Figure 6.4 provide an example of the fragment-length distribution.

Statistical methods for analyzing Cap-seq data can be broadly divided into two categories: those based on the short reads directly [424], and those

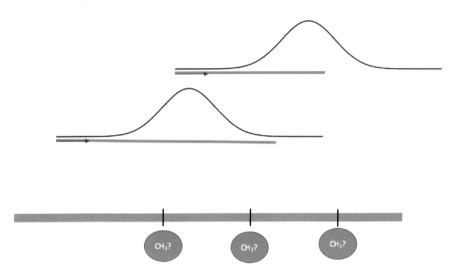

FIGURE 6.4
Schematic diagram of fragment-length library distribution and how PrRMeRCG distributing each of the reads to the methylation level of each CG site.

that infer the methylation levels at CpG sites first before launching statistical analysis to answer scientific questions (two-step methods) [120]. We describe both the direct and the two-step methods in the following two subsections.

6.5.2 Direct methods

6.5.2.1 Quantification of methylation signals

Given that Cap-seq reads are not uniquely assigned to a CG site, a common way of quantifying the signal is a "regional" based approach. The typical scheme is to divide the genome into equal-length windows/bins (e.g. 100 base pairs (bp)), and reads are then assigned to each bin according to certain rules. This scheme is depicted by the blue vertical lines in Figure 6.3, which divide the chromosomal segment into 5 bins of equal length. In Yan et al. [500], each read was first extended to its average fragment length computed from the fragment library. Each extended read is then assigned to one 100 bp window using a majority rule. Because different samples have different number of reads, the raw read counts are then normalized to "reads per million" across all the samples, leading to "counts" that can be used for cross-sample analysis. This is achieved through dividing the counts by the total number of reads for each library/sample and then multiplying by one million.

One of the problems with such a window-based method is the observation of phantom/ghost methylated windows, which refers to windows that do not contain any CpGs but have reads ascribed to them, an artifact due to binning. For example, data from 10 acute myeloid leukemia (AML) samples lead to 11, 8 and 7% of windows without CpGs but having reads assigned to them for windows of sizes 100, 200 and 500 bp, respectively [120].

Alternatively, one may consider well-annotated regions for finding DMRs, such as CpG islands, CpG shores, or gene promoter regions. Although focusing on such regions will eliminate the problem of identifying phantom DMRs, the problem of ignoring spatial correlation remains.

6.5.2.2 Detection of DMRs

For detecting DMRs within each window, a t-test can be performed [263,500] for situations in which two groups/conditions are being compared. Let X_{ij} denote the normalized read for sample j in group $i, j = 1, \cdots, n_i$ and $i = 1, 2$. Then the hypothesis of interest is that the mean methylation level in group 1, μ_1, is the same as the mean methylation level in group 2, μ_2; that is, the null and alternative hypotheses are

$$H_0 : \mu_1 = \mu_2 \text{ vs. } H_1 : \mu_1 \neq \mu_2.$$

One can then construct a t-test statistic as

$$t = \frac{\bar{x} - \bar{y}}{\sqrt{\frac{s_x^2}{n_1} + \frac{s_y^2}{n_2}}},$$

where $\bar{x}$ and $\bar{y}$ are the sample means and s_x and s_y are the sample standard deviations of groups 1 and 2, respectively.

To account for over-dispersion in sequencing read data, the negative binomial distribution has been used to model read counts, and it is adopted in MEDIPS [263]. Further, empirical Bayes can be used to borrow information across the genome (or a genomic region) to provide "moderated" variance estimates [366]. In MEDIPS, a procedure similar to Fisher's exact test [366] is used to detect DMRs.

Multiplicity adjustment procedures, such as FDR, are typically used to address the issue of multiple windows being tested. In addition to the need to address the multiple testing issue, there are other problems concerning window-based approaches. Phantom DMR as mentioned before is clearly an issue that should not be overlooked. Another issue concerns the fundamental practice of pooling reads that are mapped to a window, as such an action ignores the intrinsic correlation among reads induced by spatial correlation between methylation signals in neighboring CpG sites. Therefore, despite great cost benefit, the lack of nucleotide resolution data and the lack of adequate analysis methods both contribute to the reluctance in the use of Cap-seq techniques.

6.5.3   Two-step methods

Two-step methods address the problems in analyzing Cap-seq data head-on, with the first step devoted to extracting signals from reads to obtain "derived" nucleotide-level data, while the second step takes correlation between such data into account.

6.5.3.1   Derivation of nucleotide-level data

To infer nucleotide-level data, PrEMeR-CG [120] integrates information on the fragment lengths from each sample library with the reads. Let $L_f(> 0)$ be the random variable denoting the length of fragment f. Suppose there are a total of n fragments (which is in the order of tens of millions). The genomic location (in bp) of the beginning of the read corresponding to fragment f is denoted by $r_f, f = 1, \cdots, n$. Let $\mathcal{S}$ denote the collection of all CpG sites in the genome (or the genomic region being investigated), with g_s being the genomic location of $s \in \mathcal{S}$. For the read corresponding to fragment f, although its exact length L_f is unknown, the fragment library provides a distribution for it. That is, the survival function $P(L_f > l)$ for any $l \geq 0$ can be obtained from the distribution. As a convention, we let $P(L_f > l) = 0$ for $l < 0$, as we assume all reads are from $5'$ to $3'$ without loss of generality. Then, for a specific CG site $s \in \mathcal{S}$, its nucleotide-level methylation is assigned the following value

$$x^s = \sum_{f=1}^{n} \frac{P(L_f > g_s - r_f)}{\sum_{t \in \mathcal{S}} P(L_f > g_t - r_f)}.$$

In words, each read is distributed to CG sites in the genome, with (downstream) CG sites closer to the read getting a greater share, as the share is proportionate to the complement of the cumulative distribution function (that is, the survival function) of the fragment length random variable.

In practice, since the fragment length is finite, CG sites that obtain a share of the read are those that are "within reach" of the corresponding fragment of unknown length, as those beyond the reach are getting a share of 0. For example, suppose there are only two reads in the region spanned by 3 CpG sites as shown in Figure 6.4. Since the first CpG locus is within the reach of the only read on the left (the first read), its methylation level is its proportionate share (with the second CpG locus since it is also within the reach of the read - that is, having positive survival probability under the fragment-length distribution) according to the fragment-length distribution. On the other hand, the second CpG locus is covered by the distributions of both reads, so its methylation level will be distributed from both reads, with the majority of the contribution from the second read. Finally, the third CpG locus is only within reach of the second (but not the first) read, and will have its share of the methylation signal with the second CpG locus. However, in this hypothetical example, the fragment is in fact short and does not even cover the third CpG site, and therefore its actual methylation level would have been zero. This example

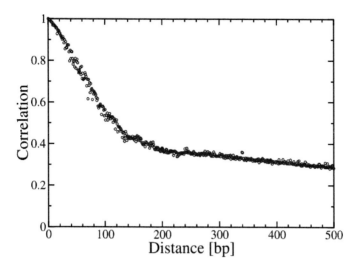

FIGURE 6.5
Correlation between neighboring CpGs decreases with distance. Correlations within the maximum length of a fragment (up to 300 bps) are likely due to the methylation signal distribution based on PrEMeR-CG; those beyond the maximum fragment length are believed to be of biological origin. (This figure is reproduced from supplementary Figure 3 of the PrEMeR-CG paper [121].)

shows the operational characteristics of the PrEMeR-CG method, which may lead to incorrect inference as it is based on probabilistic assignment. Since the number of reads that are successfully mapped to the genome are different from sample to sample, the obtained methylation values are normalized to reads per million, in the same fashion as in the window-based approach [500].

6.5.3.2 Detection of DMRs

With nucleotide-level methylation values, a choice of methods can be applied to detecting DMRs. However, methods for detecting DMCs or DMRs proposed for BS-seq data are not directly applicable, as the resulting values from the first step are not necessarily whole numbers and do not follow a binomial distribution. To detect DMRs for a predefined region, spatial correlation in the derived methylation values need to be taken into account, as such correlation can be quite strong and should not be ignored (Figure 6.5 [121]). Shown in the figure are correlations of PrEMeR-CG values between pairs of CG sites, plotted according to the distances of the pairs. As we can see from the figure, correlation decreases with the increase of distance: when the distance reaches the maximum possible length of a fragment (say 300 bps), there is still a fair amount of correlation (> 0.3). This demonstrates that, indeed, the

PrEMeR-CG derived signals between neighboring sites are correlated, due to both the shared signals from the same reads (those within 300 bps, for example) and the inherent biological correlations for methylation levels at nearby CpG sites.

For a region with m CG sites collectively denoted as $\mathcal{M}$, let $X_{ij} = (X_{ij}^s, s \in \mathcal{M})$ denote the derived methylation signals at the m CG sites for sample j in group i, $j = 1, \cdots, n_i$, and $i = 1, 2$. Let $\boldsymbol{\mu}_1$ and $\boldsymbol{\mu}_2$ denote vectors of length m of mean methylation signals for the two groups at the m CG sites, respectively. Then the null hypothesis that the region is not a DMR vs. the alternative hypothesis that the region is a DMR can be expressed concisely:

$$H_0 : \boldsymbol{\mu}_1 = \boldsymbol{\mu}_2 \text{ verus } H_1 : \boldsymbol{\mu}_1 \neq \boldsymbol{\mu}_2.$$

Since the dimension of the mean vectors, m, can easily exceed the combined sample size from the two groups, $n_1 + n_2$, traditional multivariate tests, such as the Hotelling's T^2 statistic, cannot be applied. Further, the normality assumption is clearly violated as there are excess of zero values [120]. As alternatives, MethylCapSig [16] implements other tests, including a recent method [50], which requires neither assumptions and is referred to as T_{CQ}. The test statistic is

$$T = \frac{\bar{X}_1^2 + \bar{X}_2^2 - 2\bar{X}_{1,2}}{\sqrt{2tr(\hat{\Sigma}_{X_1}^2)/[n_1(n_1-1)] + 2tr(\hat{\Sigma}_{X_2}^2)/[n_2(n_2-1)] - 4tr(\hat{\Sigma}_{X_{1,2}})}},$$

where $\bar{X}_j^2 = \sum_{i_1=1}^{n_j} \sum_{i_2=i_1+1}^{n_j} X_{i_1 j} X_{i_2 j}'/[n_j(n_1-j)], j = 1, 2, \bar{X}_{1,2} = \sum_{i_1=1}^{n_1} \sum_{i_2=1}^{n_2} X_{i_1 1} X_{i_2 2}/(n_1 n_2)$, and $tr(\cdot)$ is the trace of the covariance estimate specified within the parentheses.

In lieu of the normality assumption, only finite fourth moment is required for the distribution of the derived methylation values for each sample. Further, the total sample size does not necessarily need to be larger than the number of CG sites — the dimension of the vector. Instead, conditions are placed on the traces of the fourth powers of the covariance matrices of the distributions, essentially on the sparsity of the covariance structure [50]. With the much more relaxed assumptions, the test can be applied to detect DMRs in regions that are either sliding windows of fixed length throughout the genomes, or regions of interest, such as well-annotated gene promoters or CpG islands. The p-value is obtained by invoking the normal asymptotic property, which is formally established with the relaxed assumptions [50].

6.6 Case studies

We provide two case studies to briefly demonstrate the detection of DMRs using the two types of genome-wide methylation data based on the next generation sequencing technology. The first one demonstrates the use of BS-seq data to detect DMRs between three breast and three prostate cancer samples. The second case study focuses on the use of Cap-seq data to detect DMRs based on 10 AML patients who were stratified into two groups according to the mutation status of a gene.

6.6.1 Case study 1 – Detection of DMRs using BS-seq data

The first case study demonstrates the application of the method, BCurve, described in subsection 6.4.2. We chose to use this method as it takes between-sample variability as well as spatial correlation into account; the DMRs detected are based on a statistical inference method rather than an ad hoc, rule-based procedure. We used a publicly available RRBS dataset (accession number GSE27584) posted on the NCBI website. We utilized available samples under two conditions: three breast cancer samples and three prostate cancer samples. It is interesting to identify regions in the genome that harbor differential methylation patterns; although both may potentially be regulated by hormones, there may be a great deal of differences in how genes are regulated. To analyze the data, we first clustered the genome (22 autosomes) into non-overlapping regions, so that CpG sites being investigated within each region are not far apart. This is an important step for RRBS data as large gaps may exist [166]. This clustering step was performed chromosome by chromosome. For each of the regions that resulted from the clustering analysis, BCurve was applied to find DMRs by constructing 95% credible bands for the breast and the prostate cancer samples. Variability between the three breast samples and that between the three prostate samples are taken into account, and the variability was also allowed to be different in these two types of samples in a data-driven manner. It was found that about 10% of the CpG sites being studied were differentially methylated between these two types of cancer [329]. The DMRs identified were then analyzed further to find pathways that are enriched with genes that overlap the DMRs. One of the top pathways identified is nerve growth factor (NGF) signaling, which seems sensible biologically as NGF signaling is highly dependent on tumor origin [329].

6.6.2 Case study 2 – Detection of DMRs using Cap-seq data

In this case study, we demonstrate the application of the method described in Subsection 6.5.3, the two-step method, consisting of PrEMeR-CG in the first step and MethylCapSig in the second step. This two-step method allows for the derivation of nucleotide-level data, and the spatial correlations (existing due to biological origin or induced by PrEMeR-CG artificially) are accounted for in finding DMRs. We used Cap-seq data generated from 10 acute myeloid leukemia (AML) patients [120]. These 10 patients were divided into two groups according to whether the FLT3 gene had mutated through internal tandem duplication to be a FLT3-ITD mutant: three were mutants while the other seven were FLT3 wild types [120]. First, PrEMeR-CG was used to distribute the reads for each sample to methylation levels at each CG site. Among the 10 patients, the number of reads in the library ranged from 17.5 millions to over 26 millions, with an average of 21.6 millions. The derived methylation levels were normalized to "reads per million" to prepare for the next step of analysis.

After the normalized nucleotide-level data were obtained, the T_{CQ} test described in subsection 6.5.3.2 in the MethylCapSig package was applied to over 39,900 regions (most of them gene promoter regions) throughout the genome. As a comparison, a window-based t-test as described in Subsection 6.5.2.2 was also used to analyze the data (one t-test per region). After appropriate multiplicity adjustment, T_{CQ} identified a total of 945 DMRs (about 2.4% of all regions), compared to over 12,000 DMRs (over 30% of the regions) identified by the t-test. Less than half of those identified by T_{CQ} (401 out of 945) were also identified by the t-test to be DMRs. Given that the t-test has been shown to have extremely high type I error rate (as high as over 70%) in simulation studies [16], the majority of the DMRs identified by it are likely to be false positives. A pathway analysis showed that T_{CQ} identified three canonical gene pathways that were AML cancer-related among the top 10 pathways that were enriched by genes whose promoter regions showed differential methylation patterns between the FLT3 wild-type patients and those who have the FLT3-ITD mutation [16].

6.7   Software

There are a number of existing softwares that have been proposed specifically for analyzing sequencing-based methylation data. Table 6.2 compiles those that have been discussed in this chapter, though this is by no means an exhaustive list. Additional methods and software can be found in a review [392].

6.8   Concluding remarks and statistical challenges

In this chapter, we focus on profiling DNA methylation measured using sequencing-based technologies. We place emphases on addressing between-sample (but within-group) variability, and correlation between methylation levels among nearby CpG sites. For Cap-seq data, we have also discussed analysis strategies that derive nucleotide-level data to bring them more on-par with BS-seq data, which may make it more enticing for biological researchers, though challenges still exist [350]. On the other hand, the technology continues to evolve, including a number of single-cell BS-seq platforms and joint

TABLE 6.2

Software for detecting differentially methylated regions using
sequencing-based DNA methylation data

DataType	Software	Variability	Correlation	Type	Ref
BS-seq	MethylKit	No	No	2-Step	[3]
	DSS(-general)	Yes	No	2-Step	[107, 333]
	RADMeth	Yes	Yes	2-Step	[86]
	MethylSig	Yes	Yes	2-Step	[332]
	MOABS	Yes	No	2-Step	[421]
	Biseq	Yes	Yes	2-Step	[166]
	BSmooth	Yes	Yes	2-Step	[157]
	BCurve	Yes	Yes	1-Step	[329]
Cap-seq	MeDIPS	Yes	No	1-Step	[263]
	PeEMeR-CG*	–	–	–	[120]
	MethylCapSig	Yes	Yes	2-Step	[16]

*PrEReR-CG is for inferring nucleotide-level data from Cap-seq, not for detecting DMR
itself, and therefore the characteristics for detecting DMR are not applicable.

profiling with other genomic features. Due to cell-to-cell variability, one of
the most fundamental questions is clustering of cells based on their DNA
methylation profiles; however, the problem is challenging due to the sparsity
of single cell data and small amount of overlapping from cell to cell, even for
new methods proposed to specifically address the problem [72, 193, 221, 507].
Although our focus is on statistical methods for analyzing sequencing-based
data, numerous methods and associated software have been developed for analyzing microarray data [11, 526, 536]. An extension of BCurve discussed in
this chapter is also available for analyzing microarray data [150]. Finally, purity in DNA methylation analysis for cancer studies is a major challenge, as
tumor samples usually contain not only tumor cells, but also normal, immune,
and stromal cells. A number of methods have been proposed for taking tumor
purity into consideration in DNA methylation analysis [153, 526, 536].

7

Modeling and Analysis of Spatial Chromatin Interactions

CONTENTS

7.1   3D chromosome organization and spatial regulation

In genomes, a gene and its regulatory elements constitute a regulatory expression unit [75]. For small genomes, a gene and its regulatory elements form an uninterrupted genomic segment. In complex genomes such as those of the human and mouse, however, regulatory elements can be located hundreds of kilobases [233, 482], or even on different chromosomes [274], from the genes that they regulate. For example, most binding sites of Estrogen Receptor (ER), an important transcription factor in breast cancer, are located far away from gene promoter regions [40]. This is also the case with Androgen Receptor (AR), an important player behind prostate cancer, as many AR binding sites are placed within non-promoter regions in prostate cancer [202, 466, 510]. This long-range interaction phenomenon has been studied based on the hypothesis that spatial alignment of chromosomes brings genes and their regulatory elements close to one another, which enables communication between "far-away" genomic elements through "looping" (elements on the same chromosome) and "bridging" (elements on different chromosomes) [491]. In other words, the typical linear (one-dimensional) view of the genome from a mathematical perspective is inadequate when gene regulation is being taken into account. Rather, it is the spatial (three-dimensional (3D)) structure of the genome and proteins (togaether referred to as chromatin) that is essential in understanding gene regulation, i.e., the interplay of genes and their regulatory elements [65, 381].

Based on the interaction frequency data of chromosomes 14 and 22 produced by Hi-C, Lieberman-Aiden et al. [262] identified spatial compartmentalization and confirmed it by Fluorescence In Situ Hybridization (FISH) measurements, further supporting the notion that long-range spatial organization is a key player behind global gene regulation mechanism.

Deciphering the relationship between chromosome organization and genomic activity is a key to unlocking the black box of complex genomic processes of transcription and regulation [24, 172]. Further, chromatin interactions mediated by a particular protein of interest have captured much attention in cancer research as well as research in other complex diseases [125, 192, 400]. In particular, the presence and/or intensities of such interactions may be associated with a disease. An example is the regulation of the UBE2C oncogene, for which interactions with two distal binding sites of AR are necessary for a prostate cancer cell to progress from androgen-dependent prostate cancer (ARPC) to castration-resistant prostate cancer [400, 466]. As such, detection of differential interactions may lead to better prognosis and may provide the opportunity for designing better treatment options. Recent research has also shown that differential interactions may be associated with developmental stages of a cell. For instance, the maturation of mouse thymocytes is dependent on the long-range binding of the protein Runx1 to a silencer of the CD4

TABLE 7.1
Evolving technologies for detecting long-range chromatin interactions

Coverage	Method	Type/Resolution
Locus-specific	3C	Specific pairs of loci [76]
	4C	Large-scale interactions [401, 534]
	5C	Large-scale interactions [88]
Genome-wide	Hi-C	All genome-wide interactions (1 Mb resolution) [262]
	in-situ Hi-C	Higher resolution Hi-C assay (10 kb) [353]
	TCC	Variation of Hi-C with higher signal-to-noise ratio [215]
	(NG) Capture-C	Interactions with selected viewpoints [68, 191]
	Single-cell Hi-C	Measuring interactions in a single cell [306, 414]
Protein mediated	ChIA-PET	All genome-wide interactions involving a specific protein [125]
	HiChIP	Improves yields while requires less input [304]
	PLAC-seq	Improves efficiency and accuracy [103]

gene to regress its expression, which is found only in early stages of the mouse thymocytes development [205].

7.2 Evolving technologies for measuring long-range interaction

With recent spectacular advances in molecular technology, spatial organization of chromosomes has begun to be gradually understood. Since the introduction of the Chromosome Conformation Capture (3C) by Dekker et al. [76], a number of assays have been proposed for uncovering long-range interactions, as summarized in Table 7.1.

Using formaldehyde cross-linking and locus-specific polymerase chain reaction (PCR), 3C assay can measure physical contacts between genomic loci which, in turn, can be utilized to evaluate long-range interaction between specific pairs of loci. Using 3C data, Tolhuis et al. [436] detected long-range interactions between the β-globin and its distal control region. Other long-range regulations were also studied for various genes, including lgf2 [60, 305], TH2 cyokine [409], and α-globin [458]. To detect large-scale promoter/enhancer interactions, 4C [401, 534] and 5C [88], variants of 3C, were developed by employing inverse PCR or multiplexed ligation-mediated amplification. Even though 4C and 5C allow for higher throughput analysis, they still require the choice of a set of target loci, and thus do not permit unbiased genome-wide analysis. To address this limitation, Hi-C was introduced to enable the study of genome-wide looping and bridging by coupling with the NGS technology [262]. Hi-C aims to discover all global interactions, and has thus been the technology of choice used to recapitulate the spatial organization of chromosomes, to identify long-range interactions, and to detect chromatin domains. However, despite the great success of Hi-C, its resolution (at 1 Mb) is low, and as such, finer nuances and interactions may be missed. To further increase the capability of Hi-C to capture finer-scale biological functions, in-situ Hi-C was proposed [353], which has led to a higher resolution of interaction data at 10 Kb (or even 1 Kb if there is sufficient coverage) resolution. However, the cost of performing in-situ Hi-C is still very high and out of reach for routine analysis. Other improvements of Hi-C have also been made, including the introduction of tethered conformation capture (TCC) by performing ligations on solid substrates rather than in solution, which appears to enhance the signal-to-noise ratio [215]. Other methods that improve resolution and sensitivity have also been proposed [68, 191]. The assays discussed above are all population-based, in that the chemical treatments are applied to a population of cells so the measurement outputs need to be understood as a population average. A single cell Hi-C technology [306] was also developed several years ago. Although it was capable of capturing cell-to-cell variation of chromatin structure, it was nonetheless challenging to interpret the results due to limited sequencing depth and sparsity. However, after a few years' hiatus, there has been a recent surge in single-cell 3C-derived technologies [352, 414].

While Hi-C aims to discover all global interactions, there is also a need to focus on interactions mediated by a specific protein of interest, such as ER in breast cancer research or AR in prostate cancer studies. As such, Chromatin Interaction Analysis with Paired-End Tag sequencing (ChIA-PET) was introduced, which has been successfully applied to detect ER mediated [125], and RNA Polymerase II (Pol II) mediated [252] long-range interactions, among others. In recent years, methods that improve upon ChIA-PET have also been developed, including Hi-ChIP [304] and PLAC-seq [103].

In the following, we will describe several statistical methods for recapitulating the underlying 3D chromatin structure using Hi-C type data that capture "all-vs-all" interactions. We will also consider methods for detecting long-range gene regulations and differential interaction intensities using ChIA-PET type data that capture specific-protein-mediated interactions.

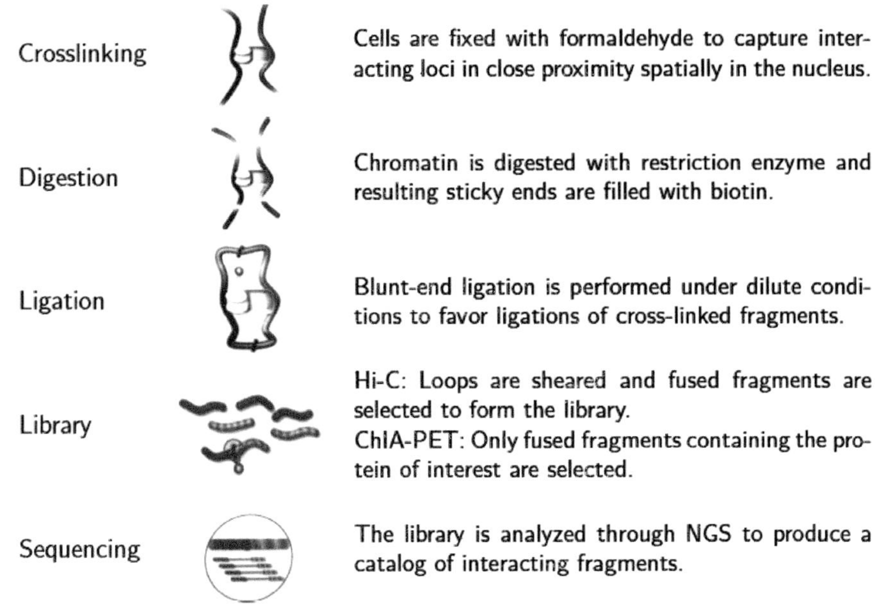

Crosslinking		Cells are fixed with formaldehyde to capture inter-acting loci in close proximity spatially in the nucleus.
Digestion		Chromatin is digested with restriction enzyme and resulting sticky ends are filled with biotin.
Ligation		Blunt-end ligation is performed under dilute condi-tions to favor ligations of cross-linked fragments.
Library		Hi-C: Loops are sheared and fused fragments are selected to form the library. ChIA-PET: Only fused fragments containing the pro-tein of interest are selected.
Sequencing		The library is analyzed through NGS to produce a catalog of interacting fragments.

FIGURE 7.1

Summaries of steps of the Hi-C and ChIA-PET technologies. Note that, schematically, the only difference between these two technologies is the library preparation step.

7.3 Methods for recapitulating 3D structures using Hi-C type data

7.3.1 Hi-C data

The Hi-C technology for generating long-range interaction data can be summarized schematically in Figure 7.1. Several steps of processing are necessary to identify long-range interactions in an unbiased, genome-wide fashion. They include cross-linking to fix interactions, restriction enzyme digestion, ligation to form loops, library preparation, and pair-end sequencing.

A typical dataset from Hi-C can be organized into a two-dimensional (2D) contact matrix (Figure 7.2), which can be divided into blocks representing intra-chormosomal interactions (diagonal blocks) and inter-chromosomal interactions (non-diagonal blocks), with n in the bottom matrix of Figure 7.2 being the number of chromosomes being considered (22 if all interactions in the autosomes are considered). The dimension of each diagonal block matrix

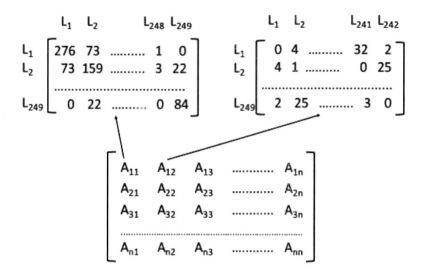

FIGURE 7.2
2D matrix layout of Hi-C data. The bottom matrix is composed of blocks
that contain either intra-chromosomal interactions (diagonal blocks) or inter-
chromosomal interactions (off-diagonal blocks). The top-left square matrix is
an example of A_{11}, intra-chromosomal interactions for chromosome 1 for Hi-C
data in 1 Mb resolution. The top-right non-square matrix shows an example
dataset for inter-chromosomal interactions between chromosome 1 and 2.

is dictated by the resolution of the data. For instance, suppose the resolution is 1 Mb, the same as the Hi-C data generated by Lieberman-Aiden [262].
Then the diagonal submatrix A_{11} is a 249 × 249 matrix, as chromosome 1
is of length 249 Mb and therefore has 249 loci (segments of length 1 Mb) as
shown in the top-left matrix of Figure 7.2. Each entry in the matrix denotes
the contact count; that is, the number of reads in the Hi-C library with the
pair-end reads mapped to the two loci. For instance, there are 276 reads with
both ends mapped to the first Mb of the chromosome, while there are 73 reads
with one end mapped to the first Mb of the chromosome and the other end
mapped to the second Mb of the chromosome. Similarly, the dimension of
each off-diagnonal block matrix (a non-square matrix) is also dictated by the
resolution. Continuing with the example of 1 Mb resolution Hi-C data, A_{12}
is a 249 × 242 matrix as it contains information for interacting chromatins
between chromosomes 1 and 2 (chromosome 2 is of length 242 Mb). Typically,
matrices denoting interchromosomal interactions are much sparser compared
to those representing intrachromosomal interactions, signified by many zeros
and much smaller numbers in most of the entries. For instance, the first Mb
segment of chromosome 1 and the first Mb segment of chromosome 2 do not

interact with one another; hence a 0 entry in the matrix (top-right matrix of Figure 7.2).

A couple review papers have compared various approaches for 3D structure recapitulations, focusing mainly on bulk data [320, 330]. One of the major categories of the existing approaches is optimization-based, which does not require the assumption of independence among the counts in a 2D matrix. On the other hand, modeling-based approaches, constituting another major category, may make strong assumptions, but can better handle bulk Hi-C data that are inherently heterogeneous. More recently, methods have been also been proposed for recapitulating single-cell 3D structures [369,544]. In the following two subsections, we will focus on discussing the two major categories of methods, optimization-based and modeling-based, respectively.

7.3.2 Non-parametric optimization-based methods

The essence of non-parametric optimization-based methods is to minimize a cost function, which typically involves the sum of differences between the distances "translated" from the contact counts and the corresponding distances obtained from the 3D structure. Such an approach is typically a two-step procedure. The first step is to translate the contact counts to distances using a biophysical model of DNA, which states an inverse relationship between contact count and distance. Then, in the second step, a consensus 3D structure (3D coordinates) is obtained by finding a set of 3D coordinates (one for each locus – a 1 Mb chromosomal segment if the data are in Mb resolution) that minimizes the cost function [22, 122, 250, 427, 532].

Suppose the genome is divided into n segments indexed with their (unknown) 3D positions (physical coordinates): $\Omega = \{\omega_1, \omega_2, \cdots, \omega_n\}$ and $\omega_i = (a_i, b_i, c_i), i = 1, 2, \cdots, n$. The 3D Euclidean distance between any two loci i and j is then $||\omega_i - \omega_j|| = \sqrt{(a_i - a_j)^2 + (b_i - b_j)^2 + (c_i - c_j)^2}$. This distance is invariant to rotation, reflection, and translation, but not scaling. We use $\boldsymbol{y} = \{y_{ij}, 1 \leq i < j \leq n\}$ to denote the $\binom{n}{2}$ $(O(n^2))$ contact counts (note that the matrix is symmetric), which are noisy and correlated observed data. There are $3n$ $(O(n))$ coordinates; these are the unknown parameters that need to be estimated. The goal of reconstructing the 3D structure is to estimate the $3n$ coordinates based on the looping intensity data (data organized into the big matrix in Figure 7.2).

We now describe the two steps of a typical optimization-based method.

Step 1. The biophysical model of DNA implies that, for a given contact count y_{ij} between loci i and j, it is related to the 3D distance of these two loci, d_{ij} through an inverse relationship:

$$y_{ij} \propto 1/d_{ij}^{\alpha},$$

where α may be pre-specified [22, 122, 427], or estimated [532] (together with Step 2).

Step 2. Define a cost function

$$C(\Omega) = \sum_{i,j \in \mathcal{S}} W_{ij}(||\omega_i - \omega_j|| - d_{ij})^2 - \eta \sum_{i,j \in \bar{\mathcal{S}}} ||\omega_i - \omega_j||^2. \qquad (7.1)$$

The goal is then to find

$$\Omega^* = \arg\min\{C(\Omega), \Omega \in R^{3n}\},$$

where R may be the entire real line or restricted to a finite region. In equation (7.1), $||\omega_i - \omega_j||$ is usually the Euclidean distance but other distance metrics, such as l_1-norm, may be used as well. The first term is the weighted sum of squared differences between the 3D distances from the hypothesized structure and the translated distances found in Step 1, and W_{ij} is the weight, which can be set in a number of ways, such as $W_{ij} = y_{ij}$, which gives more weight to those that have more frequent interactions [532]. The second term serves as a penalty, if $\bar{\mathcal{S}}$ is not an empty set. Suppose the goal is to "magnify" the distance between any two loci that do not interact with one another at all (i.e. $y_{ij} = 0$), then $\mathcal{S} = \{(i,j), y_{ij} \neq 0\}$. A positive penalty parameter η in the second term will serve the purpose of balancing out the influence of the two terms in the cost function, which can be thought of as a tuning parameter that controls the amount of penalty.

Note that when η is set to be 0, finding Ω^* boils down to obtaining the solution based on multi-dimensional scaling. Optimizing equation (7.1) can be done efficiently, such as through a semi-definite programming approach in polynomial time [532].

One example of such an approach is ChromSDE [532]. The constant parameter α in the inverse relationship is estimated instead of being treated as fixed as in some other approaches [22, 122, 427]. Its cost function has the penalty component, with those having 0 contact frequency being used in the second term of (7.1); that is, $\mathcal{S} = \{(i,j), y_{ij} = 0\}$. A variant of this general procedure is ShRec3D [250], in which the first step translates the contact frequency to distance by not only using the inverse relationship, but by also finding the shortest path connecting each pair of loci on a weighted graph. Its second step follows the general procedure without the penalty term.

7.3.3 Model-based methods

While optimization-based methods for recapitulating the 3D structure of a genome are computationally efficient and are assumption-free apart from the use of the biophysical model, the single consensus 3D structure from an optimization-based approach does not address the issue of cell heterogeneity, which is present in Hi-C data, as they are obtained from a population of millions of cells. As such, model-based methods, from which multiple 3D structures may be realized, constitute another branch of methodology for analyzing Hi-C data. Three particular features of the data, namely dependency, over-dispersion, and sparsity in the contact matrix (especially for inter-chromosomal and higher resolution data), can be directly addressed and accounted for in the model-based approaches, with details provided in the following.

Let $\boldsymbol{y} = \{y_{ij}, 1 \leq i < j \leq n\}$ be the collection of all frequency counts for a Hi-C dataset with n loci. Note that the contact frequency matrix (Figure 7.2) is symmetrical, and therefore, $\boldsymbol{y}$ only includes the contact counts in the upper triangle (not including those on the diagonal) of the matrix. Since y_{ij} is a count, it is natural to assume that it follows a Poisson distribution [182], although a normal distribution has also been used [371]. However, the contact counts are correlated as each locus may be interacting with multiple loci, and therefore assuming y_{ij} (for all $1 \leq i < j \leq n$) follows a Poisson distribution independently and identically does not reflect the nature of the data. Further, like any other data from NGS, over-dispersion is a potential concern and should be adequately addressed. Finally, the 2D data matrix can be quite sparse, especially for the blocks representing interchromosomal interaction and/or when higher resolution Hi-C data are produced and analyzed.

The following truncated Random effect EXpression model (tREX) attempts to address these three issues [328, 331]. Suppose $\boldsymbol{z}_{ij}^T$ are observed covariates, which may include fragment length, GC content, and mappability scores [182, 499]. Taking covariates into consideration, the mean contact count λ_{ij} can then be modelled as follows:

$$\log \lambda_{ij} = \beta_0 + \beta_1 \log d_{ij} + \boldsymbol{z}_{ij}^T \boldsymbol{\gamma} + X_i + X_j + U_{ij}, \tag{7.2}$$

where $\beta_1 < 0$ is imposed to reflect the biophysics property [328], as the contact count is assumed to be inversely related to the 3D Euclidean distance d_{ij} defined in Section 7.3.2. Further, $R_{ij} = X_i + X_j + U_{ij}$ is the random effect component, where

$$X_i \overset{iid}{\sim} N(0, \sigma_x^2), i = 1, \cdots, n,$$

$$U_{ij} \overset{iid}{\sim} N(0, \sigma_u^2), 1 \leq i < j \leq n,$$

and the X_i's and the U_{ij}'s are also assumed to be independent. This modeling scheme uses the X_i's to account for dependency and the U_{ij}'s to capture unknown sources of errors to allow for over-dispersion. Finally, the β_0 and $\boldsymbol{\gamma}$ in equation (7.2) are the constant offset and a vector of coefficients for the covariates, respectively.

Let $\boldsymbol{\theta}$ be the collection of all model parameters (which can be regarded as nuisance parameters). Then we can write down the likelihood for the parameters of interest Ω and $\boldsymbol{\theta}$ given the observed contact counts:

$$\log p(\boldsymbol{y}; \boldsymbol{\theta}, \Omega) \propto \sum_{(i,j) \in \mathcal{S}} \left\{ y_{ij} \log \lambda_{ij} - \log(e^{\lambda_{ij}} - I(||\mathcal{S}|| \neq ||\boldsymbol{y}||)) \right\}, \tag{7.3}$$

where $|| \cdot ||$ is the cardinality of the set and I is the usual indicator function. Suppose $\mathcal{S} = \{(i,j) : y_{ij} > 0, 1 \leq i < j \leq n\}$, then the likelihood in (7.3) addresses the data resolution issue, as the observed counts that are positive are assumed to follow a truncated Poisson distribution, and those with 0 contact counts are not taken into account in forming the likelihood. Note that if all data points, including those that are zeros, are used, then there is no truncation, and it reduces to the Poisson Random effect Architecture Model, or PRAM [328]. Suppose the random effect component R_{ij} is omitted from the model in (7.2), but there is still truncation; then, the model reduces to the truncated Poisson Architecture Model, or tPAM [326]. If there is no truncation and no random effect, then the model is simply an independent Poisson model as implemented in BACH [182] and PASTIS [456].

With model (7.2), we can see that the variance of random variable Y_{ij} (leading to observed count y_{ij}) can be greater than the mean of Y_{ij}, thus accounting for over-dispersion [328]. Dropping the subscripts of Y and λ for simplicity, we can see that

$$Var(Y) = E(Y) + Var\left(\frac{\lambda e^\lambda}{e^\lambda - 1}\right) - E\left\{\frac{\lambda^2 e^\lambda}{(e^\lambda - 1)^2}\right\} = E(Y) + \mathcal{A}.$$

One can show that a lower bound for $\mathcal{A}$ exists and is greater than 0 if $\sigma^2 = 2\sigma_x^2 + \sigma_u^2$ is large. Therefore, if there is indeed large variability in the data, then the estimated total variance will be large to accommodate for the over-dispersion.

It can also be easily seen that tREX accounts for dependencies among the contact counts [327, 328]. For $i \neq i'$ and both $i < j$ and $i' < j$,

$$
\begin{aligned}
Cov(Y_{ij}, Y_{i'j}) &= E\{E(Y_{ij}Y_{i'j'}|\lambda_{ij}, \lambda_{i'j'})\} - E\{E(Y_{ij}|\lambda_{ij})\}E\{E(Y_{i'j'}|\lambda_{i'j'})\} \\
&= Cov\left(\frac{e^{\lambda_{ij}}}{e^{\lambda_{ij}} - 1}\lambda_{ij}, \frac{e^{\lambda_{i'j'}}}{e^{\lambda_{i'j'}} - 1}\lambda_{i'j'}\right) \\
&= Cov\left(f(R_{ij}), f(R_{i'j})\right),
\end{aligned}
$$

for some real-valued function f. One can see that the last covariance is not equal to zero because $Cov(R_{ij}, R_{i'j}) = \sigma_x^2 > 0$.

To make statistical inference based on the likelihood in equation (7.3) and its variants, and to make it possible to learn about the multiple underlying 3D structures to address cell heterogeneity, Markov chain Monte Carlo (MCMC) algorithms have been derived [182, 326, 328, 371]. In other words, the 3D structures (the coordinates) drawn from the posterior distribution using MCMC algorithms can be clustered to form an ensemble of structures [371].

7.4   Methods for detecting long-range interactions using ChIA-PET type data

7.4.1   ChIA-PET data

The ChIA-PET technology for generating long-range interaction data mediated by a particular protein (P) also follows the same steps as outlined in

Interaction Frequency	Fragment 1 Region	Fragment 2 Region	Note
152	chr18:44761014-44765698	chr18:44767105-44774544	Intra-chromosomal
146	chr17:70265182-70270225	chr17:70276046-70280128	Intra-chromosomal
67	chr17:54070517-54076799	chr3: 61761763-61774548	Inter-chromosomal
43	chr17:54070517-54076799	chr20:52175586-52181065	Inter-chromosomal
10	chr16:33862882-33868962	chr2:132738254-132744384	Inter-chromosomal

FIGURE 7.3

A list showing a segment of ChIA-PET data. One end of the paired-end sequencing read is mapped to a region on one chromosome (fragment 1 region), while the other end to another chromosome (the fragment 2 region) that may or may not be the same as the chromosome harboring fragment 1 region. If the two regions are on the same chromosome, the interaction is referred to as intra-chromosomal, otherwise inter-chromosomal. Data was extracted from supplementary materials of Fullwood et al. [125].

Figure 7.1. Schematically, the only difference between Hi-C and ChIA-PET is the step for the building of the library (the fourth step). In Hi-C, all genome-wide interactions that are the results of cross-linking are selected to form the Hi-C library; whereas in ChIA-PET, only those interactions that are mediated by a particular protein are precipitated and captured to form the ChIA-PET library.

Although data from ChIA-PET may also be organized into a 2D matrix, it would be very sparse since interactions that are captured by ChIA-PET constitute only a small subset of those by Hi-C. As such, a typical dataset from ChIA-PET is a list of all interactions, with the length of the list dependent on the number of pairs of regions the interacting fragments are mapped to. Regions with special characteristics may be referred to as "anchor" regions [125]. For ease of reference, albeit in an abuse of the terminology, we refer to regions to which paired-end fragments are mapped as anchors in general, regardless of whether such regions satisfy the special properties described in Fullwood et al. [125]. Figure 7.3 shows a few items extracted from a complete list of a ChIA-PET dataset, where the interaction frequency shows the counts (number of reads) that are mapped to anchor 1 region at one end and anchor 2 region at the other end. For example, the first row shows that one DNA fragment on chromosome 18 (starting at nucleotide 44761014 and ending at nucleotide 44765698 – anchor region 1) interacts with another fragment on the same chromosome (starting at nucleotide 44767105 and ending at nucleotide 44774544 – anchor region 2) 152 times. Both the first and the second rows represent long-range interactions that are intra-chromosomal as both anchors come from the same chromosome. Rows 3–5, the last three rows, portray inter-chromosomal interactions. In particular, rows 3 and 4 depicts two inter-chromosomal interactions connecting the same anchor on 17 to two different anchors on chromosomes 3 and 20, respectively.

7.4.2   Detections of true chromatin interactions – individual pair analysis

A typical ChIA-PET experiment may generate thousands of interactions between paris of anchors, as evident in the ER-mediated ChIA-PET data (http://www.nature.com/nature/journal/v462/n7269/suppinfo/nature08497. html). However, not all interactions between a pair of anchors are due to true biological interactions mediated by the protein of interest, as "random collision" may occur between two DNA fragments during ligation due to their close proximity spatially. We refer to pairings that are due to random collision as "false pairs" to distinguish them from the "true pairs" of protein mediated interactions. Modeling the observed interaction frequencies of anchor pairs to distill the true pairs is therefore an important problem since such pairings paint the landscape of long-range gene regulations [381].

To address this, first note that the count data are inter-dependent, as those from Hi-C experiments. Let $\mathcal{S}$ denote the set of indices of pairs of anchors in which the interaction frequencies (i.e. contact counts) exceed a threshold. That is, $\mathcal{S} = \{(i,j) : y_{ij} \geq a\}$, where y_{ij} is the contact count (referred to as the joint count) between anchors i and j, and a is a positive integer. When a is set to be 1, $\mathcal{S}$ includes all pairs that interact. However, a may be set to be a larger integer to filter out pairs with low contact counts (e.g. $a = 2$ [125] or $a = 3$ [336]). For convenience, we also use $\boldsymbol{y} = \{y_{ij}, (i,j) \in \mathcal{S}\}$ to denote the collection of all joint counts with indices in $\mathcal{S}$. Their corresponding random variables will be denoted by the corresponding capital letters. The goal is then to analyze $\boldsymbol{y}$ through modeling the corresponding random variables Y_{ij}, however, this poses challenges due to data dependency. As such, one may also consider fixing various quantities, including the total reads $N = \sum_{(i,j) \in \mathcal{S}} y_{ij}$, and marginal count $n_i = \sum_{j:(i,j) \in \mathcal{S}} y_{ij}$, which is the number of reads that are connected to fragment i. Note that had the data been organized into a 2D matrix as in Hi-C, then n_i would have been the sum of row i, thus the name marginal count [314] or marginal sum [336].

Hypergeometric distribution has been used as the earliest model for analyzing ChIA-PET data as implemented in the ChIA-PET tool (CPT) [125, 252]. Assuming $a = 2$ to filter out pairs that are only connected once, Y_{ij} is assumed to follow a hypergeometric distribution with fixed total reads and marginal counts:

$$P(y_{ij} \mid 2N, n_i, n_j) = \frac{\binom{n_i}{y_{ij}}\binom{2N-n_i}{n_j-y_{ij}}}{\binom{2N}{n_j}}. \tag{7.4}$$

The idea of using the hypergeometric model for testing is to ascertain whether the observed interaction frequency between a pair of anchors is significantly greater than one would expect assuming each interaction (a read) is equally likely to happen for any pair of anchors. This is a conditional argument, i.e. Fisher's exact test, as the distribution is conditional on the assumption that the total interactions (i.e. total read counts) and the marginal counts are fixed. These assumptions are debatable though. In particular, the assumption of equally likely interactions is challenged because anchors located closer together in their linear genomic sequences are believed to have a higher probability of interacting with one another. Indeed, this property has already been incorporated into the analysis of Hi-C data [76,375]. To account for this property for the analysis of ChIA-PET data, Fisher's non-central (modified) hypergeometric distribution may be used [336]. The modification to the hypergeometric distribution is the inclusion of a weight factor (w_{ij}) to reflect the distance between the anchor pair i and j:

$$P(y_{ij} \mid 2N, n_i, n_j, w_{ij}) = \frac{\binom{n_i}{y_{ij}}\binom{2N-n_i}{n_j-y_{ij}}w_{ij}^{y_{ij}}}{\sum_{y'_{ij}} \binom{n_i}{y'_{ij}}\binom{2N-n_i}{n_j-y'_{ij}}w_{ij}^{y'_{ij}}}. \tag{7.5}$$

For a pair that is located closer than expected based on all anchors in a ChIA-PET experiment, w_{ij} will be assigned a value greater than 1 [336]. Thus, as one can see from (7.5), this will shift the weight towards those with larger observed interactions. On the other hand, a pair located farther apart than expected will be assigned a $w_{ij} < 1$, with the resulting non-central hypergeometric distribution giving more weight to those with smaller interacting counts. As a consequence, for two pairs of anchors with the same observed joint and marginal counts, the probabilities and thus the hypothesis testing result will be different, depending on the distances between the pairs. Note that if all pairs are assigned a weight $w_{ij} = 1$, then the non-central hypergeometric model given in (7.5) reduces to the standard hypergeometric model in (7.4).

Regardless of whether the standard hypergeometric distribution or the non-central hypergeometric distribution is used, the p-value for rejecting the null hypothesis that the observed interaction between pair i and j is due to random collision is $p_{ij} = P(Y_{ij} \geq y_{ij})$. Standard methods for multiple testing, such as Bonferroni or False Discovery Rate, may be applied [125, 336].

7.4.3 Detections of true chromatin interactions – joint analysis of all pairs

As we can see from the above, each pair is analyzed individually to ascertain whether each observed count is significantly greater than what one would expect from a random collision. However, the counts among pairs are interdependent, and as such, it may be more sensible to analyze them jointly. One such approach is MDM, which is based on modeling the counts as coming from a mixture of distributions with two components, one depicting the distribution

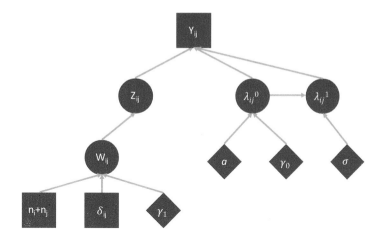

FIGURE 7.4

Schematic flowchart of the mixture hierarchical model. In the diagram, a square (□) denotes observed data; a circle (○) denotes a random variable that is dependent on hyper-parameters with their own distributions; a diamond (◇) represents a hyper-parameter whose distribution has fixed parameter values.

for counts due to random collision and one for those that truly represent chromatin interactions [314]. The schematic diagram of the hierarchical modeling framework is given in Figure 7.4. Mathematically, we introduce a Bernoulli random variable for each pair. Specifically, for anchors i and j, define

$$Z_{ij} = \begin{cases} 1 & \text{if the joint count represents true interaction,} \\ 0 & \text{if the joint count is due to random collision.} \end{cases}$$

Then

$$\begin{aligned} Y_{ij} \mid Z_{ij} &= 1 \sim \text{tPoi}(\lambda_{ij}^1, a) \text{ and} \\ Y_{ij} \mid Z_{ij} &= 0 \sim \text{tPoi}(\lambda_{ij}^0, a), \end{aligned}$$

where tPoi denotes a Poisson distribution truncated at a, the threshold as given in the definition of $\mathcal{S}$, and λ_{ij}^1 and λ_{ij}^0 are the Poisson intensity parameters. These two parameters may be modeled further with hyper-parameters (Figure 7.4) with the constraint $\lambda_{ij}^1 > \lambda_{ij}^0$ to reflect the property that the count from a truly interacting anchor pair is expected to be larger than that due to random collision. A key component is the modeling of the prior probability that a pair of anchors is truly an interacting pair. Specifically, let $P(Z_{ij} = 1) = W_{ij}$ where W_{ij} may be influenced by the marginal counts ($n_i + n_j$, treated as fixed), but also by a hyperparameter (γ_1). Furthermore, because the interactions targeted by ChIA-PET are those that represent long-range gene regulation, the genomic features of the regions where the anchors reside may also contain information that can be utilized to help separate the true interacting pairs from random collisions. This integrative approach leads to the incorporation of δ_{ij}, which is the minimum total linear genomic distance between one of the anchors and the nearest transcription factor binding site of protein P, and between the other anchor and the nearest gene promoter. Since a smaller δ_{ij} and a larger $n_i + n_j$ are indicative of a greater probability of (i, j) being a true interacting pair, the mean of W_{ij} can be set to be proportional to $n_i + n_j$, but inversely proportional to δ_{ij}, as follows:

$$W_{ij} \sim \text{beta} \left(\frac{\gamma_1 (n_i + n_j)}{\delta_{ij}}, \frac{\gamma_1}{||\mathcal{S}||} \sum_{(i,j) \in \mathcal{S}} \frac{n_i + n_j}{\delta_{ij}} \right).$$

This will lead to a mean value of greater than $1/2$ for a pair that has large marginal counts and/or whose anchors are in functional genomic regions. This modeling approach may be considered as a Bayesian version of the mixture of experts model, as the weights of the two mixture components are dependent on the values of other variables [198].

To make inference about whether (i, j) is a true interacting pair, we find the posterior probability that Z_{ij} is equal to 1. This probability can be calculated based on samples drawn from an MCMC algorithm [314].

In the approach described above, the genomic distance between two anchor regions is not taken into account, which may lead to a less efficient algorithm. In contract, Mango [341] and MICC [164] are two software packages that use information on the genomic distance, but not the distances from the anchors to the protein of interest. Although some comparisons have been carried out [314], the data generation process of using a Poisson distribution [336] may not reflect real data well. Other methods have assessed their performance by comparing with Hi-C or 5C results [164, 341]. Recently, a comprehensive comparison between CPT, MDM, MICC, and Mango was carried out to assess their relative merits [275].

7.4.4 Detections of pairs with differential chromatin interactions

In the above subsection, the focus is on detecting whether a joint count for a pair of anchors truly represents long-range chromatin interactions. Considering differential interacting pairs and differential looping intensities under different conditions is also of great importance, as such differences can be associated with a disease or the progression of a disease. A recently proposed analysis pipeline for detecting differential chromatin interactions using edgeR [364], diffloop, also includes visualization of loops [244]. In the following, we focus on describing the MDM method [313].

TABLE 7.2

Six categories of paris according to chromatin interaction status under two conditions.

	Chromatin Interaction		
Category	Condition 1	Condition 2	Change*
1	Yes	Yes	Increase
2	Yes	Yes	Same
3	Yes	Yes	decrease
4	Yes	No	Decrease
5	No	Yes	Increase
6	No	No	NA

*The change in interaction intensity from Condition 1 to Condition 2.

In the situation where two conditions are considered, there are 6 possible categories of outcomes, as listed in Table 7.2. To classify pairs of anchors into these six categories, a 6-component mixture model can be used [313]. Let $\mathcal{S} = \{(i,j) : x_{ij} \geq a \text{ or } y_{ij} \geq a\}$, where x_{ij} and y_{ij} are the contact counts between anchor i and j under Condition 1 (C1) and Condition 2 (C2), respectively. As in the previous subsection, $a = 1$ will lead to the inclusion of all anchor pairs that have non-zero contact counts in at least one of the two conditions. Note that $\mathcal{S}$ is revised from the previous subsection to reflect the two-conditions setting. The six-component mixture model for the corresponding random variables X_{ij} and Y_{ij} is

$$P_{X_{ij}, Y_{ij}}(x, y \mid \boldsymbol{\theta}) = \sum_{k=1}^{6} w_{ij}^{(k)} P_k(x, y \mid \lambda_{ij}^{(k)}, \eta_{ij}^{(k)}),$$

where $\boldsymbol{\theta}$ is the collection of all model parameters, and the usual constraint on the component weights (sum to 1) as well as other constraints are imposed on the parameters for identifiability [313].

Each component is the joint distribution of two independent Poissons conditional on the criterion that one of the contact frequencies needs to be at least a:

$$P_k(x, y \mid \lambda_{ij}^{(k)}, \eta_{ij}^{(k)}) = \frac{\left(\lambda_{ij}^{(k)}\right)^x \left(\eta_{ij}^{(k)}\right)^y}{x! y! \left(e^{\lambda_{ij}^{(k)} + \eta_{ij}^{(k)}} - \sum_{x < a} \frac{\left(\lambda_{ij}^{(k)}\right)^x}{x!} \sum_{y < a} \frac{\left(\eta_{ij}^{(k)}\right)^y}{y!}\right)}.$$

As in subsection 7.4.3, one of the most critical steps is integrating other genomic information for setting up the prior distribution for the component weights. For detecting true chromatin interactions in 7.4.3, the component weight (a single weight parameter suffices for a two-component mixture) is

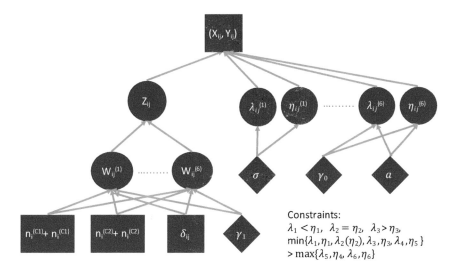

FIGURE 7.5
Schematic flowchart of the mixture hierarchical model. In the diagram, a square ($\square$) denotes observed data; a circle ($\circ$) denotes a random variable that is dependent on hyper-parameters with their own distributions; a diamond ($\diamond$) represents a hyper-parameter whose distribution has fixed parameter values.

assumed to follow a beta distribution with its parameters dependent on the marginal counts and features of the genomic regions where the two anchors reside. The same idea applies for detecting differential interaction intensities, albeit the beta distribution is replaced by its multivariate counterpart, the Dirichlet distribution, for a six-component mixture. The schematic is given in Figure 7.5, while the details can be found in Niu et al [313].

7.5 Case studies

The methods described in this chapter are demonstrated briefly in three case studies. The first one is the analysis of a human lymphoblasstoid Hi-C dataset generated by Lieberman-Aiden et al. [262] to study the underlying 3D chromosomal organization. The second is the analysis of a ChIA-PET dataset of the MCF-7 cell line [125] to identify true chromatin interactions mediated by Estrogen Receptor α (ERα). The last case study investigates differential chromatin interactions between the MCF-7 and the K562 cell lines using ChIA-PET data mediated by Polymerase II (Pol II) [253].

7.5.1   Case study 1 – Reconstruction of 3D structure from Hi-C data

The first study demonstrates an application of the methods described in Section 7.3 to analyze a Hi-C dataset generated from a karyotypical normal human lymphoblastoid cell line [262]. This dataset is in fact the combination of data from two experiments on the same cell line; the pooling of the data is justifiable given their high producibility [262, 326–328, 331]. This case study focuses on data from two chromosomes, 14 and 22, as Fluoresence In Situ Hybridization (FISH) data are available for 9 pairs of loci (5 on chromosome 14 and 4 on chromosome 22), which allows for empirical evaluations of the accuracy of the reconstructed 3D strcutures from different methods. The most interesting features from the FISH data for experimental validation are that (1) the spatial distance between loci 2 and 4 (on chromosome 14), L24, is in fact shorter than that between loci 2 and 3, L23, which are closer together in terms of their linear genomic distance, and (2) similarly, the spatial distance between loci 6 and 8 (on chromosome 22), L68, is shorter than that between loci 6 and 7, L67, despite their shorter linear genomic distance. The data are processed at 1 Mb resolution, leading to 89 1-Mb segments (loci) on chromosome 14 and 36 loci on chromsome 22. As such, the data matrix is a submatrix of that displayed in the bottom row of Figure 7.2: two diagnoal blocks of sizes 89×89 (for chromosome 14) and 36×36 (for chromosome 22), respectively, and two off-diagonal blocks (one is the transpose of the other) of sizes 89×36 (or 36×89) depicting interchromosomal interactions between chromsomes 14 and 22. Focusing on these two chromosomes leads to a total of $n = 125$ loci (7750 data points) for reconstructing the structures for these two chromosomes. The analysis results, presented in Table 7.3, show that the estimated distances (after standardization [331]) for L24 from all six methods are indeed shorter than their corresponding distances for L23, consistent with the FISH results. The average of the differences between the estimates and the FISH measurements across all nine pairs of loci for each of the methods is also calculated, with ShRec3D appearing to have a much larger difference than the rest of the methods. The correlations between the estimated distances and the FISH measures are all quite high according to three measures, attesting to the adequate properties of the reconstructed 3D structures, although those from ShRec3D are a bit lower than the other methods.

7.5.2   Case study 2 – Detection of true chromatin interactions from ChIA-PET data

In this case study, we show an application of the methods described in Subsections 7.4.2 and 7.4.3 for analyzing a dataset generated from a ChIA-PET

TABLE 7.3

Results from analyzing a Hi-C dataset

	Chrom	**ShRec**	**PASTIS**	**BACH**	**tPAM**	**tREX**	**FISH**
L23[1]	1.150	1.313	1.388	1.135	1.112	0.940	0.959
L24	0.525	1.122	0.810	0.706	0.457	0.498	0.674
L67	0.408	0.248	0.522	0.589	0.620	0.602	0.703
L68	0.283	0.141	0.364	0.435	0.412	0.411	0.567
Ave diff[2]	0.184	0.307	0.143	0.096	0.143	0.082	–
Pearson[3]	0.930	0.701	0.851	0.923	0.867	0.971	–
Kendall	0.667	0.555	0.667	0.667	0.833	0.944	–
Spearman	0.850	0.583	0.800	0.833	0.933	0.983	–

[1] For Chrom (ChromSDE), ShRec (ShRec3D), and PASTIS, L23, L24, L67, and L68 are the single consensus estimates, whereas for BACH, tPAM, and tREX, they are the medians of 10,000 MCMC estimates. The corresponding entries for FISH are the median of 100 measurements. All measurements are standardized [331].

[2] For each method, the average difference is between its median estimate and the corresponding median FISH measurement over all nine pairs of loci.

[3] The three correlations, Pearson's correlation coefficient, Kendall's tau, and Spearman's rank correlation, measure the relationships between the median estimates and the median FISH measurements for each method.

experiment to study ERα mediated long-range gene regulations in the MCF-7 cell line (IHM001F) [125]. This dataset contains 1451 intra-chromosomal and 838 inter-chromosomal interactions, resulting in a total of 2289 joint counts, ranging from 2 to 83. Note that singleton interactions were filtered out before the data were made publicly available [125]. The hypergeometric (HG) model from the ChIA-PET Tool software [252] and the hierarchical mixture model from the MDM software [314] were applied to analyze the data. For the HG analysis, a pair with a joint count of at least 2 is declared to represent a true chromatin interaction if the FDR (adjusted from the raw p-values) is smaller than 0.05. For the MDM analysis, the additional information on the genomic features of the anchors were obtained from the UCSC Genome Browser (hg18), while the ERα protein binding sites for the MCF-7 cell were taken from the information provided with the ChIA-PET experiment [314]. Based on the estimated model parameters, if the posterior probability that a pair with a joint count of a least 2 is greater than 0.5, then it is declared to be a true chromatin interacting pair by invoking the Bayes rule. The cross classification of the results are given in Table 7.4. One can see that the result from HG is quite consistent with that from MDM for the pairs identified to be true by MDM — only 9 of the pairs identified to be true by MDM were declared to be false by HG. However, the results are quite discrepant for those pairs declared to be random collisions by MDM, as a super majority of them (88%) were identified to be true pairs by HG. This vast difference in the results

TABLE 7.4

Cross classification of the results from HG and MDM based on a
hierarchical mixture model

		Hypergeometric	
		True Pair	False Pair
Mixture Model	True Pair	423	9
	False Pair	1641	216

could be due to the additional information provided by the genomic features
of the anchor regions for MDM [314]. That is, pairs with anchors that are
far away from regions with the desired genomic features will be more likely
to be labeled false pairs by MDM compared to HG. On the other hand, true
chromatin interacting pairs with small joint counts may be "rescued" by the
additional genomic feature information: three of the nine pairs classified as
true by MDM but false by HG are all anchored in a region on chromosome 17
that contain ERα binding sites [314].

7.5.3   Case study 3 – Detection of differential chromatin interaction intensities from ChIA-PET data

Our last case study illustrates an application of the method described in Subsection 7.4.4 to analyze two ChIA-PET datasets and identify differential chromatin interactions and differential intensities. The first ChIA-PET dataset
was obtained from the MCF-7 cell line, whereas the second was from the
K562 leukemia cell line. Both cell lines were treated by Polymerase II (Pol II)
before the ChIA-PET experiments [253]. To focus on identifying chromatin
interactions that are unique to the MFC-7 cell lines, the corresponding ChIA-
PET data were filtered through a threshold ($a = 2$), which discarded singleton
reads, resulting in a total of 9739 pairs [313]. Note that among these pairs,
the joint counts may be less than 2 for K562, as no filtering was made for this
ChIA-PET dataset. The results, with the pairs classified into 6 categories,
are provided in Table 7.5. One can see that the number of pairs for which
there is chromatin interaction in MCF-7 but not in K562 (category 4) is much
larger (it is in fact 100 times larger) than its counterpart (category 5). This
is not surprising given that the filtering was applied to the MCF-7 data only.
Indeed, these results are in direct response to the goal of the analysis, which
aims to identify interactions that only exist in MCF-7 but not in K562 [313].
A particular interaction that stands out is the one that centers around the
gene GREB1 (with a count of 15 and 1 in MCF-7 and K562, respectively), as
the gene is known to be involved in long-ranged regulation in MCF-7 but not
in K562 [253].

TABLE 7.5

Characterization of differences in chromatin interactions between MCF-7 and K562

Category[1]	1	2	3	4	5	6
Number	393	1180	86	977	5	7098
Percentage(%)	4.0	12.1	0.9	10.0	0.1	72.9

[1]The six categories of chromatin interactions between two conditions (C1: MCF-7; C2: K562) are as given in Table 7.2.

7.6 Software

A number of softwares have been proposed for analyzing Hi-C data to recapitulate the underlying 3D chromatin structures of a genome. There are also a couple softwares for detecting true interactions and differential intensity of chromatin interactions based on ChIA-PET data. Table 7.6 compiles those that have been discussed in this chapter. R packages for simulating realistic Hi-C and ChIA-PET data are also available [275, 537]. This is by no means an exhaustive list: for example, there are recent methods proposed specifically for analyzing HiChIP data, such as hichipper [245].

TABLE 7.6

Software for analyzing chromatin interaction data

DataType	Software	Model	Additional Data	Ref
Hi-C[1]	MCMC5C	Normal	No	[371]
	BACH	Poisson	No	[182]
	PASTIS	Poisson	No	[456]
	tPAM	truncated Poisson	No	[326]
	tREX	truncated Poisson	No	[331]
	ChromSDE	–	No	[532]
	ShRec3D	–	No	[250]
ChIA-PET[2]	ChIA-PET Tool	Hypergeometric (HG)	No	[252]
	ChiaSig	Non-central HG	No	[336]
	MDM	Mixture	Yes[3]	[313, 314]
	Mango	Binomial	No	[341]
	MICC	Mixture	No	[164]
	diffloop	Neg. Binomial	No	[244]

[1]The softwares listed are for 3D structure reconstruction.
[2]These softwares are for detecting true interacting anchor pairs or for detecting differential intensities of interactions.
[3]The additional data used are genomic features of the region in which the anchors of ChIA-PET reside.

7.7 Concluding remarks, and statistical and computational challenges

In this chapter, we focus on addressing two problems in analyzing chromatin interactions. Based on Hi-C data, we describe model-based and optimization-based approaches for recapitulating the underlying 3D structure of chromatins. Understanding the chromosomal organization in its natural 3D space is extremely important for correlating with genomic functions, such as for studying the relationship between chromatin packing density and important genomic marks (e.g. histones) [262]. However, there are an array of other biological problems, including normalization, identification of typological domains and their boundaries, detection of interaction points genomewide, and integrative analysis of Hi-C data with other data types to enhance the accuracy of estimations and predictions. Solving such problems presents a multitude of statistical and computational challenges due to data dependency, resolution, and many others beyond the sheer size of a typical dataset [350]. Recall that the Hi-C protocol considers a population of cells; therefore methods for analyzing bulk Hi-C data discussed in this chapter are not necessarily directly applicable to single-cell data. In particular, the sparsity in single cell data could be due to either structural zero or sampling zero, also referred to as dropouts, making the analysis even more challenging (this will be a topic of the next chapter). Furthermore, due to computational intensity, some methods as they currently stand may not be scalable to analyzing genome-wide Hi-C data; modifications are needed for such a feat [389]. For ChIA-PET type data, we elucidated the problem of detecting true interacting anchors and those that exhibit differential interaction intensities under two different conditions. We discussed one way of integrating genomic features of the anchor regions into the analysis, but clearly other data may also be integrated to improve the sensitivity and specificity of finding true long-range gene regulation, as demonstrated in a recent study [275]. Finally, detecting changes in gene-regulation overtime as a disease progresses is particularly relevant, but analyzing time-course data is challenging and remains understudied.

8

Digital Improvement of Single Cell Hi-C Data

CONTENTS

DOI: 10.1201/9781315153728-8

8.1   Sparsity of single cell Hi-C data

As discussed in Chapter 7, the past decade has seen tremendous progress in understanding chromosome 3D structures, which play an essential role in transcription regulation both in normal cells and in various stages of disease cells [76, 78, 115, 125, 262, 304, 353, 498]. In particular, high-throughput single-cell technologies developed in recent years [73, 83, 116, 306, 307, 352, 415] (collectively referred to as single cell Hi-C – scHi-C) have greatly enhanced our understanding of cell-to-cell variability, and especially cell-type heterogeneity. However, work toward comprehensive analysis of 3D structures within and between different cell types, especially in discovery of subpopulations/subtypes, still has yet to reach maturity and continues to be developed. One of the main reasons is the sparsity of scHi-C data relative to bulk Hi-C (bHi-C; topic of Chapter 7), typically composed of millions of cells. Using the K562 data [116] as an example, Figure 8.1(a) is a heatmap representation of the first five chromosomes of a bHi-C that has more than 2,000,000 reads organized into a two-dimensional (2D) contact matrix (map), in which each position displays the interaction intensity (count) between two evenly divided segments (referred to as loci) of the genome. On the other hand, Figure 8.1(b) is the heatmap of an scHi-C 2D matrix with about 300,000 reads (representative of a small proportion of scHi-C in the K562 dataset); there is reduced intensity when compared to the bHi-C but the patterns remain similar. However, for the majority of the single cells in the K562 dataset, each has only about 1000 reads and shows much reduced resemblance to bHi-C, as demonstrated in Figure 8.1(c).

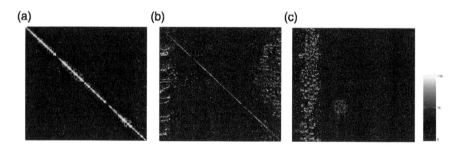

FIGURE 8.1
Heatmaps of the first 5 chromosomes of one bHi-C and two scHi-C data from a K562 dataset (accession number: GSE80006) with discrepant sequencing depths. (a) bHi-C (bulk A), depth: 2,036,405 reads; (b) scHi-C (sc79B), depth: 356,255 and (c) scHi-C (sc201A), depth: 1057.

8.1.1 Digital improvement of data quality

Sparsity is one of the major challenges genomic researchers face today in analyzing single cell data, and particularly scHi-C, as sparsity is an order of magnitude more severe compared to most of other types of single-cell data [541]. Since Hi-C data are represented as 2D contact matrices, the coverage of scHi-C (0.25–1%) is much smaller than that of single cell RNA-seq (scRNA; 5–10%) [541]. As such, data quality improvement is a necessary step in other to perform meaningful downstream analysis [511, 541, 545]. In fact, digital quality improvement as an intermediate step using statistical and machine learning methods have been applied even to bHi-C data [177, 450, 506, 530], with the same goal of obtaining more meaningful downstream analysis results. Regardless of whether the improvement is for scHi-C or bHi-C data, the available improvement methods to date are largely embedded as an intermediate step for the eventual goal of answering biological questions such as enhancing data resolution, constructing 3D structure, and clustering single cells [450, 506, 530, 541, 545]. Of course, the problem for scHi-C is much more challenging, and methods proposed for data quality improvements have achieved varying degree of success [499].

8.1.2 Separating structural zeros from dropouts

In addition to sparsity, a further complication for analyzing single cell Hi-C data is that not all zeros are created equal. Among observed zeros in an scHi-C contact matrix, some are true zeros (i.e. structural zeros) because the corresponding pairs in that particular cell do not interact with each other at all due to the underlying biological function, whereas others are sampling zeros (i.e., dropouts) as a consequence of low sequencing depth. Differentiating between structural zeros and dropouts is important since correct inference would improve downstream analyses such as clustering and 3D structure recapitulation. For example, a 3D structure recapitulation method may include a penalty term in an objective function to position two loci in the 3D space as far as possible if they truly do not interact [369, 531]. However, if an observed zeros due to insufficient sequencing depth is incorrectly inferred as a structural zero, then applying the method would lead to an artificial separation of two loci that in fact have coordinated effects on certain biological functions. Conversely, if an observed zero is incorrectly treated as a dropoout and imputed as currently practiced [506], then two loci that are in different compartments or domains may be artificially placed close to one another.

In scRNA research, the concepts of structural zero and dropout are well understood and have received considerable attention, with a number of methods developed to identify SZs and impute DOs [48, 258, 295, 339, 454, 524]. In contrast, the concepts of structural zero and dropuout have not been widely disseminated and recognized yet, as they have only emerged recently in scHi-C research [153]. Methodologically, methods developed for scRNA-seq may be adapted and repurposed for identification of structural zeros and imputation of dropouts, and for quality improvement of scHi-C data more generally. Nevertheless, it is easily seen that special features such as spatial dependencies among interreations in the 2D matrix are not accounted for in straightforward adaptation [153].

8.2 Adaptation of scRNA imputation methods for improving scHi-C data quality

Since the issue of sparsity is only addressed in a rather limited manner in scHi-C research [230, 260, 525], yet a large number of methods have been developed for scRNA data, in this section, we describe an adaptation of an scRNA method for inference on structural zeros and imputation on dropouts for scHi-C data. Methods that have been adapted for scHi-C data include those that may be classified as machine learning or statistical modeling based (Table 8.1), with scImpute [258] and McImpute [295] as the representative of these two categories of methods, respectively. The methods shown in Table 8.1 and adapted in [153] were shown to work well for scRNA data [48, 524] and are reasonable alternatives for scHi-C data [153].

8.2.1 Preparation of scHi-C data for using scRNA methods

Since scHi-C chromatin interaction data are represented as a 2D contact matrix, whereas scRNA data is a vector of gene expression levels for each single cell, one needs to first vectorize the scHi-C matrix for each single cell by concatenating the columns, but only using the lower (or upper) triangular values from the scHi-C 2D matrix. Then one combines vectors from multiple single cells into a new scHi-C data matrix with rows representing locations (pairs

TABLE 8.1

scRNA-seq imputation methods adapted for scHi-C data by Han et al. [153]

Category	Method	Aim
Statistical modeling	ScImpute [258]	Imputation & Structural zeros inference
Machine learning	MAGIC [454]	Imputation
	McImpute [295]	Imputation & Structural zeros inference
	SCRABBLE [339]	Imputation
	scRMD [48]	Imputation & Structural zeros inference

of interacting loci in the genome) and columns representing cells. In this way, imputation methods for scRNA can be directly applied to scHi-C data.

8.2.2 Machine-learning algorithms for imputation

Matrix decomposition algorithms [48, 181], low-rank matrix completion algorithms [235, 295, 339], and data diffusion methods [454] are among popular approaches for developing imputation methods for scRNA data. A representative is McImpute [295], which was found to be competitive among several methods recently adapted for structural zero identification and quality improvement of scHi-C data [153]. Like other methods for scRNA data analysis, McImpute makes use of information from similar single cells, thereby effectively increases the information for imputation, as described in the following.

Based on the low-rank matrix completion algorithm similar to solving the Netflix movie recommendations problem [235], McImpute uses the nuclear norm minimization algorithm to recover the low-rank matrix. The observed data matrix, after data preparation (Section 8.2.1) to combine scHi-C data from similar single cells, is normalized in the following manner: dividing by the library size in each cell (column sum) and then multiplying by the median of column sums. Further, a $\log_2$-transformation is applied to each pseudo-count (adding 1 to the count). McImpute solves the matrix completion problem to obtain the predicted matrix by minimizing an objective function which involves two terms. The first term is the sum of the squared differences between the pseudo-count matrix and the predicted matrix only for the observed non-zero positions. The second term represents a penalty on the rank of the predicted matrix, which includes not only the predicted for observed non-zeros but also values for the zero positions.

Inference for structural zeros is made post imputation using the following rule, which is shared by a number of other imputation methods [153]. All the imputed values smaller than 0.5 are replaced with a 0, effectively declaring them structural zeros. One advantage of McImpute is that it does not make any assumption on the data distribution, and values in the entire matrix are taken into consideration for imputation.

8.2.3 A block algorithm to account for spatial correlations

As described above, in order to use methods designed for scRNA for analyzing scHi-C data, one needs to vectorize each scHi-C 2D matrix. However, doing so results in the loss of the most important feature of a contact matrix, which is the spatial correlation within neighborhood regions inherent in such data. To alleviate this problem, an initial CUT step can be executed first to take dependencies among contacts in neighboring positions into consideration to enrich information and reduce noise. This step essentially entails a partition of

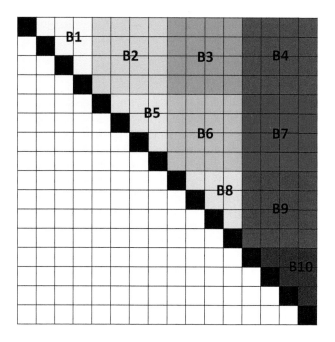

FIGURE 8.2
Schematic diagram of the result of a CUT Step. There are a total of 10 blocks,
B1, B2, ..., B10.

the whole scHi-C matrix into blocks so that imputations can be carried out for
each block separately. A schematic diagram depicting how to partition the 2D
contact matrix into blocks is shown in Figure 8.2, in which we divide the upper
triangular part of the matrix above the main diagonal into 10 blocks. The
partition of blocks allows McImpute and other scRNA imputation methods
to take advantage of information contained in neighboring positions of a 2D
matrix. Each block is then analyzed one at a time using McImpute as before
by vectorizing data contained in the block and combining across multiple cells
into an scRNA-like matrix. Since some positions in the original 2D matrix
may be overlapped from performing the CUT step, the imputed value for an
overlapped position is taken as the average from all involved blocks [153].

In Han et al. [153], the authors demonstrated the feasibility of adapting
scRNA imputation methods for scHi-C data, but weaknesses were also identified. In particular, as discussed above, straightforward adaptations of scRNA
methods ignore the spatially correlated feature of Hi-C data represented in
a 2D matrix. Although adding the CUT step to the methods alleviates the
concern to some extent, it is ad hoc, and the problem of selecting an optimal
block size has not been resolved [153]. As such, adaptations and modifications of existing scRNA methods for identification of structural zeros and

imputation of dropouts, although adequate, may not be the best choice for analyzing scHi-C data.

8.3 Methods designed for improving Hi-C data quality

Approaches for addressing sparsity to improve scHi-C data quality all aim to "smooth" the data by borrowing information from spatially correlated neighbors and may be classified into three categories depending on the underlying methodology: (1) kernel smoothing, (2) random walk and (3) neural network. A list of representatives in each category can be found in Table 8.2. For kernel smoothing, the types of kernels that have been used in the literature are uniform kernels or 2D Gaussian kernels. For example, HiCRep [506], which aims to assess the reproducibility of Hi-C data, applies a uniform kernel (or referred to as 2D mean filters — 2DMF) by replacing each entry in the 2D contact matrix with the mean count of all contacts in a neighborhood. Another method, scHiCluster [541], has proposed the use of a method in its first step that may also be classified into this category: it uses a filter that is equivalent to taking the average of the genomic neighbors, although the filter may also incorporate different weights during imputation. While a uniform kernel (2D MF) takes the average of the genomic neighbors with equal weights, a 2D Gaussian kernel (2DGK) uses a weighted average of neighboring counts according to a 2D Gaussian distribution: the farther away a neighbor is from the entry that is being imputed, the smaller the weight. For instance, SCL [545] applies a 2D

TABLE 8.2
Three categories of methods for improving Hi-C data quality and the biological questions of interesst

Category	Method	Aim
Kernel smoothing	HiCRep [506]	Reproducibility[a]
	SCL [545]	3D Structure
	scHiCluster [541]	Clustering
Random Walk	scHiCluster [541]	Clustering
	GenomeDISCO [450]	Reproducibility[a]
	SnapHi-C [511]	Chromatin contacts
Neural network	HiCPlus [530]	Data Resolution[a]
	DeepHiC [177]	Data Resolution[a]
	Higashi [525]	TAD domain boundaries

[a]The biological questions of interest are addressed with respect to bulk Hi-C data.

Gaussian function to impute scHi-C contact matrices before inferring the 3D chromosomal structure.

Methods referred to as random walks have also been proposed as a way to smooth out an observed 2D matrix for improving data quality [450, 511, 535, 541]. The idea of a "random walk" process is to borrow information from neighbors in a fashion different from the "neighborhood" idea in kernel smoothing. Any position that is on the same row or column as the entry being imputed will contribute to the "smoothed" count in each step of the random walk. In GenomeDISCO [450], it is found that taking three steps of the random walk (RW3S) would lead to the best results in the problems investigated therein. Random walk with restart is another popular approach [511, 535, 541], in which a Markov chain has some probability of going back to the starting position in every step of the chain.

Another way to improve data quality is through applying neural network, a set of deep learning methodology commonly applied to analyzing imaging data; HiCPlus [530], DeepHiC [177], scHiC-Rep [535] and Higachi [525], are such supervised learning techniques for improving data quality. For example, HiCPlus aims to infer high-resolution Hi-C matrix from low-resolution Hi-C matrix, with the assumption that Hi-C data matrices are composed of repeated local patterns and are spatially correlated in a neighborhood. On the other hand, Higashi represents the scHi-C dataset as a hypergraph and train a neural network based on the hypergraphs to capture high-order interaction patterns as a way of improving data quality.

8.4   Self representation smoothing for imputation and structural zeros inference

The methods presented in Section 8.3 are for improving data quality in low-resolution bulk Hi-C or sparse scHi-C data. Such methods do not aim to separate observed zeros into structural zeros and sampling zeros (dropouts), although they may be adapted to for such a purpose (Section 8.6). Further, all the methods described there do not make use of information from similar single cells, missing the opportunity to utilize data that can further help improve data quality. In this section, we describe scHiCSRS [496], a package that implements a self-representation smoothing (SRS) method and a mixture model approach to identify structural zeros in addition to data quality improvement.

8.4.1   SRS model and estimation

Suppose we have scHiC data from K single cells (SCs) that are similar (e.g. of the same known cell types) and sequencing-depth normalized, as assumed throughout this section. Let count y_{ijk} represent the observed interaction frequency (appropriately log-transformed if necessary) between loci i and j $(1 \leq j \leq i \leq n)$ (genomic segments of equal sizes; Chapter 7) for SC $k \in \{1, \ldots, K\}$, and n is the total number of loci considered. In a nutshell, scHiCSRS is a two-step approach, wherein the first step is SRS, which imputes the interaction frequencies in all positions in the 2D data matrix for each SC. This imputation step sets up a model that accounts for spatial correlation in a 2D matrix, while the contributions from neighbors are not fixed based on a specific kernel prior to imputation (i.e, different from the kernel-based data-quality improvement methods [450, 541, 543]). Rather, the contributions from the neighbors, the weights, are estimated using a self-representation formulation [98]. As the formulation unfolds in the following, we will see that information from similar SCs are also incorporated for learning about the weights, a feature missing from existing data-quality improvement methods discussed in Section 8.3.

Specifically, the SRS step uses the following model to incorporate information from neighbors and similar single cells:

$$y_{ijk} = \sum_{i'j' \in \delta(ij)} \beta_{(ij)(i'j')} y_{i'j'k} + \sum_{k' \neq k} \gamma_{kk'} y_{ijk'} + \varepsilon_{ijk}, \quad 1 \leq j \leq i \leq n, \ k = 1, \cdots, K,$$

(8.1)

where $\delta(ij)$ is the neighborhood of position (ij) in the 2D data matrix, $\beta_{(ij)(i'j')}$ is the contribution (weight) from a neighboring position $(i'j')$ on the interaction frequency at position (ij), $\gamma_{kk'}$ is the contribution from SC k' at the same position, and ε_{ijk} is a random error term assumed to have mean 0. The neighborhood can be specified flexibly as the problem dictates; typically, it may be set as $\delta(ij) = \{(i'j') : j' \leq i', 0 < |i' - i| \leq a, 0 < |j' - j| \leq a\}$ for a small integer a. For example, if we set $a = 1$, then the neighborhood is simply the eight (or up to eight since only upper triangular data are considered) immediate neighbors of (ij) but not including (ij) itself. Note that the β's are shared among all SCs, whereas the γ's are shared among all the positions within the same SC.

For efficient estimation of the β's and γ's, equation (8.1) is rewritten in a matrix representation by mapping each position (ij) in the 2D matrix to a single unique index $p = i(i - 1)/2 + j, 1 \leq p \leq N = n(n + 1)/2$; therefore, $y_{ijk} \equiv y_{pk}$. The collection of data for K single cells can then be organized into a single matrix of dimension $N \times K : Y = (y_{pk})$. This reorganization of the data is similar to the preparation step for using the adapted scRNA methods to analyze scHi-C data (Section 8.2.1). Similarly, all the β's are collected into a single sparse matrix of dimension $N \times N : B = (\beta_{pp'})$, and all the γ's into a single matrix of dimension $K \times K : \Gamma = (\gamma_{kk'})$ with all diagonal elements set to 0. The B is sparse because for each row $p \equiv (ij)$, the entries are zeros except for those that are in $\delta(ij)$, the neighborhood of (ij). Finally, the error terms are also written into a single matrix (in the same way as representing the 2D data) of dimension $N \times K : E = (\varepsilon_{pk})$. With these matrix representations, one can re-express (8.1) as $Y = BY + Y\Gamma + E$. The estimates for the B and Γ matrices are obtained by solving a Lasso-type least-squares optimization problem:

$$(\hat{B}, \hat{\Gamma}) = \arg\min_{B, \Gamma}\{\|Y - BY - Y\Gamma\|_F + \lambda\|\Gamma\|_1\}, \tag{8.2}$$

where all the entries in $\hat{B}$ and $\hat{\Gamma}$ are constrained to be non-negative; $\|\cdot\|_F$ represents the Frobenius norm, $\|\cdot\|_1$ denotes the L_1 norm (as in Lasso) for each element in the matrix, and λ is the amount of penalty. We note that in (8.2), the entries in B are not penalized as the matrix is already sparse by taking spatial correlation into account. The penalty λ is a tuning parameter controlling the sparsity level of Γ (larger λ leads to a smaller set of most similar SCs retained in SRS). Using Gordon's Theorem [206, 459, 533], λ is estimated as follows before optimizing (8.2): $\hat{\lambda} = sd(Y)$, where $sd(Y)$ is the standard deviation of entries in data matrix Y. This way of selecting the tuning parameter is advantageous as it performs well in genomic applications [206] and avoids the computational intensity of cross-validation or similar procedures. A coordinate descent algorithm is then applied to solve the optimization problem in (8.2) by iterating between minimizing the objective function with respect to B while keeping Γ fixed or vice versa.

With the estimated $\hat{B}$ and $\hat{\Gamma}$, one obtains

$$\tilde{y}_{ijk} = \sum_{i'j'\in\delta(ij)} \hat{\beta}_{(ij)(i'j')}y_{i'j'k} + \sum_{k'\neq k} \hat{\gamma}_{kk'}y_{ijk'}.$$

To account for the uncertainty in the estimated $\tilde{y}_{ijk}$, a weighted average between $\tilde{y}_{ijk}$ and the observed y_{ijk} is used to obtain the final imputation value $\hat{y}_{ijk} = (1-w)y_{ijk} + w\tilde{y}_{ijk}$, with the weight w roughly inversely proportional to the variability of the estimated frequency for the position across SCs. Mathematically, this weighted average is the Bayesian estimate (posterior mean), where the observed contact frequency is assumed to follow a Poisson distribution [186, 206]. Its unknown intensity parameter is further supposed to distribute as a Gamma with mean related to $\tilde{y}_{ijk}$ and variance – related to the weight – estimated across SCs.

8.4.2   Inference for structural zeros

For identification of structural zeros, it is possible to simply use a thresholding approach: any position (ij) in SC k will be identified as SZ if $\hat{y}_{ijk} < 0.5$ as in some of the scRNA analysis methods [153, 295] (assuming the SCs are appropriately depth-normalized). A different strategy is based on probabilistic modeling. For each position (ij) in the 2D matrix, a Gaussian mixture model will be fitted to the log-transformed imputed data, $z_{ijk} = \log_1 0(\hat{y}_{ijk} + 1.01)$:

$$z_{ijk} \sim \sum_{q=1}^{Q} \eta_{ij}^q N(\mu_{ij}^q, \sigma_{ij}^q),$$

where $\sum_{q=1}^{Q} \eta_{ij}^q = 1$, η_{ij}^q is the weight for normal component $N(\mu_{ij}^q, \sigma_{ij}^q)$ with μ_{ij}^q and σ_{ij}^q being the mean and variance, respectively. This modeling-based

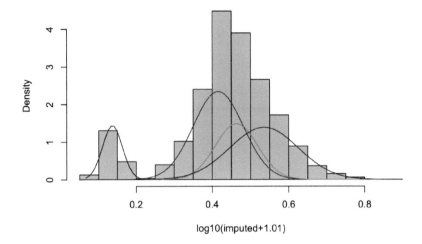

FIGURE 8.3
Distribution of imputed data with superimposed component densities of a mixture of normal distributions. The component with the smallest mean (blue component) was identified to have captured the structural zeros.

approach allows for flexibility without assuming an empirical relationship between dropouts and the mean of contact frequencies, as does the simple thresholding, rule-based approach. Among the components, those with much smaller means are treated as capturing structural zeros whereas the rest are assumed to model dropouts with larger imputed values. For a given fixed number of component, Q, the parameter in this mixture model is estimated using the Expectation-Maximization (EM) algorithm [77]. The number of mixture components, Q is determined based on the BIC criterion [388]. Figure 8.3 is an example of the individual mixture components fitted to a position in a simulated dataset mimicking the K562 real data analyzed (Section 8.6.2).

With a slight abuse of notation to avoid cluttering, assume the parameters in the model represent those that are estimated and selected based on the EM algorithm and the BIC criterion. Now assume the components means are ordered such that $\mu_{ij}^1 < \mu_{ij}^2 < \cdots < \mu_{ij}^Q$. Starting from the component with the smallest mean value, each successive component q is labelled as capturing structural zeros until the following stopping rule is satisfied: when $q = Q - 1$ or when $\mu_{ij}^q - \mu_{ij}^{q-1} > m(\mu_{ij}^{q+1} - \mu_{ij}^q)$, where m is a fractional number preset to be small enough to avoid over detection of structural zeros and is specified to be $1/10$ in scHiCSRS [496]. In other words, the components with small means clustered together will be treated as capturing structural zeros.

Suppose the first component is the only one that represents structural zeros among the observed zeros (Figure 8.3), then the probability that an observed zero is a structural zero is calculated as

$$P_{ijk}^{SZ} = \frac{\eta_{ij}^1 f_1(z_{ijk}; \mu_{ij}^1, \sigma_{ij}^1)}{\sum_{q=1}^{Q} \eta_{ij}^q f_q(z_{ijk}; \mu_{ij}^q, \sigma_{ij}^q)},$$

where $f_1, \cdots, f_Q$ are the normal density functions. An observed zero at position (ij) for SC k is labeled an SZ if the posterior probability $P_{ijk}^{SZ} \geq 0.5$, following the Bayes rule. Other threshold values may be considered as so desired.

8.5   Bayesian modeling for identifying structural zeros and imputing dropouts

In the above section, we see that scHiCSRS contains a post-processing step to separate structurl zeros from sampling zeros. In this section, we describe HiCImpute, a method for which structural zeros inference is embedded into the procedure; in other words, HiCImpute treats inference for structural zeros as an integral part of the procedure. Another feature that sets HiCImpute apart from the rest described in this chapter is that it also utilizes information from bulk data (if such are available), in addition to borrowing information from neighbors are similar single cells, to provide a powerful method for correct identification of structural zeros and accurate imputation of dropped-out values.

8.5.1   The HiCImpute model

Suppose we have scHi-C data from K single cells (SCs) that are deemed homogeneous and a relevant bulk Hi-C dataset related to the SCs. Using the same notation as in Section 8.4, let Y_{ijk} and Y_{ij}^b be the counts that represent the observed interaction frequency between loci i and j $(i < j)$ for single cell (SC) $k, k = 1, \cdots, K$, and the bulk data, respectively. For the observed counts thta are zeros, some are true zeros since the two loci are never interacting in this particular cell and therefore are structural zeros, whereas others are sampling zeros since they interact in reality since they interact in reality but drop out from the sample as their interaction is not observed due to insufficient sequencing depth. Thus, this zero-inflating problem is complicated since not all zeros are created equal. For each $Y_{ijk} = 0$, our goal is to make statistical inference to tease out those that are structural zeros from those that are dropouts. Since Y_{ijk} represents count data, it can be reasonably modeled by a Poisson distribution. Let $T_k = \sum_{i<j} Y_{ijk}$ denote the sequencing depth of SC k. Let μ_{ij}^k be the parameter representing the intensity of SC k if the SC were depth-normalized to a desired sequencing depth T, which may be

the maximum among the SCs, that is, $T = \max T_k$, or to an intended level appropriate for downstream analysis, say 300,000 (the level of the best SCs observed in the K562 data in one of the case studies (Section 8.6.2)). Let $\lambda^k = T_k/T$ is the proportional sequencing depth of SC k relative to the intended one. However, note that this is simply a construct since the observed, not any normalized, count is used in the modeling; the reason for having this construct will become clear as the formulation unfolds. Let $Y_{ijk} I_{Y_{ijk}>0} \sim \text{Poisson}(\lambda^k \mu_{ij}^k)$, where $I_{Y_{ijk}>0}$ is the usual indicator function. Define an indicator random variable S_{ij}, which is equal to 1 if loci i and j do not interact; otherwise, it is 0. Further, let $S_{ij} \sim \text{Bernoulli}(\pi_{ij})$, with unknown parameter π_{ij}. Let $Y_{ijk}|S_{ij} = 0 \ \text{Pois}(\lambda_k \mu_{ij}^k)$, and $P(Y_{ijk} = 0|S_{ij} = 1) = 1$. Then Y_{ijk} follows a mixture of a point-mass distribution at 0 and a Poisson distribution with mixing proportions π_{ij} and $1 - \pi_{ij}$, respectively. Noting that $P(Y_{ijk} > 0|S_{ij} = 1) = 0$, we have

$$
\begin{aligned}
P(Y_{ijk}) &= P(Y_{ijk}|S_{ij} = 1)P(S_{ij} = 1) + P(Y_{ijk}|S_{ij} = 0)P(S_{ij} = 0) \\
&= \begin{cases} \pi_{ij} + (1 - \pi_{ij})\text{Poisson}(0; \lambda^k \mu_{ij}^k) & \text{if } Y_{ijk} = 0, \\ (1 - \pi_{ij})\text{Poisson}(Y_{ijk}; \lambda^k \mu_{ij}^k) & \text{if } Y_{ijk} > 0. \end{cases}
\end{aligned}
$$

Let π_{ij} follow a beta distribution so that the mean of the beta distribution is governed by the proportion of zeros across the SCs in that position. The idea is that if there is a large proportion of zeros in that position, it is more likely to be a structural zero, and any non-zero observations could be due to random collision [125], or cell-to-cell variability.

Recall that μ_{ij}^k is the intensity parameter for a common sequencing depth, but to allow for cell-to-cell variability, an additional hierarchy is used to model this parameter as follows: $\mu_{ij}^k \sim \text{Normal}^+(\mu_{ij}, \sigma^2)$, where σ^2 is set to control the degree of variability from cell to cell, and μ_{ij} is further assumed to follow a gamma distribution whose mean uses information from the bulk data and data from the neighborhood of the SCs. Specifically, let $Y_{ij}^{(nSC)} = \sum_k Y_{ijk} / \sum_k \lambda^k$, which is the weighted average of the "normalized" (to sequencing depth T) count of contacts between i and j over all the SCs analyzed together, with the weight being proportional to the sequencing depth of each SC. We denote the overall mean across all locations of the 2D matrix as $\bar{Y}^{(nSC)} = \sum_{i<j} Y_{ij}^{(nSC)} / [n(n-1)/2]$, where n is the dimension of the contact matrix. Similarly, let $Y_{ij}^{(nB)} = Y_{ij}^B (T / \sum_{i<j} Y_{ij}^B)$ and $\bar{Y}^{(nB)} = \sum_{i<j} Y_{ij}^{(nB)} / [(n(n-1)/2]$ be the "normalized" contact counts and overall mean for the bulk data, respectively. Then the mean of the gamma distribution is specified as

$$
\left(\frac{\sum_{(i,j)\in\Omega} Y_{ij}^{(nSC)}}{\|\Omega\|} \right) \left(\frac{\sum_{(i,j)\in\Omega} Y_{ij}^{(nB)}}{\|\Omega\| \bar{Y}^{(nB)}} \right),
$$

where Ω is the neighborhood, which is set to include the eight immediate neighbors as in scHiCSRS [496], and $\|G\|$ is its cardinality.

8.5.2   Statistical inference

The hierarchical modeling procedure naturally leads to a Bayesian inference framework. To fully specify the parameters of interest and nuisance parameter, additional non-informative priors for the nuisance parameters are used [498]. With the posterior distribution specified up to a normalizing constant, MCMC methods are used to sample the parameters from the posterior distribution and statistical inference is made based on the posterior samples. In particular, using samples generated by MCMC from the posterior distribution of π for a particular pair that have an observed zero count in an SC, a natural decision based on the Bayes rule is to declare a zero for an SC to be a SZ if the corresponding π_{ij} is estimated by the posterior sample mean to be greater than 0.5. Otherwise, the imputed value for the dropped out value of the observed zero is taken to be $\hat{\lambda}^k \hat{\mu}_{ij}^k$ for each single cell k, where $\hat{\lambda}^k$ and $\hat{\mu}_{ij}^k$ are the MCMC estimates of the parameters λ^k and μ_{ij}^k, respectively. For data quality improvement, all non-zero observed counts are also imputed using the same procedure as for the dropout values.

This HiCImpute procedure has the advantage that information from bulk data, if available, are incorporated to help enhance the quality of the imputed data. Further, observed counts are directly used without having to be sequencing-depth normalized; it is the intensity parameter that get "normalized" and thus the use the Poisson distribution is appropriate. Furthermore, potential over-dispersion is taken care of by the hierarchical modeling of the mean of the Poisson distribution.

8.6   Case studies

The methods described in this chapter are demonstrated briefly in three case studies. The first one is to illustrate the improvement of cell clustering using scHi-C data enhanced from scRNA imputation method discussed in this chapter, McImpute. The second is the counterpart of the first case study, but using methods specifically proposed for scHi-C data. The last case study is to illustrate the utility of scHi-C analysis methods for cell subtype discovery.

8.6.1   Case study 1 – improved cell clustering with enhanced scHi-C data from scRNA imputation method

This case study is based on information from a scHi-C dataset (https://www.ncbi.nlm.nih.gov/geo/query/acc.cgi?acc=GSE117876) composed of single cells from two cell lines: 14 lymphoblastoid (GM) and 18 peripheral blood mononuclear cell (PBMC). This case study demonstrates that enhanced scHi-C data from McImpute leads to improved separation of the 32 cells

TABLE 8.3

GSE117876 data summaries and clustering results before and after imputation

(A) Data summary

	Mean	**Min**	**Q1**	**Median**	**Q3**	**Max**
Seq Depth	1837	2252	2674	2829	3113	4818
% Zeros	48.5	54.4	58.4	57.6	60.8	68.1

(B) Clustering results

	K-means clustering				
	Observed			**McImpute**	
Cell type	**Cluster 1**	**Cluster 2**		**Cluster 1**	**Cluster 2**
GM	13	1		14	0
PBMC	7	11		2	16
Misclas Rate	0.25			0.06	
ARI	0.23			0.76	

according to their respective cell types. Specifically, 30 consecutive loci on chromosome 1 that have a large percentage of observed zeros (Table 8.3(A)) were selected for use in clustering the cells before and after imputation with McImpute [153]. With the observed data (before imputation), 1 GM cell was classified with the PBMC cluster, while 7 PBMC cells were classified with the GM cluster (Table 8.3(B)). The clustering results from the data enhanced with McImpute improved to only having two misclassifications, with 2 PBMC cells remain clustered with the GM cells. This reduces the misclassification rate and increases the adjusted rand index (ARI) (Table 8.3(B)).

8.6.2   Case study 2 – cell clustering with enhanced data from methods proposed specifically for scHi-C analysis

This case study considers three real scHi-C datasets. The five methods discussed in this chapter, 2DMF, 2DGK, WR3S, scHiCSRS, and HiCImpute, are applied to these data to demonstrate the improvement (or the lack thereof) for clustering cells.

GM/PBMC data

The first scHi-C dataset (GSE117874) is the same as the one used in the first case study (Section 8.6.1). Using the same sub-2D matrix (consisting of 30 consecutive loci on chromosome 1), the results show that data enhanced by both HiCImpute and scHiCSRS lead to the same improved clustering result, reducing the misclassification rate while increasing the adjusted rand index (Table 8.4(A)). Specifically, for the data improved by both methods, all GM cells are now correctly classified, and there are only three misclassified PBMC cells. In contrast, using data imputed by 2DMF and 2DGK does not result in any improvement, whereas using the WR3S imputed data in fact leads to more misclassifications on the GM and PBMC cells than directly using the observed data. Comparing the clustering results with those presented in Case Study 1, it can be seen that McImpute-enhanced data in fact has one fewer misclassified cell than the best performer in this case study.

K562 data

The second dataset contains a mixture of bulk and single cell K562 Hi-C data (https://www.ncbi.nlm.nih.gov/geo/query/acc.cgi?acc=GSE80006). Specifically, there are two bulk Hi-C data – K562A (bulk A) and one K562B (bulk B) – and 19 scHi-C data of K562A cells and 15 K562B cells [116]. However, only 10 of the 34 single cells have a sequencing depth of over 5K; most of the remaining ones have sequencing depth of only 1K. Since it is unrealistic to expect that 3D structure can be sufficiently recovered from scHiC data with very low sequencing depth, this case study only considered the 10 cells that have sequencing depth of at least 5K [495, 498]. The two bulk data are incorporated by HiCImpute to further improve imputation. Using the observed data on these 10 single cells, K-means led to one of the two K562A cells being clustered with the eight K562B cells (Table 8.4(B)). Clustering using imputed data from 2DMF, 2DGK, or RW3S did not yield any improvement over the outcome from simply using the observed data. In contrast, clustering using improved data from HiCImpute and scHiCSRS both led to the correction of the single misclassified K562A cell, resulting in perfect separation of all K562A and K562B cells. It is noted that, while scHiCSRS did not incorporate information from the bulk A and bulk B data, as did HiCImpute, it did analyzed single cells together, thus enriching the information for delineating between these two subtypes of single cells.

Prefrontal cortex L4 and L5 subtypes

The third scHi-C dataset (https://github.com/dixonlab/scm3C-seq) consists of prefrontal cortex cells of subtypes L4 (131 cells) and L5 (180 cells) produced by a recent single-nucleus methyl-3C sequencing technique that simultaneously captures both DNA methylation and 3C data [248]. There are 14 cell subtypes of the prefrontal cortex cells, including eight neuronal subtypes; however, based on the observed scHi-C data, cells from these subtypes were all clustered together [248]. Among the eight are L4 and L5, two excitatory neuronal subtypes located on different cortical layers. The K-means analysis using the originally observed (not imputed) L4 and L5 scHi-C data for 311 cells confirms the finding that these two subtypes are indeed mixed together (Table 8.4(C)). Applying 2DMF, 2DGK, and RW3S did not improve the results at all. The misclassification rates for these methods are close to 0.5 with

TABLE 8.4

Clustering results for three datasets based on observed data and data improved by Hi-C analysis methods*

(A) GSE117876

Cells	KM	Obs	Bayes	SRS	2DMF	2DGK	RW3S
GM	C1	13	14	14	13	13	11
	C2	1	0	0	1	1	3
PBMC	C1	7	3	3	7	7	8
	C2	11	15	15	11	11	10
MisRate		0.25	0.09	0.09	0.25	0.25	0.34
ARI		0.23	0.65	0.65	0.23	0.23	0.07

(B) GSE80006

Cells	KM	Obs	Bayes	SRS	2DMF	2DGK	RW3S
K562A	C1	1	2	2	1	1	1
	C2	1	0	0	1	1	1
K562B	C1	0	0	0	0	0	0
	C2	8	8	8	8	8	8
MisRate		0.10	0.00	0.00	0.10	0.10	0.10
ARI		0.52	1.00	1.00	0.52	0.52	0.52

(C) scm3C-seq

Cells	KM	Obs	Bayes	SRS	2DMF	2DGK	RW3S
L4	C1	76	131	131	77	77	76
	C2	55	0	0	54	54	55
L5	C1	105	0	6	105	104	105
	C2	75	180	174	75	76	75
MisRate		0.49	0.00	0.02	0.49	0.49	0.49
ARI		0.00	1.00	0.92	0.00	0.00	0.00

*KM = K-means; Obs = Observed; Bayes = HiCImpute; SRS = scHiCSRS; MisRate = Misclassification Rate; ARI = Adjusted Rand Index

ARI being practically 0, indicating random mixing of the L4 and L5 cells. Thus, the problem appears to be much more challenging compared to the first two datasets: not only are the two types of cells mixed together, but the dataset is also much larger (hundreds compared to tens of cells). Nevertheless, using data improved by HiCImpute led to perfect separation of the two subtypes. Although the results with improved data from scHiCSRS is not perfect, there is a marked improvement, with only 6 of the L5 cells misclassified with the L4 cells, leading to a very small misclassification rate and an ARI of over 0.9.

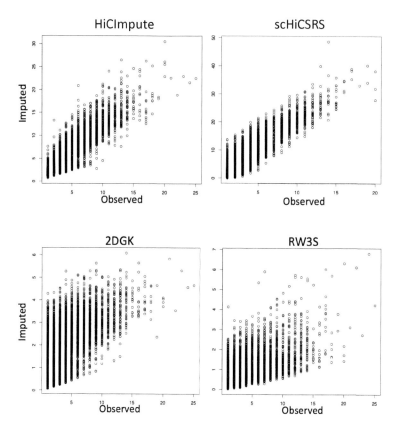

FIGURE 8.4
Scatterplots of observed vs. imputed for a representative L4 cell.

Plotting the imputed vs. the observed data for a representative single cell from L4 shows that HiCImpute is the best in preserving the original scale and having a high correlation with the observed data while improving the quality (Figure 8.4). For scHiCSRS, the point cloud also scatters tightly around a straight line, although the imputed values are larger compared to the observed due to data normalization across single cells. On the other hand, the plots for 2DGK and RW3S yield point clouds that scatter much more loosely, and there is a great reduction in the values due to smoothing (averaging). Note that for the purpose of clustering, shrinking or magnifying the observed counts usually has a negligible impact on the outcome. The results for 2DMF are not shown in the figure for the sake of brevity, since they are almost identical to those for 2DGK. Considering all single cells, the correlations between the observed and imputed are 0.91, 0.82, 0.65, 0.70, and 0.72, for HiCImpute, scHiCSRS, 2DMF, 2DGK, and RW3S, respectively.

8.6.3 Case study 3 – discovery of subtypes based on improved data

Cell to cell variability is a driving force behind the development of single cell technologies [426]. Based on single cell RNA-seq data, subtypes of L4 and L5 have been discovered. For example, two L4 subtypes, Exc L4-5 FEZF2 SCN4B and Exc L4-6 FEZF2 IL26, were found to be highly distinctive as they occupied separate branches of a dendrogram [175]. On the other hand, cells of the L4–IT–VISp–Rspo1 subtypes were shown to exhibit heterogeneity along the first principal component of scRNA-seq data [428]. Similarly, two subtypes of L5, Exc L5-6 THEMIS C1QL3 and L5-6 THEMIS DCSTAMP, were also found to be on two separate branches of a dendrogram [175], while there was also research that further classified L5 cells into L5a and L5b subtypes [428]. Other works have also found subclusters of excitatory neurons including L4 and L5 [110, 294, 345].

Given the excellent performance of HiCImpute for separating the L4 and L5 cells, it is of interest to further consider whether the improved data can also lead to detection of subtypes. To provide a baseline result for comparison, the observed data (without any imputation) are visualized using t-SNE [453] and the embedded data are clustered using K-means. Consistent with the results in Table 8.4(C), the best number of cluster is determined to be 2 (based on K-means by identifying where the total within cluster sum of squares elbow occurs, or the adjusted rand index, as both give the same results), although each of these two clusters contains a mixture of L4 and L5 cells (Figure 8.5(A)). Following the same procedure as above, but using the HiCImpute-improved data, four well-separated clusters emerge, two of each of the L4 or the L5 cell types without any mixing (Figure 8.5(B)). Although the lack of information on individual cell identities prevents the mapping of each of the subtypes discovered using HiCImpute-improved data with those already given in the literature using scRNA data, these results demonstrate that scHi-C, like scRNA, contains information that can be used to study cell-to-cell variability.

8.7 Software

A number of softwares have been proposed specifically for improving scHi-C data quality through inference of zeros and imputation of dropout values. The software packages discussed and used in this chapter, including HiCRep [506], SCL [545], and GenomeDISCO [450], HiCImpute [498], scHiCSRS [496], and those listed in Table 8.2, are all freely available. Software that have been proposed for imputation of scRNA data and have been adapted for improving

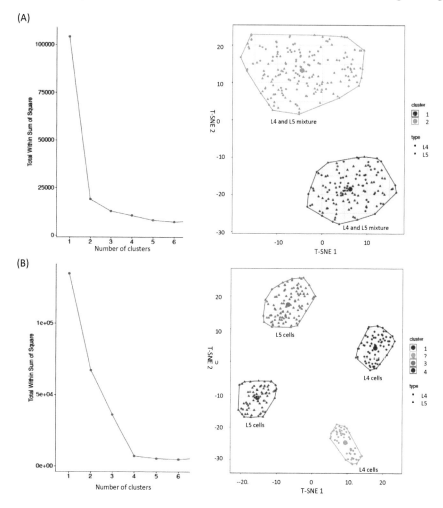

FIGURE 8.5

Discovery of further subtypes of L4 and L5 cells by HiCImpute. (a) Results from the originally observed data; (b) Results from HiCImpute-improved data.

data quality for scHiC data are listed in Table 8.1. These methods, either explicitly discussed in this chapter, or listed in the tables, are by no means the only ones available. Particularly, the field of scHi-C research is fast evolving, and new software are becoming available at a breathtaking pace.

8.8 Concluding remarks, and statisitcal and computational challenges

This chapter addresses the problem of data quality improvement for single cell Hi-C. Because of the limited sequencing depth, sparsity is a major concern, and the necessity to impute observed zeros to improve single cell Hi-C data quality has been amply discussed and numerous methods, algorithms, and software have been proposed. However, imputation is only half of the story for addressing the issue of sparsity, as some of the observed zeros are in fact structural zeros and should stay zeros. Of the two methods that have the best performance compared to the other methods proposed for scHi-C data quality improvement, both have the capability to identify structural zeros and impute dropouts. Although the two methods are based on different types of approaches, they share the common feature of utilizing similar single cells, which is not shared by the other methods. This pooling of information is seen to help improve the data quality sufficiently for separating cells of known types, as demonstrated in one example . It is further noted that one of the two methods, HiCImpute, can also utilize information from bulk data if such are available. Despite the observation that the methods discussed in this chapter can indeed improve scHi-C data quality and lead to more meaningful clustering results, the impact of the improved data on other biological questions of interest, such as identification of TAD-like domain boundaries and recapitulation of 3D structures, is not addressed in this chapter. Related topics may be found in the literature [152, 523].

9

Metabolomics Data Preprocessing

CONTENTS

9.1   Introduction

Metabolomics is the study of small molecule metabolites that reflect chemical cellular processes in a biological system. The metabolome, the comprehensive collection of low molecular weight molecules in a tissue or biofluid, reflects cellular activity downstream from the genome. The metabolome encompasses intermediate and end products of multiple enzymatic reactions in the cell, influenced not only by genetics but environmental influences as well (Figure 9.1). Hence, metabolomics investigations offer a unique integrative perspective on both genetic and environmental contributions to cellular metabolism. Since small molecule metabolites are basic machinery for physiological function, circulating metabolites can be viewed as intermediaries between genetic and environmental influences and downstream phenotypes. Metabolic profiling is therefore a promising strategy for beginning to untangle cellular mechanisms underlying associations between exposures and phenotypic outcomes.

DOI: 10.1201/9781315153728-9

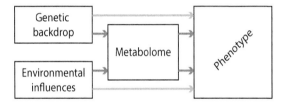

FIGURE 9.1
The metabolome reflects both genetic and environmental influences on cellular organisms and, at least in part, is thought to mediate effects of these factors on phenotypes.

The total size of the metabolome is unknown and estimates vary widely depending on the organism under study. A total of approximately 600 metabolites have been estimated for *Saccharomyces cerevisiae* [119] and up to 200,000 metabolites have been estimated for plants [112]. Estimates for total numbers of metabolites do not necessarily match the number of unique entries in metabolite databases. For example, databases for metabolites detected in human urine or blood serum generally contain reliable mass spectrometry profiles for 10,000–15,000 metabolites [429]. Estimated total numbers of metabolites and reliable annotations will very likely converge over time with large scale efforts to catalog data describing metabolites.

Specimens used for metabolomics investigations depend largely on experimental design and scientific hypotheses being evaluated. Plant metabolomics, where much pioneering metabolomics work was done, is conducted using a variety of plant tissues. Blood serum and urine are typically used for human metabolomics investigations and both have their strengths and drawbacks depending on the platform being used. Human metabolomics studies have been conducted for a wide range of applications including obesity and diabetes [189, 308], cancer [10, 487] and the altered metabolic state during pregnancy [385, 386]. Exacting sample preparation protocols are published for all types of samples and are designed to control variability in observed data attributable to sample preparation to the fullest extent possible [229, 282, 390, 445, 475, 514].

Metabolomics investigations often deploy a series of analytic technologies since the chemical diversity of small molecule metabolites precludes reliable, comprehensive assay using only one approach at this time. Generally, multiple assays including both targeted and non-targeted techniques are applied. Targeted technologies produce stable, calibrated measurements of metabolite abundance but typically focus only on one or a few chemical classes of metabolites. Non-targeted technologies are more comprehensive in nature, but require far more pre-processing and curation and tend to be more prone to technical variability. While similar in nature to small molecule metabolomics, lipidomics has emerged as a somewhat distinct field of study due to unique complexities of this chemical class [408, 481]. Lipidomics will not be emphasized in the current chapter.

In this chapter, we will discuss a range of technology platforms and approaches that are typically used for metabolomics investigations. We will describe data formats that often accompany these technologies and are readily accomodated by analytic software. For non-targeted assays that rely on mass spectrometry, peak identification algorithms are paramount and will be described. Normalization techniques and experimental design strategies to control for various sources of technical variability will also be discussed.

9.2 Data sets

9.2.1 HAPO metabolomics study

Metabolomics data from women who participated in the Hyperglycemia and Adverse Pregnancy Outcome (HAPO) Metabolomics Study will be used to illustrate statistical analyses in this chapter and the next. HAPO was an international, epidemiologic study designed to identify associations between maternal glucose during pregnancy and adverse pregnancy outcomes in both mothers and their newborns [23]. HAPO included more than 23,000 participants from 15 field centers around the world, and included standardized protocols and rigorous training of research staff to ensure collection of high quality phenotypic data [142]. During HAPO, pregnant women underwent a 75g 2-hour oral glucose tolerance test (OGTT) between 24 and 32 weeks gestation after an overnight fast. Fasting blood samples prior to the start of the OGTT were among the blood samples obtained. Many primary measurements were made, and additional blood samples were collected and stored at -80C. The HAPO Metabolomics Study was designed to investigate the metabolic milieu underlying associations between maternal glucose during pregnancy and newborn size outcomes. Both targeted and non-targeted gas chromatography-mass spectrometry (GC-MS) measurements were made on stored samples from the HAPO OGTT using techniques as described [385]. For illustration purposes, we will focus on metabolomics data for 400 fasting samples from HAPO mothers and cord serum samples from their newborns. Women were sampled such that maternal glucose and body mass index (BMI) as well as newborn birthweight spanned the range of measures observed in the full HAPO study. Table 9.1 includes descriptive statistics for these HAPO mothers and their newborns.

TABLE 9.1

Summary statistics for 400 HAPO mothers and newborns

	N(%)
Field center	
Belfast, UK	188 (47.0)
Brisbane, Australia	136 (34.0)
Newcastle, Australia	76 (19.0)
Parity	
First child	203 (50.8)
Second or later child	197 (49.2)
Newborn sex	
Male	209 (52.2)
Female	191 (47.8)

	mean (sd)
Maternal characteristics	
Age at OGTT (yrs)	29.4 (5.1)
BMI at OGTT (kg/m^2)	29.0 (4.9)
Mean arterial pressure (mmHg)	83.1 (6.9)
Fasting plasma glucose (mg/dL)	131.7 (27.2)
Sample storage time (yrs)	9.9 (1.3)
Newborn characteristics	
Gestational age at OGTT (wks)	28.6 (1.4)
Gestational age at delivery (wks)	40.2 (1.2)
Birthweight (g)	3667.6 (495.3)
Sum of skinfolds (mm)	12.8 (2.7)

9.3   Technology platforms

Metabolomics technologies were largely developed subsequent to earlier waves of transcriptomics and other DNA- or RNA-based "omics" assay techniques, within initial efforts published in the late 1990s [312, 319] and consistent development since that time. Metabolomics experiments often employ a two-pronged approach, relying on both targeted and non-targeted metabolomics assays. The full metabolome of small molecules in a cell includes multiple chemical classes with varying physical properties, as well as a very large dynamic range of metabolite concentrations. Targeted assays are designed to specifically assay compounds of a certain chemical class and typically incorporate stable standards for reliable quantification of metabolites in a given sample. Non-targeted assays are intended to comprehensively characterize all metabolites in a sample across the full dynamic range, including those with unknown identities. Targeted and non-targeted approaches are often deployed together to provide focused and reliable measurements of certain metabolites that are most relevant to the system under study, complemented by a more comprehensive view that is generally understood to require validation.

Targeted panels are developed to specifically assay certain classes of compounds with reliable biochemical annotations, for example amino acids, acylcarnitines, fatty acids and intermediary metabolites [290,308]. Detailed protocols exist to ensure standardized sample preparation and technology parameter specification for a variety of different assay types. Extraction methods are generally tuned to the chemical class under study and inclusion of internal, stable standards ensure these are calibrated quantitative assays with specific units of measurement [363]. Data from targeted metabolomics panels are generally of modest dimension, with numbers of metabolites ranging 20–50 depending on the chemical class being studied. Stable internal standards used to calibrate results minimize technical variability and in general batch effects or other similar potential sources of technical noise are of minimal concern. Data values may be missing for targeted measurements, and in general this missingness indicates lack of detectability given instrumentation thresholds. For the most part, data generated using targeted metabolomics technologies require very little pre-processing prior to statistical analysis.

In contrast to targeted metabolomics, non-targeted techniques are designed to comprehensively assay the full set of metabolites in a sample. The most commonly used approaches include nuclear magnetic resonance (NMR) spectroscopy, liquid chromatography-mass spectrometry (LC-MS) and GC-MS. NMR is a spectroscopic technique that relies on atomic nuclei energy absorption and re-emission as external magnetic fields vary. NMR not only allows for quantification of metabolite abundance, but can also be used to investigate chemical structure. Mass spectrometry-based approaches gather data in the form of mass-to-charge (m/z) ratios and retention times, with peak areas at these parameters used to estimate relative metabolite abundances.

While non-targeted approaches are more comprehensive, they also demand significantly more pre-processing prior to statistical analysis to address scientific hypotheses. Non-targeted methods that use mass spectrometry require complex computational peak detection and alignment, as well as chemometric approaches to match identified peaks to available libraries for annotation. These strategies will be described in the next section. Non-targeted metabolomics data are sometimes described as "semi-quantitative" since the abundance measures that result from these approaches are not strictly calibrated to internal standards with well defined units of measurement, but rather use peak areas to measure relative metabolite abundance across samples. It is often the case that identified peaks are not included in available libraries, in which case experimental investigation including analytical chemistry must be performed to identify unannotated compounds of interest. Non-targeted assays can also be prone to bias, for example only reliably detecting metabolites within a certain abundance range depending on tuning of the instrumentation. Despite these complications, non-targeted metabolomics techniques do offer opportunity for novel metabolite discovery, unlike targeted assay counterparts.

9.4 Data formats

For non-targeted data in particular, recording results according to consistent and computationally tractable standards is key. Similar to other high-throughput array-based data reporting constructs, several initiatives provide standard approaches for non-targeted metabolomics. The netCDF (network common data form) binary data format developed and maintained by Unidata (http://www.unidata.ucar.edu/software/netcdf/) [359] has frequently been used as a non-targeted mass spectrometry-based data output format. mzXML [267,337] and mzML [80] are two related mark-up language approaches specifically developed for proteomics mass spectrometry data that are often harnessed for metabolomics data. Many technologies include conversion tools to transform raw output into these data formats. Strategies such as mzAPI also exist for converting one format into another [13]. Similar to other omics data types, efforts at standardization have led to substantial improvements in data reporting and eased computation since pre-processing algorithms are often programmed to act explicitly on these formats. Efforts to increase generalizability and sophistication of these data models continue.

9.5 Peak identification and metabolite quantification for non-targeted data

A unique challenge in the analysis of non-targeted metabolomics data is the substantial amount of pre-processing required to identify metabolites present in the samples under study and then quantify their relative abundance for downstream data analyses. For mass-spectrometry based approaches, this requires peak detection, alignment and deconvolution, followed by compound identification using the mass spectrometry output. NMR pre-processing involves similar steps. The order of these pre-processing components depends on the approach being used, but the general idea is to identify high-quality peaks within each spectra for a sample and then align peaks across samples that likely represent the same metabolite. Areas under these peaks are then used as representative measurements of relative metabolite abundance in each sample.

9.5.1 Peak detection

Peak detection methods generally incorporate initial smoothing of mass spectra followed by filtering of peaks using different strategies to identify those representing metabolites in the sample and not simply background noise. XCMS is a commonly used preprocessing algorithm that was originally developed for LC/MS data that includes peak detection as a first step [405]. Observed data are first sliced into mass units, for example 0.1 m/z, and intensity maxima at each retention time within the slices are filtered and background subtracted using a second-order Gaussian derivative as a model for the peak shape. Other methods have also been proposed that incorporate different functions for smoothing mass spectra peaks, for example for example moving averages and Gaussian windows [501], wavelet transforms [90], local maxima and recursive thresholding [223,224]. Each approach has its strengths depending on the nature of the application and metabolomics pre-processing software pipelines like MZMine2 [344] generally offer a variety of options for the user to choose. After smoothing, filters are generally applied to select peaks with areas above a background noise threshold and to remove low signal artifacts, for example using signal-to-noise ratios in detected peaks as in XCMS.

9.5.2 Peak alignment

Peak alignment is another crucial pre-processing step that is performed either prior to or following peak detection, depending on the algorithm. Experimental noise often results in a computed mass peak at a retention time that is slightly different from the true value, challenging proper alignment and often resulting in multiple candidates with a reasonable mass match. The idea in peak alignment is to identify and control for systematic deviations in retention times that may occur during chromatography for a series of samples using LC/MS or GC/MS. Figure 9.2 demonstrates peak detection for lactic acid for two different studies using GC/MS technology. The upper panel contains mass spectrometry peaks run on an older GC/MS machine and illustrates the need for peak alignment to reliably quantify abundance in the assayed samples. The lower panel contains mass spectrometry peaks run on a newer GC/MS machine and depicts more accurate alignment.

Original alignment methods were based on the concept of warping in which non-linear transformations were used to maximize correlations among spectra, for example correlation optimized warping (COW) and dynamic time warping (DTW) [4,437]. COW aligns each spectrum to a reference spectrum and DTW maintains equivalence of connected data points across spectra.

More recently, spectral segmenting alignment methods have gained popularity and demonstrate favorable performance. In these methods, a constant shift is used for spectral points either for the full spectra or after splitting the spectra into segments. Icoshift is a commonly used approach that relies on a reference spectrum and aligns spectra to maximize spectral

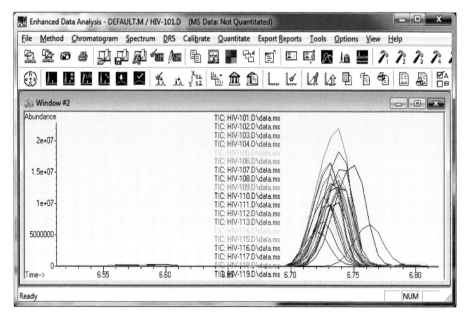

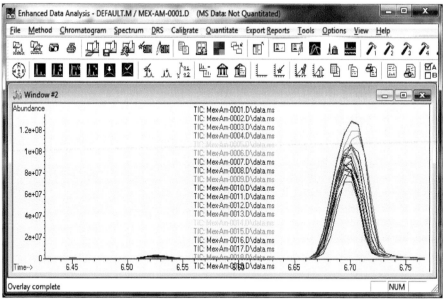

FIGURE 9.2

Lactic acid peaks in 24 injections of preparations of human serum from two
clinical studies, analyzed by GC/MS. Data for the lower plot, acquired on
a new 7890B GC-5977B MS (Agilent Technologies, Santa Clara, CA), show
stability of peak retention time at 6.7 minutes. Data in the upper plot were
acquired on an Agilent 6890N GC-5975 MS system with an optional ProSep
inlet after a decade of use. Retention times are less stable in the older instrument, necessitating peak alignment for annotation of this feature as lactic
acid.

correlations [377]. Computation time is decreased by computing correlations using the fast Fourier transform (FFT) and, if desired, by combining with automatic segmentation methods to optimize segmentation splits [460]. Using a reference panel for peak alignment may introduce bias, particularly if the reference spectrum is not representative of the spectra being aligned. At the expense of increased computation time, methods such as FOCUS have been described that rely on iterative maximization of correlations only over the spectra being analyzed [5]. Peak alignment methods developed in NMR have shown adequate performance for mass spectrometry-based data and vice versa, hence methods are often used interchangeably for either data type.

In XCMS, peaks are aligned after they are detected. XCMS identifies peak coordinates within m/z bins for alignment purposes and identifies retention time boundaries using kernel density estimation to define boundaries within which observed peaks across samples likely represent the same metabolite. Peaks that are not represented in at least half of the samples, or depending on the application, at least half of the samples within an experimental condition or sample type, are eliminated to reduce background noise and unmeaningful peaks. When multiple peaks are represented within these boundaries for a sample, tie-breaking strategies are employed to identify the peak to be used for analysis. Following peak alignment, retention times are then aligned using median values for reliable peaks that are well represented across samples and for which the retention time ranges for the most part only have one identified peak. Importantly, XCMS does not require use of a reference spectrum and therefore avoids related biases.

A variety of other methodological refinements have been proposed, and are implemented in algorithms and pipelines such as msInspect [25], MZmine [223], OpenMS [416], XAlign [527], apLCMS [512], xMSAnalyzer [447] and MetaBox [2]. These algorithms are similar in spirit, but employ different statistical modeling approaches for identifying peak maxima and peak boundaries. A variety of approaches have been developed to address various challenges in peak identification, including approaches that consider metabolite derivatives or fragments [279].

9.5.3 Peak deconvolution

It is not uncommon that peaks from multiple metabolites overlap, thus complicating abundance estimation for these metabolites. To resolve this difficulty, peak deconvolution methods have been developed. These techniques estimate the relative area attributable to each overlapping peak. Peak deconvolution methods exist for both NMR and mass spectrometry-based data. AMDIS (Automated Mass Spectrometry Deconvolution and Identification System) was one of the original approaches to peak deconvolution and is still frequently used in practice [91, 149, 413]. AMDIS consists of three major components: the first identifies noise across segments of the chromatogram that may be indicative of multiple proximal peaks; the second called "component perception" identifies individual chromatographic components distinct enough to explain the noise detected; and the third models the shape of each purported peak. The strength of the match of detected peaks to retention time libraries of known peaks is typically summarized by AMDIS using a "Reverse" score. This general approach to deconvolution is used by other algorithms as well, for example MetaboliteDetector [173] and ADAP-GC [311], although specific analytical details differ [91].

Initial pre-processing algorithms for NMR data used binning approaches to identify peaks, although peak detection methods initially developed for mass spectrometry data have demonstrated superior performance in NMR data as well [463, 484]. Peak alignment is also an issue for NMR data with observed variability in parts-per-million, or ppm, being similar in nature to retention time shifts in mass spectrometry data. BATMAN is an open-source software that is frequently used for NMR data that relies on a Bayesian model selection strategy [14, 158, 159]. This approach incorporates data on compounds with known spectral properties catalogued in databases (by default Human Metabolome Database [486]) to construct prior distributions. Observations for unannotated metabolites are then modeled as a linear combination of wavelet basis functions. Peak shifts are also included in the specified model. An MCMC approach is used to estimate metabolite abundance for compounds both with and without catalogued spectral data.

9.5.4   Metabolite identification

High-dimensional non-targeted metabolomics data must be annotated for meaningful interpretation. For some algorithms, such as BATMAN for NMR as just described, spectral data for known compounds are used to inform peak pre-processing. In many cases, however, peak annotation follows detection or is used in iterative fashion to refine peak alignments. Metabolite identification generally relies on matching of observed data to libraries of values for known metabolites, for example expected mass-to-charge ratios (m/z) and retention times for mass spectrometry analysis of known metabolites. Constructing these libraries is an expensive, labor intensive process that involves mass spectrometry analysis of known compounds and addition of the observed peaks to the library. Often times, existing publicly available libraries are expanded upon at high volume metabolomics analysis centers to address the needs of a variety of studies. Large-scale centers often have libraries including spectra for several thousand compounds.

Pre-processed data using the classic AMDIS approach have often been aligned with the Fiehn spectral library [232] for annotation. The Fiehn library (FiehnLib) currently contains mass spectral and retention time data for over 1000 known compounds. FiehnLib is often used as the base library for labs who wish to expand an existing library to include their own metabolites of specific interest. A variety of other publicly available databases detail m/z ratio and a variety of other data for small molecule metabolites. Some resources are devoted to data from human samples, for example the Human Metabolome Database (HMDB) [485]. Others can be linked specifically to output from pre-processing algorithms in an informatics pipeline; for example, METLIN is tightly integrated with the XCMS algorithm [404]. Other resources include the MassBank public repository [178], ChemSpider [338] and WEIZMASS [395].

Most databases used for metabolite annotation include not only metabolite names, but also standardized identifiers that link to additional information. For example, FiehnLib includes maps to PubChem IDs. PubChem is a component of the various NCBI (National Center for Biotechnology Information) resources that is specifically devoted to cataloging detailed information on chemical information, chemical structures and experimental biological activity data for small molecule metabolites. Metabolites are also frequently mapped to compound IDs from the KEGG pathway database [218, 219]. KEGG provides data on the position and role of metabolites within large-scale, well annotated metabolic pathways. CAS Registry Numbers from the Chemical Abstracts Service (https://www.cas.org/content/chemical-substances) are also often used as stable chemical compound identifiers to link across various data resources. Table 9.2 catalogs several existing compound identification systems and related resources. Investigation of these resources can enhance interpretation of experimental findings based on published literature, and also support analytic pipelines including pathway enrichment analyses as will be described in the next chapter.

Despite ever increasing efforts to build and expand retention time libraries, it is very often the case that only a subset of identified peaks are reliably

TABLE 9.2

List of databases with compound identifiers often used for metabolomics investigations

ID Name	Resource
BioCyc	BioCyc Database [41] https://biocyc.org/
ChEBI	Chemical Entities of Biological Interest Database [162] http://www.ebi.ac.uk/chebi/
HMBD	Human Metabolome Database [485] http://www.hmdb.ca/
KEGG	Kyoto Encyclopedia of Genes and Genomes [218] http://www.genome.jp/kegg/
MetaCyc	MetaCyc Metabolic Pathway Database [42] https://metacyc.org/
PubChem	PubChem Open Chemistry Database [231] https://pubchem.ncbi.nlm.nih.gov/

annotated. For example, it is not uncommon to identify GC/MS peaks numbering in the low thousands for studies of human samples with only a few hundred reliably annotated. Depending on the scientific question, it may be reasonable for statistical analysis to first focus only on annotated peaks, followed by discovery of trends for unannotated peaks. While there is significant possibility for new discovery by analyzing unannotated peaks, extensive work including analytical chemistry and possibly targeted assay development is likely to be required to confirm and validate observations.

9.6   Normalization

When raw non-targeted metabolomics data have been successfully run through peak detection, alignment, deconvolution and annotation pipelines, estimated peak areas are the final output that are used as measures of metabolite abundance. Peak areas are generally $\log_2$ transformed for both numeric stability and improvement of normality of observed data. When applied on a large scale, non-targeted metabolomics assays typically require batching of sample assays and can be run over the course of days, weeks and sometimes months. Substantial variations in estimated metabolite abundance levels attributable to batch, and even the run order of samples within a batch, are well-documented in particular for GC-MS data [92, 423]. In addition, due to the chemical diversity of metabolites being assayed, batch and run order effects can vary substantially for different metabolites.

Data from internal controls confirm batch effects with widely varying trends across different analytes [92, 217]. This phenomenon was observed for the HAPO Metabolomics Study. In this study, small aliquots from all fasting maternal samples of analytical interest were combined to create a quality control (QC) pool. Aliquots from this QC pool were run at the beginning, middle and end of 50 batches that were used for the HAPO Metabolomics samples. Since these aliquots come from identical QC pools, observations should be consistent across batches; however, trends attributable to batch are clearly observable in the data. In the example in Figure 9.3, observed values for alanine demonstrate a sharp increase starting at batch 30 and decrease until roughly batch 40. In contrast, values observed for glucose show a sharp increase from roughly batch 30 to batch 40. Three points are included for each batch, with small, medium and large points corresponding to QCs from the beginning, middle and end of each batch. For alanine in particular, it is clear based on QCs that observations from the beginning of the batch are lower than those at the end of the batch. This illustrates effects of run order within batch that must be controlled during statistical analyses. The dotted line in the plot for serine is used to indicate undetectable levels of this metabolite for some of the batch and run order combinations. Serine is a relatively low abundance compound and was undetectable in QC samples for some batches, presumably due to varying thresholds of detectability across batches. This truncation

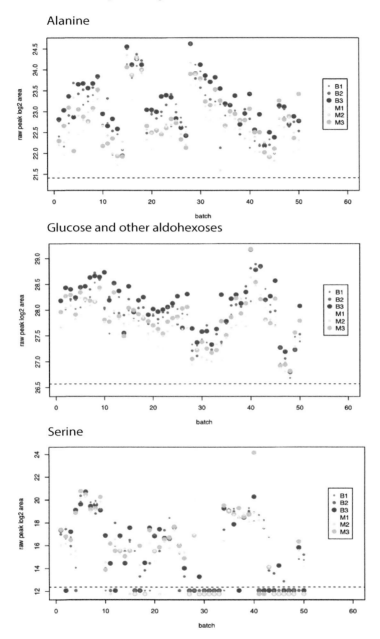

FIGURE 9.3

Plots of metabolite levels for peaks identified as alanine, glucose and other aldohexoses, and serine for identical baby (pink) and mom (blue) QC aliquots run at the beginning (B1, M1), middle (B2, M2) and end (B3, M3) of 50 HAPO Metabolomics batches. Since QC aliquots are identical, variability is attributable strictly to batch and run order within batch for each sample type. Values below the dotted line were undetectable for the indicated batches.

effect is common for GC-MS data and should be accounted for in normalization, as well as downstream statistical analyses. This simple example for only three metabolites from HAPO Metabolomics QC samples illustrates that, prior to statistical analyses pertaining to scientific study hypotheses, it is crucial to examine batch and run order variability for non-targeted metabolomics data and normalize with regard to these factors to control technical variability to the fullest extent possible.

Many approaches for controlling technical variability, i.e. normalization, have been described and some are included in existing metabolomics data analysis pipelines [131, 190, 226]. Figure 9.4 illustrates alanine data from HAPO Metabolomics after the application of several normalization methods that will be described, demonstrating the ability to successfully address technical variability using a variety of approaches. Straightforward approaches calculate scaling factors to be applied uniformly to all metabolites measured in a given sample and do not specifically rely on QC samples [419, 461]. For example, mean-centering of metabolites may be performed by calculating the mean of a given metabolite level across all samples in a batch and then subtracting that mean from observed metabolite levels in the batch. This is generally performed in metabolite-specific fashion and will result in a mean value of 0 for all metabolites across all batches. If desired, a constant, for example the mean of the metabolite across all batches prior to mean centering, can be added to mean-centered values so that data resemble the original scale. Adding this constant may be most relevant if calculating the relative standard deviation (RSD), the ratio of the metabolite's standard deviation to its sample mean, is of interest for reporting on assay quality. Specifically, let y_{im} represent the observed value of metabolite m for individual i ($i = 1, ..., n$). Let $x_i = b$ if the sample for individual i was included in batch b, with $b = 1, ..., B$. Letting $\mathbf{1}$ represent the indicator function, mean-centering is often performed as follows:

Mean-centering:

$$\left(y_{im} - \frac{\sum_{j=1}^{n} y_{jm} \mathbf{1}(x_j = x_i)}{\sum_{j=1}^{n} \mathbf{1}(x_j = x_i)} \right) + \frac{1}{n} \sum_{i=1}^{n} y_{im}. \tag{9.1}$$

Median scaling is similar in spirit to mean scaling. While there are variations in median scaling [468], one simple approach is to divide observed values for a metabolite in a given batch by the batch-specific median for that metabolite, and then multiply by the median for the metabolite across all samples in all batches to return the data to their approximate original scale. For example, median scaling may be performed as follows:

Median-scaling:

$$\left(\frac{y_{im}}{\underset{j:x_j = x_i}{\text{median}}\{y_{jm}\}} \right) \underset{i:i=1,...,n}{\text{median}}\{y_{im}\}. \tag{9.2}$$

Normalization strategies developed for gene expression microarray studies have also been applied to metabolomics data with varying success. Quantile normalization is based on the idea of equating distributions of metabolite abundance levels across all samples [31]. Metabolite abundance levels are sorted for each sample, means of the ranked metabolites are calculated across all samples, and the calculated means are substituted for the original observed metabolite levels according to their rank. Quantile normalization can be used in conjunction with empirical Bayes batch effect correction (ComBat) [209]. ComBat was designed to be used as a pairing to other normalization algorithms for batches of small sample size (<25) and uses metabolite-specific mean and variance estimates to correct for batch effects, while maintaining treatment or phenotype effects. EigenMS was adapted from surrogate variable analysis that was originally developed for gene expression data [249], but modified for use with non-targeted metabolomics data. Categorical "treatment" effects are estimated using ANOVA and then singular value decomposition is applied to the matrix of residuals to remove additional bias trends from the data to produce a normalized data set. This approach is most appropriately suited for experimental designs that are evaluating categorical treatments or phenotypes rather than metabolite associations with continuous measures. Variance stabilizing normalization (VSN) [188] was also developed for gene expression data and addresses the fact that variance of observed values depends on overall signal intensity. In VSN, a smooth transformation is applied that mimics linear scaling for low intensity values and a log transformation for high intensity values such that variance is stabilized across the full range of observations.

The main difficulty in applying normalization methods developed for gene expression data to non-targeted metabolomics data pertains to assumptions underlying the approaches. In general, these methods assume that gene expression is largely unchanged across the full range experimental conditions being assayed with only a relative few genes demonstrating biologically relevant changes. Depending on the method, other assumptions are that a roughly equal number of genes show increased and decreased expression across samples and that batch effects in general affect observed gene expression values in similar ways. These assumptions can easily be violated for metabolomics data [461]. Large-scale metabolic fluctuations are often observed across experimental conditions, rendering the assumption of relative stability of metabolite abundance unsuitable. Chemical diversity of metabolites and related differential batch and run order effects, often going in opposite directions, also violates the assumptions of similar batch effects across metabolite features.

To overcome these difficulties in adapting gene expression normalization methods to metabolomics data, more recently developed approaches advocate use of multiple internal standard compounds to the samples under study for normalization purposes [423]. The NOMIS (Normalization using Optimal selection of Multiple Internal Standards) method applies a multiplicative model for true metabolite abundance and batch variation that can be estimated using the internal standards. This approach is different from the use of QC samples. QC samples are aliquots from identical pools that are included as their own sample in a GC-MS batch. In the NOMIS approach, internal standards are compounds that are added directly to the analytical samples under study. NOMIS was developed using LC-MS data and specifies the following for log transformed peak areas:

NOMIS:

$$y_{im} = \mu_m + \rho_{im}(\Omega) + \epsilon_{im} \tag{9.3}$$

where

$$\rho_{im} = \sum_s \beta_{ms} \left(\Omega_{si} - \frac{1}{N} \sum_{i=1}^{N} \Omega_{si} \right) \text{ and } \epsilon_{im} \sim N(0, \sigma_m^2) \tag{9.4}$$

In NOMIS, Ω_{si} is the observed log transformed intensity for individual i for internal standard s, with $s = 1, ...S$. This model formulation estimates required corrections for each sample based on deviations of internal standards in a given sample from the mean of the same internal standards across all other samples. Maximum likelihood is used to estimate model parameters and their estimates are then used as correction factors for observed data. All equations used for maximization are described in detail by Sysi-Aho et al. [423].

The MeltDB 2.0 metabolomics pipeline incorporates normalization to a single internal standard, with ribitol being the default normalization standard unless otherwise specified [226]. Another approach is a priori identification of at least a few metabolites expected not to change in the experimental conditions of interest for modeling of batch effects [70]. After pre-selecting a set of metabolites that are assumed not to change, factor analysis is used to estimate sources of technical variability according to the following model:

Factor analysis using non-changing metabolite sets:

$$y_{im} = x_i \beta_m + \hat{w}_i \alpha_m + \epsilon_{im} \tag{9.5}$$

where x_i and $\hat{w}_i$ are row vectors for observed factors of interest and estimated factors of technical variation, respectively, and β_m and α_m are corresponding estimated parameter vectors.

While approaches that rely on addition of internal standards to samples can be useful, selecting appropriate standards and identifying non-changing sets can vary substantially depending on sample type and metabolite classes of interest. Furthermore, for sample types that are relatively new to the investigator, it is quite likely that selected internal standard or non-changing compounds may not be representative of retention times and aligned mass spectrometry peaks for metabolites observed in analytical samples [217].

Therefore, to more comprehensively address issues pertaining to chemical diversity of batch and run-order effects, the difficulty in selected representative internal standards and unsatisfied assumptions for other omics-based normalization algorithms, repeated assay of quality control (QC) samples is increasingly applied in many large-scale metabolomics studies [92,93,217,467]. This involves creation of standard control pool, often created from small aliquots from all samples to be assayed in an experiment, and inclusion of samples from this pool with all batches over the course of an experiment. Repeated assay of QC samples within a batch is also merited since metabolite abundance can be affected by run order within batch (Figure 9.3).

Normalization methods have been developed specifically to incorporate data from QCs included with GC/MS batches, for example, Batch Normalizer [467]. This regression-based algorithm incorporates total abundance of each sample to estimate and correct for batch and run order effects. The median total abundance level across all metabolites for all QC and analytical samples is identified as $\text{median}(TA_*)$. Metabolite levels are scaled by the ratio of this $\text{median}(TA_*)$ to the sum of all observed metabolite levels for the jth QC sample to which it belongs, TA_j. A linear regression model on the scaled metabolite abundances for control samples only is then used to estimate a batch-specific intercept and run order effects as follows:

Batch Normalizer:

$$\frac{\text{median}(TA_*)}{TA_j} \times y_{jm} \sim \sum_{b=1}^{B} (\beta_{mb}\mathbf{1}(x_j = b) + \alpha_{mb}r_j) \qquad (9.6)$$

where $x_j = b$ if the jth QC sample was in batch b and r_j is the run order position of the jth QC sample in the batch. The model parameters β_{mb} and α_{mb} are estimated using QC sample data only. Observed data for analytical samples of interest are then rescaled by the ratio of the median abundance for the metabolite of interest in the control samples to the sum of the batch-specific intercept and estimated run order effect for that sample's run order position, specifically for sample i and metabolite m:

$$\frac{y_{im} \times \frac{\text{median}(TA_*)}{TA_i} \times \text{median}(\{y_{jm} : j = 1, ...J\})}{\sum_{b=1}^{B} (\hat{\beta}_{mb}\mathbf{1}(x_j = b) + \hat{\alpha}_{mb}r_j)} \qquad (9.7)$$

where J is the total number of QC samples. Additional formulaic and estimation details are included in the original description of the algorithm [467]. While this approach has merit, it should be noted that its reliance on total metabolite abundance within a sample requires normalization for all peaks identified in mass spectrometry output. If investigators are interested in emphasizing only the annotated subset for statistical analysis, Batch Normalizer should be run first on the full set of observed peaks, including the unannotated subset.

Mixnorm is another mixture model-based approach for GC/MS data normalization that explicitly incorporates QC data for normalization purposes and also models batch-specific detection thresholds and resulting data truncation [356, 385]. In addition to batch effects observed for metabolite abundance, detection thresholds also vary across batches. This is evidenced by the detection of low abundance compounds from identical QC samples in some batches but not others (Figure 9.3) [141,180]. Many normalization algorithms either eliminate undetected metabolite levels or require imputation of these undetected values. Mixture modeling offers an alternative to imputation and jointly models the probability of a metabolite being absent in a sample due to degradation or presence below detectability along with observed abundance levels. Across a range of frequencies of undetected values, mixture modeling estimates batch and run-order effects with reasonable accuracy. Mixnorm is applied separately for each metabolite feature in a data set, thus accommodating variations in batch and run order effects for compounds of different chemical classes. Batch and run order corrections are first estimated using data for QC samples only, and estimated corrections are then applied to analytical samples of interest prior to statistical analysis. The likelihood contribution to the mixnorm normalization model for a given metabolite in the jth QC sample is specified as follows:

Mixnorm:

$$((1 - p_j) + p_j \Phi\left[(T_j - \mu_j)/\sigma\right])^{(1-\delta_j)} \left(p_j \exp\left[-(y_j - \mu_j)^2/2\sigma^2\right]/\sqrt{2\pi\sigma}\right)^{\delta_j}$$

where $\delta_j = 1$ if the metabolite is detected in the QC sample and 0 otherwise, T_j is the batch-specific threshold of detectability (often taken to be the minimum observed value in a given batch), p_j is the probability of metabolite presence in the sample, μ_j is the expected mean abundance level for the metabolite and Φ is the standard normal cumulative distribution function. A logistic model is used for p_j, specifically $\log(p_j/(1 - p_j)) = x'_j\beta$ and a linear model is specified for μ_j, specifically $\mu_j = z'_j\alpha$. Covariate vectors x_j and z_j are used to specify batch and, if desired, run order within batch as covariates for each component of the model. Model parameters are then estimated using computational likelihood maximization techniques and location shifts are applied to analytical samples of interest based on the batch and run order value for each sample to remove technical variability.

9.7   Experimental design

Like any other experimental investigation, experimental design principles are paramount to the successful execution of metabolomics research (Figure 9.4).

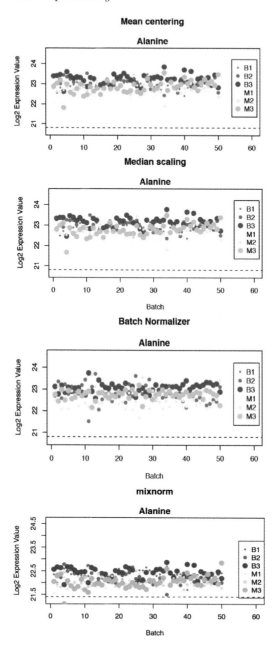

FIGURE 9.4

Plots of alanine levels for baby (pink) and mom (blue) QC aliquots run at the beginning (B1, M1), middle (B2, M2) and end (B3, M3) of 50 HAPO Metabolomics batches after application of mean centering, median scaling, Batch Normalizer and mixnorm normalization methods. While minor variations are evident, for alanine, all four methods worked reasonably well to stabilize batch effects previously observed in Figure 9.3.

Depending on the nature of the investigation, study teams should carefully consider whether targeted panels of specific compounds, more comprehensive non-targeted approaches with the possibility of new compound discovery or multiple panels used together are most appropriate for the scientific questions at hand. Careful selection of the biospecimen to be used for assay is also crucial and investigators should ensure that observations culled from valuable resources are indeed relevant to answering carefully developed hypotheses. Metabolomics technology experts can play a crucial role in multidisciplinary research teams to help make decisions about which technologies and biospecimens are most appropriate for a variety of investigations. They can also speak to subtleties in sample preparation that should be standardized to yield the most reliable data. Several helpful commentaries and guidelines exist on technology and sample selection issues for designing metabolomics experiments [79, 143].

From a statistical perspective, standard power calculation strategies can be naturally applied to metabolomics experiments to guide the number of independent biological replicates that are reasonable to answer proposed scientific questions. Careful consideration should be given to whether metabolomics values will be treated as predictors or outcomes and whether individual metabolite associations are of interest or if metabolites will be analyzed jointly. Since metabolomics technologies by nature measure panels of metabolites or attempt comprehensive assay, multiple comparisons adjustments should factor into power and sample size calculations. Non-targeted metabolomics data in particular can also be prone to unique missingness mechanisms, often due to absence of a metabolite in a sample or presence below a threshold of detectability. Depending on the goals of the study, this may also have power and sample size implications. Multiple comparisons adjustment and strategies for handling missing data will be addressed in the next chapter.

As discussed, non-targeted mass spectrometry-based metabolomics data are prone to batch and run order variability. Careful experimental planning is highly recommended to strategically construct batches to allow optimal control of technical noise so that biologically relevant associations can be detected. If paired samples are to be used in the study, it is recommended that these be placed in the same batch. If possible, samples should be allocated to batches such that experimental conditions or phenotypes of interest and any likely confounding variables are balanced across batches to the fullest extent possible. It is also strongly recommended that internal standards from a source representative of the samples under study be included in the batching scheme. If possible these standards should be included at the beginning, middle and end of each batch, or in some other carefully designed sequence, so that effects attributable to run order within a batch can be estimated and accounted for prior to downstream statistical modeling.

10

Metabolomics Data Analysis

CONTENTS

Once crucial data pre-processing analyses are complete as described in the previous chapter, a rich and varied collection of statistical approaches are often applied for metabolomics data analysis. Initial analyses often begin with per-metabolite investigations to identify the strongest associations of metabolites with phenotypes or experimental conditions. Since small molecule metabolites contribute to cellular function in a coordinated functional system and not as independent agents, it is well recognized that multivariate modeling along with pathway and network analyses are crucial tools to understanding joint metabolite activity. Several considerations in statistical analyses of metabolomics data mirror concerns in other omics realms, for example gene expression data, but often with unique adaptations to address subtleties of metabolomics data types. The techniques described in this chapter will be illustrated using the HAPO Metabolomics data set of targeted and non-targeted measurements for 400 fasting samples from mothers and 400 cord serum samples from newborns discussed in the last chapter on metabolomics data pre-processing. Mixnorm was applied to these data for normalization purposes [356].

DOI: 10.1201/9781315153728-10

10.1   Per-metabolite analyses

Per-metabolite analyses are conducted for metabolomics data in very similar fashion to gene expression data. Just as gene expression microarray data are often filtered to focus analyses only on genes exhibiting reasonable variability across experimental conditions or passing other quality assurance criteria, it is advisable to explore metabolomics data in descriptive fashion and determine analysis priorities for metabolites and identified peaks at the outset of investigation. Of particular importance for non-targeted mass spectrometry data is the summary of missingness and determination of whether missingness is likely attributable to true absence from the sample or presence below detectability, or if absence may be an artifact of misalignment of peaks during pre-processing.

As in most high-throughput investigations, multiple comparisons techniques are also warranted to correct for the multiplicity of hypothesis tests conducted for per-metabolite analyses. The specific approach to multiple comparisons adjustment can depend on the scientific question at hand and should ideally adjust for the fact that metabolite levels are typically not independent but in fact demonstrate fairly high correlations with each other.

10.1.1   Redundant metabolite measurements

Prior to statistical analyses, it can be appropriate to identify which components of metabolomics investigations are priority for evaluating scientific hypotheses. For example, in the HAPO Metabolomics data set, GC-MS non-targeted metabolomics assays were performed along with conventional metabolite assays and targeted panels of amino acids and acylcarnitines. In several cases multiple non-targeted GC-MS peaks were mapped to the same compound in the AMDIS peak detection and annotation pipeline. Curation of co-eluting compounds was required for identification of the most reliable peaks and only those with highest Reverse scores were deemed optimal matches for a given metabolite and were included for analysis (see Figure 10.1). This focused analyses to GC-MS data deemed most interpretable and reduced the multiple testing burden for likely redundant peaks. Also, only GC-MS peaks that could be annotated were included for analysis in this pilot study with examination of unannotated peaks reserved for larger projects to support more in-depth efforts for novel metabolite discovery.

Of note, targeted amino acid assays were performed for the HAPO Metabolomics study, but GC-MS also yielded non-targeted data for multiple amino acids included in the targeted panel. For this study, the investigators chose to analyze amino acid data from the targeted panel and exclude observations for amino acids from non-targeted GC-MS (Figure 10.2). For HAPO Metabolomics the investigators elected to prioritize the more stable, calibrated

Annotation	RT	Reverse
1-Eicosanol	21.633	89.1
1-Hexadecanol/pentadecanoic acid 1	18.0614	90.1
1-Hexadecanol/pentadecanoic acid 2	17.9326	87
1,5-Anhydroglucitol	16.9252	92
2-Aminobutanoic acid	8.4632	94
2-Hydroxybutyric acid	7.7756	97.9
2-Hydroxypyridine	6.4891	81.6
2-Hydroxyvaleric acid	8.3317	66.4
2-Ketoleucine 1	9.0459	90.9
2-Ketoleucine 2	8.5416	82.7
2-Ketoleucine 3	8.8374	79.4
2-Ketovaline	7.5084	81.7
3-Indoleacetic acid	18.0968	82.9

...

Serine 1	11.0942	98.8
Serine 2	9.7049	83.2
Stearic acid	20.705	98.4
Succinic acid	10.4874	95.5

...

Urea	9.5729	99.8
Uric acid	19.3464	95.6
Valine 1	9.0984	98.9
Valine 2	7.2541	88.8

FIGURE 10.1
Retention times and Reverse scores for annotated peaks using AMDIS pipeline for HAPO Metabolomics data. Of 162 total annotated peaks, 23 were deemed likely redundant with other peaks and were removed from further consideration. Higher Reverse scores were used to identify the peak used for further statistical analysis.

targeted measurements to reduce multiple testing burden and streamline multivariate, pathway and network analyses such that each metabolite was only represented once. In other settings, it may be warranted to include both targeted and non-targeted data for the same metabolites to confirm observations using multiple approaches. Given the overlap in metabolite representation that can arise for targeted and non-targeted approaches, it is advisable to decide prior to statistical analyses how redundancies will be handled. This will avoid

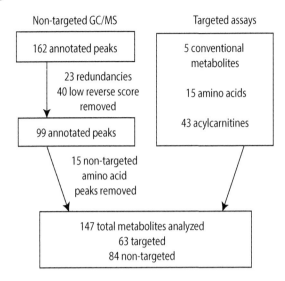

FIGURE 10.2
HAPO Metabolomics non-targeted GC/MS data first yielded 162 annotated peaks. Removal of redundant peaks and peaks with low Reverse scores yielded 99 annotated peaks. Of these 15 amino acids were represented on a targeted amino acid panel, so these non-targeted data were removed from further consideration. This yielded a total of 147 metabolites for analysis in HAPO Metabolomics, of which 63 were targeted and 84 were non-targeted measurements.

post-hoc determination of results that favor scientific hypotheses and inform selection of multiple comparisons procedures.

10.1.2   Descriptive statistics

In addition to identifying redundant measurements of metabolites, it is also advisable to perform initial descriptive examinations of metabolites. While many features are assayed and this can be a time consuming task, some visualization of distributions is far more accomplishable for the low hundreds of annotated metabolites that typically result from metabolomics assays as opposed to gene expression investigations that assay tens of thousands of genes. Simple histograms and/or boxplots can confirm whether typical $\log_2$ transformations for peak areas indeed satisfy normality and if any outliers exist in the data. If noted outliers are all attributable to a single sample, it may be reasonable to exclude all data for that individual. In addition to visual examination, it can be advisable to implement automated methods of outlier removal, for example eliminating single metabolite observations that are 5 standard deviations from the mean for that metabolite in either direction. Other outlier detection methods can be implemented as well. The spirit of outlier detection is to reduce spurious detection of associations, especially given the multiplicity of tests being conducted.

Another frequent descriptive summary of metabolomics data is the relative standard deviation, or RSD. In the statistical literature, this is the same descriptive summary known as the coefficient of variation, although RSD tends to be more popular in metabolomics literature. For a given metabolite m, the RSD is calculated as follows using observed data from individuals $i = 1, ..., n$.

$$RSD_m = \frac{s_m}{\bar{y}_m} = \frac{\sqrt{\frac{1}{n-1} \sum_{i=1}^{n} (y_{im} - \bar{y}_m)^2}}{\bar{y}_m} \tag{10.1}$$

where $\bar{y}_m$ and s_m are the sample mean and standard deviation for metabolite m, respectively. Practical guidelines for evaluating RSD have been described across a variety of technologies and sample types [334].

Notably, summarizing missingness relative to observed abundance for detected metabolites can also be key to understanding the nature of observed metabolite data. The simple scatterplot in Figure 10.3 compares the 25th percentile of observed abundance levels to the frequency of undetected values for non-targeted metabolites in HAPO Metabolomics maternal fasting samples. If too many observations are missing, data for the metabolite may not be reliably analyzed and the metabolite may be excluded from analysis. For example, metabolites corresponding to gray points were excluded from analysis since over half of the values were not observed. Simple frequencies of missingness can help determine whether missingness is frequent enough to warrant statistical modeling that accomodates joint analysis of presence vs. absence of a metabolite in a sample along with analyses of observed quantitative values. It is also advisable to summarize missingness frequency and determine whether missingness is attributable to abundance below detectability. If missingness is common and observed values are close to the lower limit of detection for the technology, then missingness may believably be attributed to low abundance or presence below detectability. Metabolites corresponding to green points were analyzed using mixture models, attributing undetected values to either true absence from the sample or presence below detectability. If instead missingness is relatively uncommon and the observed values indicate higher overall abundance, this type of missingness may be attributed to peak misalignment for that sample, for example the metabolites corresponding to pink points in Figure 10.3. In this case, it may be worth manual review of proximal peaks to identify the correct observed value if possible. Metabolites with no or minimal missingness, for example those corresponding to blue points, can typically be treated as regular continuous values and simple exclusion of infrequent undetected values may not substantially alter analysis results.

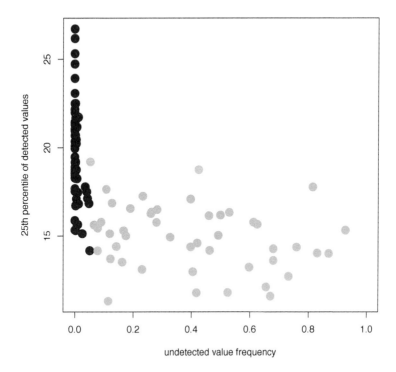

FIGURE 10.3
Plot of 25th percentile of detected abundance values vs. missingness frequency
in HAPO Metabolomics fasting maternal samples. Blue points were treated as
continuous variables and infrequent missing values were excluded from analysis. Gray points were excluded from analysis since over half of the values
were not observed for a given metabolite. Metabolites corresponding to green
points were analyzed using mixture models to jointly model detected and undetected values. Pink points of reasonably high missingness and overall abundance merit further examination for sources of missingness, including possible
peak misalignment.

10.1.3   Individual tests of metabolite association

Individual tests of metabolite associations typically proceed in a regression
framework and are generally performed separately for each metabolite. Depending on whether metabolites are viewed as predictors or outcomes, and
whether phenotypes or experimental conditions are categorical or continuous,
either logistic or linear regression can be employed. Covariate adjustments
should be specified in a consistent manner for all metabolites and may be

determined by prior literature or investigation of associations with metabolites and/or outcomes prior to formal per-metabolite modeling. Units of measurement vary for targeted panels and are very different in nature for targeted vs. non-targeted assays. Even for targeted metabolite measures with common units, a one-unit change in the level of one metabolite can have very different physiological implications compared to a one-unit change in another. For this reason, along with typical beta and standard error estimates from regression models, it can be helpful to report statistics that are neutral to the scale of measurement, for example partial correlations. This can help identify and compare varying strengths of association across metabolites.

10.1.4 Missing data

Missing data provide a unique challenge in metabolomics investigations, especially for non-targeted GC-MS assays. Contrary to missingness that results from failure to collect data, for example in a classic longitudinal study, missing GC-MS observations often result from data points that are assayed but not observed with quantifiable abundance in a given sample. This missingness can be attributed to multiple different sources. First, a metabolite may be truly absent from a sample under study. In this case, the data point isn't in fact missing, but the quantitative value to use for analysis isn't entirely clear. Normalized $\log_2$ transformed peak areas fall in a numeric range determined by GC-MS technology tuning parameters, but that range is not necessarily one for which "0" would accurately correspond to an absent metabolite. Similarly, a metabolite may be present in a sample but at very low abundance below a threshold of detectability. It is not generally possible to distinguish presence below detectability from true absence, and again, the correct analytical value to use in this case in unclear. In many applications, missing data of this nature will either be excluded from analysis, or imputed at the minimum or half-minimum observed value either within a metabolite or across all observations in the study [180, 226, 393]. If missing data are not too prevalent this may be an adequate strategy to represent low abundance, but there are obvious drawbacks in terms of bias for both beta and standard error estimates for widespread imputations of the same numeric value.

An alternative to excluding or imputing missing observations is statistical modeling of true absence from the sample or presence below detectability. Mixture models have been adapted for this purpose with the likelihood specified to allow for logistic modeling of presence vs. absence in a sample and linear modeling of quantifiable abundance [302, 315]. For the ith observation in a study, the mixture model likelihood can be specified as follows:

$$\left((1 - p_i) + p_i \theta \left[(T_i - \mu_i)/\sigma\right]\right)^{(1-\delta_i)} \left(p_i \exp\left[-(y_i - \mu_i)^2/(2\sigma^2)\right]/\sqrt{2\pi}\sigma\right)^{\delta_i},$$

where p_i represents the probability of metabolite presence in the ith sample, T_i is the threshold of detectability for the ith sample, and δ_i is an indicator equal to 1 if the metabolite is detected and 0 otherwise. A logistic model is specified for p_i, specifically $\log(p_i/(1 - p_i)) = x_i'\beta$, where x_i and β are covariate and parameter vectors, respectively. A linear model is specified for the mean of the observed metabolite level μ_i, with $\mu_i = z_i'\alpha$, where z_i and α are covariate and parameter vectors, respectively. The first component of the likelihood corresponds to undetected metabolite values and includes probability of absence as well as presence below detectability. The second likelihood component is simply a linear model for quantifiable abundance. Settings for T_i may be informed by technology settings or can be based on data, for example the minimum detected value for samples in the same batch. An adaptation of mixture modeling for normalization purposes was explained in the previous chapter using data from QC samples and specific covariates defined by batch and run order. The more general mixture model proposed here can accommodate main predictors of interest, experimental design parameters and any other relevant covariate adjustments.

Of note, the mixture modeling approach just described treats metabolites as outcomes in statistical analysis. If instead it is more reasonable to treat metabolites as predictors and missingness is reasonably substantial, it may be reasonable to categorize metabolites for analysis. A category including all missing values can be interpreted as a group of observations for which that metabolite is either absent or present with particularly low abundance. This can provide for meaningful interpretation especially when compared against quantile-based categories for metabolites with observed values.

10.1.5 Multiple comparisons adjustment

Multiple comparisons considerations are similar to those encountered in other high-dimensional omics data analysis settings, with both family-wise error rate (FWER) and false discovery rate (FDR) methods frequently applied. We refer the reader to the rich published literature about multiple comparisons adjustment procedures. We will point out, however, that special consideration should be given to the use of both targeted and non-targeted assays. Targeted panels may be included for specific consideration of well-formulated scientific hypotheses. In this case, it may be reasonable to adopt more conservative FWER methods for stronger control of multiple comparisons and rigorous justification of findings from targeted panels. Given their more exploratory nature, FDR methods may be best suited for non-targeted panels since discoveries made with these approaches typically result in further experimental follow-up, particularly for unannotated peaks. Multiple comparisons adjustment within targeted / non-targeted technology type, or across all statistical tests being conducted, should also be considered depending on the scientific aims of the study.

10.2 Multivariate approaches

Multivariate approaches in metabolomics data analysis largely echo techniques employed for gene expression data analysis as described in earlier chapters. Arguably the most popular method for multivariate metabolomics data analysis is principle component analysis (PCA) [35,490]. PCA transforms the full metabolomics data set into a series of orthogonal variables, calculated as linear transformations of the original variables, that are known as principle components. The first principle component is constructed to explain the maximum amount of variance in the data. Successive components are then constructed such that they maximize remaining variance explained for the data, conditional on their being orthogonal to all previously constructed components. PCA is sensitive to data transformation and variables are often mean-centered and scaled by their standard deviation prior to analysis. PCA results in a set of loading vectors and score vectors. Loading vectors are series of coefficients for contributions of each of the original variables, in this case metabolites, to each principle component. Score vectors are the values that result after applying each loading vector to the data. Plotting scores for the first few principle components can be a helpful descriptive technique for summarizing metabolomics data. Clustering of scores can be indicative of similarity among samples, which can either correspond well with experimental conditions, or may reveal sources of bias or confounding in a data set.

In some cases, PCA scores have been used as predictors or outcomes for subsequent analyses, depending on the application [308, 394]. While on the one hand this can be viewed as a helpful summary of broad scale associations, interpretation of these results is not always clear. PCA ultimately produces linear combinations of the original variables. Interpretation of the strongest contributing variables is not always clear, nor is the interpretation of the resultant scores. PCA is also complicated by truncation features of non-targeted mass spectrometry data since PCA was not originally developed to handle missing data. In most cases, imputed data are used for unobserved metabolite levels to perform PCA [82].

PCA is considered an "unsupervised" multivariate method since it is performed, in our case, only on the metabolomics data set without regard to a phenotype or outcome. In contrast, one of the most common "supervised" methods that does incorporate phenotypes or outcomes hypothesized to be related the metabolites is partial least squares (PLS) [117]. PLS has adaptations for quantitative outcome variables (regression) or for categorical outcomes, typically referred to as PLS discriminant analysis (PLS-DA). PLS components maximize the covariance between the metabolic features and the outcome variable of interest, hence there is a different interpretation of the resultant vectors than for PCA. Various modifications to PLS-DA have been proposed with one of the most notable being orthogonal PLS (O-PLS). O-PLS has been proposed as an improvement upon original PLS methods to ensure that variables that are correlated with the phenotype contribute most strongly to the component estimation.

Other multivariate approaches can also be applied to metabolomics data in the same way they are applied to other omics data. For example, unsupervised approaches including hierarchical clustering and k-means clustering have been reported in the literature [176, 259], although these approaches can be prone to lack of reproducibility in independent data sets [137, 203]. It has also been noted that supervised multivariate techniques can be subject to overfitting [36]. For prediction purposes, random forest methodology has also demonstrated utility, often identifying metabolic features that help predict outcomes above and beyond other known factors [51, 385].

10.3   Pathway enrichment analyses

Pathway enrichment analyses are a familiar concept from gene expression studies, but also particularly relevant to metabolomics since metabolites rarely act alone, but rather act in concert with other metabolites through enzymatic pathway reactions. A variety of approaches and resources exist for metabolic pathway analyses, and most harness statistical concepts originally developed for gene expression data. Many pathway analysis approaches are designed to identify pathway enrichment, that is, statistically significant evidence of coordinated associations of a set of metabolites in the same pathway with a given experimental condition or phenotype. The original *gene set enrichment analysis* (GSEA) concept was published using a Komolgorov-Smirnov test statistic to identify whether a given set of genes previously annotated to a common "set" were represented at the extremes of a distribution of test statistics from a gene expression microarray experiment [297, 418]. While the specific statistical Komolgorov-Smirnov test statistic has been criticized for lack of power and difficulty of interpretation, the main contribution of this work to analyze high-throughput data in sets, rather than as individual features, made major impact and has carried over to other omics studies, including metabolomics. Two primary approaches are mostly applied for pathway enrichment analyses using metabolomics data, generally referred to as "competitive" and "self-contained" enrichment analyses as defined for gene expression data [134].

Regardless of approach, pathway enrichment analyses begin with metabolic pathways identified a priori. Databases such as KEGG [218, 219] and Unipathway [299] provide computationally tractable pathway annotation files that are often used in analyses of this type. Open source software tools including IM-PaLA [216] and MetaboAnalyst 3.0 [493] also have built in functionality for identifying metabolic pathways of interest from public databases for enrichment analysis. Enrichment analyses require a mapping of metabolites to the pathways with which they are associated. While this concept seems simple, constructing this mapping is often complicated by the complexity of experimentally observed metabolites and database identifiers; for example, isomers are in general mapped to the same KEGG ID. Chemometrics experts can provide crucial insight into confirming metabolite-pathway mappings prior to conducting pathway enrichment analyses. In many applications, analyses are performed without regard to the order of metabolic reactions in a given pathway, which can be viewed as a disadvantage of enrichment approaches.

Enrichment testing also require per-metabolite test statistics for association of each metabolite with the experimental condition or phenotype of interest. Competitive testing requires explicit designation of statistical significance or lack thereof, for example p<0.05. A hypergeometric test is then used to evaluate whether, given the total number of metabolites in the experiment, the number of metabolites from a pathway that are also statistically significantly associated with the phenotype or outcome is higher than would be expected by random chance. Similar to the gene expression context in which competitive testing was developed, this approach is complicated by designation of "statistically significant" association as well as determination of the full set of metabolites in an experiment. In particular, satisfactory handling of similar and possibly redundant metabolite peaks and unannotated peaks is unclear.

A more interpretable approach that avoids the difficulties of competitive testing is self-contained enrichment analysis. This framework still requires a prior identification of metabolic pathway members and per-metabolite test statistics for association with phenotypes. However, enrichment is evaluated only using test statistics for the pathway members, with various approaches to evaluate whether a summary test statistic for all pathway members, for example the mean test statistic, is more extreme than would be expected by random chance. Depending on the test statistic being used, theoretical distributional properties can apply to determine significance of the gene set, or permutation techniques can be used. The "globaltest" strategy is one reliable self-contained approach that is often used for metabolic pathway enrichment analyses [135]. Score test statistics are computed for each metabolic feature represented in the pathway, and the mean over all score test statistics is viewed as the summary test statistic. The overall mean is asymptotically normal, but can also be evaluated according to a scaled χ^2 distribution or through permutation testing.

Table 10.1 represents pathway enrichment test results with at least two metabolites annotated to the pathway and FDR-adjusted $p < 0.001$ using the HAPO Metabolomics maternal fasting data and the self-contained "globaltest" approach. The analyses were conducted in MetaboAnalyst 3.0, which provides a user-friendly interface for conducting these types of analyses. Notably, the number of metabolites represented in the HAPO Metabolomics data set is far lower than the number of metabolites annotated to these pathways. Given the reasonably low representation, it is advisable to confirm pathway

TABLE 10.1

Subset of MetaboAnalyst self-contained set enrichment analyses for HAPO Metabolomics fasting maternal data

Pathway name	No. of metabolites annotated to pathway	No. of measured metabolites	FDR-adjusted p-value
Galactose metabolism	41	4	4.5e-07
Glycolysis or Gluconeogenesis	31	3	4.5e-07
Alanine, aspartate, glutamate metabolism	24	5	4.7e-05
Taurine and hypotaurine metabolism	20	3	4.7e-05
Pyruvate metabolism	32	2	3.1e-05
Pentose phosphate pathway	32	4	9.2e-05
Fatty acid biosynthesis	49	2	1.5e-04

enrichment analysis results using these techniques with other methods of analysis.

Some pathway enrichment analysis techniques integrate existing pathway annotation with data driven network analyses. For example, Metabolomics Pathway Analysis (MetPA) [494] integrates statistical pathway enrichment approaches with metabolic network data to identify topological features of pathways that are particularly relevant to a phenotype under study. In MetPA, edges model direct metabolic cascades that comprise an annotated pathway. Summary statistics such as "relative betweenness centrality" can be calculated to describe features for metabolites within the network. Additional network considerations of particular use in metabolomics will be developed in the next section.

10.4   Network analyses

Similar to many other omics studies, a variety of network analyses have proven useful for metabolomics data analyses. Network approaches offer an integrated view of joint activity among metabolites and can be very helpful for placing identified associations into context. In network analyses, metabolites are typically represented by nodes and relationships among nodes are represented by edges. Edges can be determined by a variety of criteria, depending on the software being used or the scientific question under study. A comprehensive treatment of network data analysis is beyond the scope of this book, but some of the most commonly used tools will be described.

10.4.1 Constructing and describing networks

Network modeling can rely exclusively on experimental data, not necessarily requiring external annotation as is required for pathway enrichment analyses. While this can add some challenge to interpreting output, there is also greater potential for novel observations and discovery. A simple approach for constructing a metabolic network is to let nodes represent metabolites and then connect nodes with an edge if the Pearson correlation, or partial correlation after adjustment for a set of defined covariates, for the metabolite pair exceeds a certain threshold. This can be a helpful descriptive approach for understanding dependencies among metabolites in a data set.

Figure 10.4 illustrates results from a network analysis of maternal fasting HAPO Metabolomics data. Nodes represent metabolites and are colored according to chemical class. Edges represent partial correlations >0.25 or <-0.25 after adjustment for study site, gestational age, maternal age and mean arterial pressure at the oral glucose tolerance test, newborn sex and sample storage time. The maternal fasting network for this data set is densely connected with many metabolites correlated with each other, particularly among a given class of metabolites.

10.4.2 Metabolic subnetwork optimization

Network modeling of pairwise correlations or partial correlations can assist in understanding metabolite dependencies, however, this type of network modeling does not capture association with phenotype. Several approaches have been explored for handling phenotype data. For example, subnetwork optimization techniques can be used to identify subsets of connected nodes from an overall network that collectively demonstrate high association with a phenotype or set of phenotypes. In the algorithm proposed by Dittrich et al. [85], a network is first constructed according to criteria appropriate to the study at hand but independent of phenotype, and node scores are then assigned according to p-values for association with a phenotype. A unique and useful component of this approach is the ability to jointly consider multiple phenotypes when identifying subnetworks. Multiple phenotypes can be incorporated using an aggregate approach based on order statistics for the p-values associations with each phenotype. Let $P_1, P_2, ...P_w$ be random variables for the p-values resulting from tests of association with w phenotypes. Since p-values are uniformly distributed under the null hypothesis of no association between the metabolite and phenotypes, then the distribution of the ith order statistic, $P_{(i)}$ follows a beta distribution with parameters i and $w - i + 1$, or

$$P_{(i)} \sim \frac{w!}{(w-i)!(i-1)!} p^{i-1} (1-p)^{w-i}, 0 \le p \le 1.$$

Using the beta distribution, an overall aggregate p-value p_{aggr} can then be assigned to the node using the desired order statistic, for example the minimum

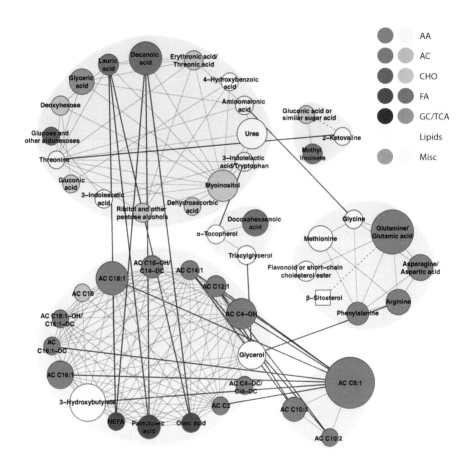

FIGURE 10.4

Network analysis of HAPO Metabolomics maternal fasting data. Nodes represent metabolites and are colored according to chemical class. Edges represent partial correlations >0.25 or <−0.25 after adjustment for study site, gestational age, maternal age and mean arterial pressure at the oral glucose tolerance test, newborn sex and sample storage time. The subnetwork shown here is the optimal subcomponent whose nodes demonstrate collective association with maternal BMI at the time of the oral glucose tolerance test. Gray shading corresponds to spinglass clusters in the optimized subnetwork. AA = amino acid, AC = acylcarnitine, CHO = carbohydrate, FA = fatty acid, GC/TCA = glycolysis/tricarboxylic acid. Bold (light) shading represents positive (negative) association with BMI.

or maximum. Aggregate p-values are then converted to node scores. Dittrich et al. suggest a node score based on a mixture model of signal-to-noise with the beta distribution representing signal and a uniform distribution representing noise. The original Dittrich et al. algorithm was proposed for gene expression data of far higher dimension than metabolomics data and the assumption of substantial noise in the p-value distribution (i.e. lack of association with phenotype) is in general far more applicable than in metabolomics data. For smaller dimensional metabolomics data, a simpler transformation of the aggregate p-value to a node score works well, for example $S = -log(p_{aggr}) + log(c)$, where c is a constant between 0 and 1 below which p_{aggr} should result in a positive node score.

After node scores are calculated, an optimal subnetwork is then identified either using a heuristic algorithm or by optimization of a scoring function using integer-linear programming. The intent is to identify the subnetwork with the highest overall sum of node scores and lowest "cost" according to edge weights. Edges are given higher weight if their adjacent nodes have negative scores and low degree, where degree is defined as the number of incident edges. Subnetwork optimization of this type is intended to focus attention on components of a metabolic network that are most relevant to the phenotype under study. An example of the optimized subnetwork for maternal fasting metabolites related to BMI during pregnancy is depicted in Figure 10.4. A maternal fasting metabolic network optimized for both for maternal BMI as well as newborn birthweight in the HAPO Metabolomics data is illustrated in Figure 10.5. As might be expected, a subset of the nodes included in the network for maternal BMI only are also included in the network when both the maternal BMI and newborn birthweight phenotypes are considered. In general, the network related to both phenotypes is somewhat smaller, suggesting that not all maternal metabolites associated with maternal BMI have direct implications for newborn birthweight. Somewhat surprising at first glance, some metabolites are included in the joint phenotype network that were not included in the maternal BMI only network. This is explained in part by the aggregate p-value component of the algorithm. P-values that suggested only trends of association for either phenotype individually confirm consistent association when considered jointly.

10.4.3   Local community detection

Review of the HAPO Metabolomics correlation network also suggests distinct communities within the network that are more interconnected than others. Several techniques exist for local community detection and many can be implemented using the *igraph* package in R [62]. Figures 10.4 and 10.5 demonstrate application of "spinglass" clustering with nodes colored according to membership in distinct, non-overlapping communities. These communities tend to identify correlations among metabolites of similar chemical classes, for example amino acids, acylcarnitines and fatty acids. Many other local community

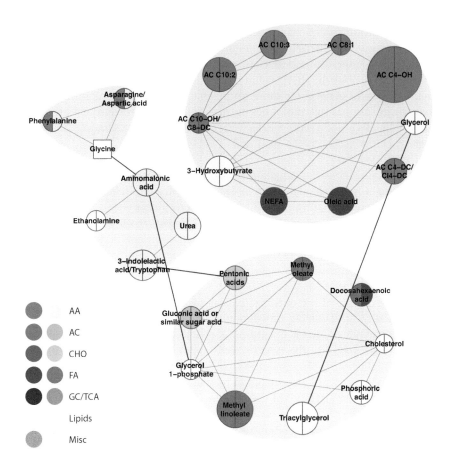

FIGURE 10.5

Network analysis of HAPO Metabolomics maternal fasting data. Nodes represent metabolites and are colored according to chemical class. Edges represent partial correlations >0.25 or <−0.25 after adjustment for study site, gestational age, maternal age and mean arterial pressure at the oral glucose tolerance test, newborn sex and sample storage time. The subnetwork shown here is the optimal subcomponent whose nodes demonstrate collective association with maternal BMI at the time of the oral glucose tolerance test as well as newborn birthweight. Gray shading corresponds to spinglass clusters in the optimized subnetwork. AA = amino acid, AC = acylcarnitine, CHO = carbohydrate, FA = fatty acid, GC/TCA = glycolysis/tricarboxylic acid. Bold (light) shading represents positive (negative) association with BMI. The lefthand (righthand) side of the node corresponds to the metabolite's association with maternal BMI (newborn birthweight).

detection options can be implemented in R using the *igraph* R package, including edge betweenness, fast greedy, label propagation, leading eigenvector, Louvain, optimal, spinglass and walktrap.

10.4.4 Differential metabolic networks

Another promising approach to network analysis is differential network analysis. Rather than constructing a metabolic network based on metabolite observations across a full range of phenotypes and then identifying associated nodes, differential network analysis examines differences in network structures according to phenotype. For example, for a categorical phenotype, two different partial correlation graphs could be constructed and their edge topologies compared. An approach called Differential Network Analysis in Genomics (DINGO) includes formal statistical testing to identify edges exhibiting differential correlation under one condition compared to another [145]. DINGO starts with a Gaussian graphical model to define edges among metabolites. In Gaussian graphical models, nodes represent metabolites and edges represent partial correlations among metabolites after adjusting for all other metabolites as well as any other relevant covariates. These models identify metabolites whose pairwise associations are strongest among all metabolites in an experiment and have shown high overlap with metabolic reactions that are documented in the literature and in simulations for various reaction model systems [20,240,241]. Once the Gaussian graphical model is identified, DINGO then specifies a model including local components for edges unique to either group and a global component including edges relevant to both groups. Full development is explained in detail in the manuscript, but ultimately differential edges are evaluated with respect to a differential score δ_{ab}^{12}, defined as:

$$\delta_{ab}^{12} = \frac{\hat{\phi}_{ab}^{(1)} - \hat{\phi}_{ab}^{(2)}}{s_{ab}^{B}}$$

where $\hat{\phi}_{ab}^{(1)}$ and $\hat{\phi}_{ab}^{(2)}$ are Fisher's Z transformation for the partial correlation estimates for metabolites a and b in the two groups under study and s_{ab}^{B} is a bootstrap estimate of the standard error. A cutoff value k is used to determine edges in the local or global networks, such that edges with $|\delta_{ab}^{12}| > k$ are part of the differential network and other edges are included in the global component. While k is a user specified constant, the default value is 2 since this corresponds to the Wald-type test statistic cutoff for significance.

An example of differential network analysis results for HAPO Metabolomics data is included in Figure 10.6. Edges in this network are pairs of maternal metabolites exhibiting differential partial correlations in mothers of boys compared to mothers of girls. This network suggests that not only are individual metabolites associated with sex of the baby, but the ways in which they jointly correlate also differ according to sex.

The current formulation of DINGO compares partial correlation networks for two distinct groups, but extensions to multiple groups follows fairly

EU
Maternal fasting metabolites
Differential network for boy vs. girl moms
50 nodes, 170 edges

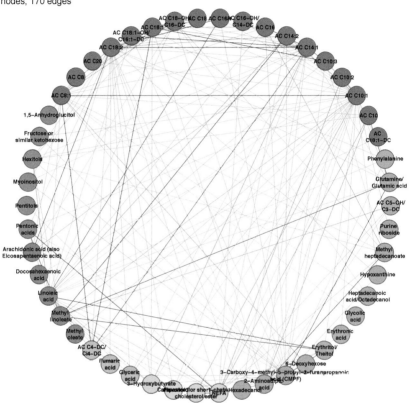

FIGURE 10.6

Results from differential network analysis using DINGO, comparing the maternal fasting metabolic network for mothers of boys compared to mothers of girls for HAPO Metabolomics. Green edges demonstrate positive partial correlation in mothers of boys and girls, but differ in magnitude. Red edges demonstrate negative correlation in mothers of boys and girls but differ in magnitude. Yellow edges demonstrate partial correlation in opposite directions for mothers of boys compared to mothers of girls.

naturally (see [145]). Extensions to evaluate differential networks across a continuous phenotype are also possible, but await further methodological development. The field of differential network analysis is ripe for methodology contributions. For metabolomics in particular, understanding coordinated differences in metabolic network behavior across phenotypes or experimental conditions of interest will undoubtedly be relevant for many applications.

10.5 Case studies on data integration

Data integration efforts for metabolomics and other omics data types are underway with comprehensive modeling efforts still under development. Many initial efforts have focused on integration of genetic variant data or transcriptomics with metabolic profiles. A specific challenge in data integration is that metabolomics and genomics cannot be directly linked in terms of a common gene, unlike proteomics and genetic data. This adds a layer of complexity, both biological and computational, that must be accommodated for when jointly modeling these data types. Still, integration has demonstrated some success. Several case studies are discussed below.

10.5.1 Metabolomics and genetic variants

"Metabotyping," or identifying metabolite levels associated with genetic variants, is a promising strategy to unite these two data types. Early efforts adopted a classic GWAS framework, testing associations of individual SNPs with metabolites treated as outcomes using an additive model for allele counts [133]. Extensions to this basic approach have included analysis of metabolite ratios [194] and emphasis on SNPs or metabolites associated with other phenotypes [211, 420]. Enrichment analysis strategies have also been adapted to jointly analyze metabolite sets using a phenotype set enrichment analysis (PSEA) approach [362]. In PSEA, multiple metabolites are treated as phenotypes and genetic variants are evaluated for their collective association with multiple phenotypes. Similar in concept to GSEA but with the "sets" in some sense reversed to evaluate sets of outcomes rather than sets of predictors, PSEA can improve power for detecting subtle associations.

The promise of jointly analyzing metabolomics and genetic variant data is supported by larger reported effect sizes for metabotypes than GWAS for disease. Since metabolic profiling is "closer" to underlying genetics in terms of cellular activity than clinical phenotypes, it is logical than genetic influence on disease mechanisms might be more noticeable for metabolites than a specific disease outcome. Furthermore, heritability of metabolite levels is greater than clinical phenotypes [227] and more metabolite variability is explained by genetic than clinical factors [360]. All of these efforts support further integration efforts as understanding genetic linkages to metabolite activity may ultimately explain phenotypic development.

10.5.2   Metabolomics and transcriptomics

Rather than evaluating genetic sequence variants, several studies have also demonstrated promising joint analyses of gene expression data, or transcriptomics, and metabolomics data. One simple approach to integrating metabolomics and transcriptomics is to identify pairwise correlations among metabolite and gene features. Classic Pearson and Spearman correlations have been applied and for several studies, more statistically significant correlations were identified than would be expected by random chance using this approach [106, 449]. Pairwise correlation methods are potentially less useful when studying metabolic flux through pathways since the timing of transcript and metabolic changes may not align well depending on the investigation. In these settings, data alignment through algorithms such as Dynamic Time Warping may be critical for informative analysis [44, 303, 430]. Another critical factor that complicates application of simple correlation approaches pertains to sample sources for transcriptomics and metabolomics. If experiments are investigated using different tissues, or experimental conditions are not well aligned, there is a distinct possibility that correlations will be low for the two data sets. Cavill et al. [43] provide a helpful review of experimental design conditions and analysis strategies for integrating metabolomics and transcriptomics data.

Other approaches for metabolomics and transcriptomics data analysis harness classic data multivariate techniques such as PCA and PLS and integrate data in various ways. For example, in some applications, metabolomics profiles are treated as outcomes and transcriptomics as predictors and PLS is used to study joint transcriptomic changes that are associated with variations in specific metabolites [261, 342, 473]. Sparse regression models have also been used in a similar spirit [199]. In other settings, PCA is performed separately for metabolomics and transcriptomics data in what are called multi block analyses, and relationships among the resulting loadings and scores are summarized through quantitative measures called "super weights" [167] or through various visual summaries [161]. Hierarchical modeling strategies have been used for integrating transcriptomics and metabolomics with additional data sources, for example proteomics. PCA or other dimension reduction techniques are typically applied first to each data set individually, followed by hierarchical model building to integrate the data. While interpretation is complicated, this approach has demonstrated improved class separations in some settings [213].

Pathway analysis resources can also be applied to integrate transcriptomics and metabolomics data. For example, Metscape 2.0 provides flexible functionality for combining gene set enrichment analysis results using transcriptomics data with metabolic pathways and experimental metabolomics data for visualization and interpretation of significant pathways reflecting both data types [222]. Integrated omics data analysis strategies are most informative when strategically aligned with scientific hypotheses and will undoubtedly be an area of continued, robust methodology development.

11

Appendix

CONTENTS

11.1 Basics of probability

Probability theory is a mathematically consistent and coherent way of putting a numerical value to uncertainty. To understand it, we must learn to speak the language first. In the context of probability theory, the word "experimen" will mean any kind of activity that produces results or outcomes. For example, throwing a ball and tossing a coin are experiments; so is eating dinner or listening to a lecture. There are basically two types of experiments: those with a perfectly predictable outcome and those with many possible outcomes so that one is never sure which one is going to occur. Experiments of the first kind are called *deterministic* and those of the second kind are *random* or *stochastic*. An illustration of a deterministic experiment is to mix together an acid and a base (or alkali), because they will invariably produce a specific salt along with some water. Another example would be to heat a bowl of water to 100 degrees Celsius (or 212 degrees Fahrenheit) under normal atmospheric pressure, at which point it will surely start boiling. However, things are not so predictable when you flip a fair coin. It can either land heads up or tails up. So this is a random experiment. So is throwing a dart onto a dartboard with your eyes closed or rolling a cubic die – it is difficult to predict exactly where the dart will land or exactly what number will turn up on the die.

So it should be clear that probability theory deals primarily with a random experiment because there is some uncertainty associated with its outcomes. The first step towards quantifying this uncertainty is a complete enumeration of all the outcomes. Such a complete list of all possible outcomes of a random experiment is known as its *sample space*. Depending on the experiment, the

---

sample space can be finite or countably infinite. For example, when a coin is flipped, the sample space is S = {Heads up, Tails up} or simply $S = \{H, T\}$. When a regular cubic die is rolled, $S = \{1, 2, 3, 4, 5, 6\}$. However, if our experiment is counting how many flips of a coin was needed to see the first "H," then S consists of all positive integers and is infinite. Once again, the sample space of a random experiment is determined solely by the description of the experiment. For example, the simple experiment of drawing a card from a (well-shuffled) deck of 52 cards with our eyes closed can have different sample spaces depending on the extent of details we allow. If we draw a card, blindfolded, from a deck of 52 cards and just watch its color, the sample space is simply {red, black}. However, if we do the same experiment and watch only the suit it is coming from, S will be {diamonds, spades, hearts, clubs}. Finally, in the context of the same experiment, if we are interested in all the features of a card (its color, suit as well as the number or picture on it), then S will be a list of 52 distinct cards.

Having written down the sample space of a random experiment, we now must link it to the concept of an *event*. Any subcollection of outcomes listed in the sample space is called an event. In real life, we are more often interested in such collections of outcomes than in individual outcomes. In the language of set theory (the branch of mathematics that deals with the relationships among, and the properties of, various subcollections of a bigger collection), the sample space is the *universal set* and the events are its *subsets*. So, following set-theoretic notations, events are usually denoted by A, B, C, etc. For example, in the context of rolling a cubic die (what was the sample space for it?), A could be the event that an odd number turned up. In other words, $A = \{1, 3, 5\}$. Similarly, B could be the event that a number ≤ 2 turned up, i.e., $B = \{1, 2\}$. In the context of drawing a card from a deck of 52 cards, A could be the event that the card drawn is an ace, B could be the event that it is black, C could be the event of it being a face-card and D, the event of it being a diamond. These are examples of simple events. Given two such simple events, one can construct all sorts of compound events using the set-theoretic operations *union* ($\bigcup$), *intersection* ($\bigcap$) and *complementation*. Given two events (or sets) A and B, their union is defined as the bigger collection containing all the outcomes in A as well as those in B. So, using the notation "$\in$" to denote that an outcome x belongs to an event A, the event $A \bigcup B = \{x : x \in A \text{ or } x \in B\}$. Notice that this "or" is not an exclusive or, i.e., it includes the possibility that an outcome x may belong to both A and B. The event $A \bigcap B$ is defined as the subcollection containing only those outcomes that are common to both A and B. In other words, it is $\{x : x \in A \text{ and } x \in B\}$. For an event A, its complement (A^c or $\bar{A}$) is the collection of all outcomes in the sample space that do not belong to A, i.e., $A^c = \{x : x \notin A\}$. Applying the set-theoretic operations on these, one can create compound events such as $A \bigcap B^c = \{x : x \in A \text{ and } x \notin B\}$ (also called $A - B$), $A^c \bigcap B = \{x : x \notin A \text{ and } x \in B\}$ (also called $B - A$), $A \Delta B = (A - B) \bigcup (B - A)$ (also called the symmetric difference), etc. Let us point out, for future reference, that the operations of union and intersection are *distributive*, i.e., for any three sets A, B and C, $A \cup (B \cap C) = (A \cup B) \cap (A \cup C)$ and $A \cap (B \cup C) = (A \cap B) \cup (A \cap C)$. All these can be nicely depicted by a simple diagram called a *Venn diagram*, which consists of an outer rectangle representing the universal set (or sample space) and a circle representing each simple event $A, B, \ldots$ inside it. If A and B have some elements in common, the two circles representing them should be overlapping. If A and B do not overlap as events (in which case, they are called *mutually exclusive* or *disjoint*), the circles should be drawn as separated too. In the former case, the total area covered by the double circle represents $A \cup B$, the overlap between the two circles represents $A \cap B$, the crescent-shaped portion of the circle A that is outside the overlap region stands for $A - B$ and the corresponding portion of the other circle stands for $B - A$. If one covers up the circle A, the rest of the rectangular box that is still visible represents A^c. Similarly, for B^c. Notice that the union of any event and its complement is the entire rectangular box. Now the question is: what should we call the "outside border" region of the rectangular box that is still visible when we cover up the entire double circle (i.e., $A \cup B$)? Since, in general, "outside" means "complement," one correctly concludes that it should be called $(A \cup B)^c$. But is there another name for it?

Result 11.1. $(A \cup B)^c$ is the same as $A^c \cap B^c$. Similarly, $(A \cap B)^c$ is the same as $A^c \cup B^c$.

These are the famous *De Morgan's rules*. It is easy to verify them. For example, if x is an outcome in $(A \cup B)^c$, that means $x \notin A \cup B$, which, in turn, means that x is neither in A nor in B. In other words, $x \in A^c$ and $x \in B^c$. But this is equivalent to saying that $x \in A^c \cap B^c$. So every outcome in $(A \cup B)^c$ also belongs to $A^c \cap B^c$. Similarly, by retracing this line of reasoning, one can conclude that the converse is true too. This shows that the two collections are identical. The other De Morgan's rule can be similarly established. But what are these useful for? Here is an example.

Example 11.1. Suppose that, in a two-set Venn diagram, $A - B$ has 15 outcomes (which is often denoted by $n(A - B) = 15$ where "n" means *cardinality*), $(A \cap B)^c$ has 43 outcomes, B itself has 20 and $A^c \cap B^c$ has 15. Can we find out $n(A)$ and $n(S)$? Notice first of all that $A \cup B = (A - B) \cup B$, so that $A \cup B$ in this case has 15+20=35 outcomes. Now, by De Morgan's first rule, $(A \cup B)^c = A^c \cap B^c$, so in this case $(A \cup B)^c$ has 15. Also, since the union of any event and its complement gives us the whole sample space S, $n(S)$ must be $n(A \cup B) + n(A \cup B)^c = 35 + 15 = 50$. Finally, since $A \cap B$ and $(A \cap B)^c$ are complements of each other, their union is the whole sample space. So $n(A \cap B) = n(S) - n(A \cap B)^c = 50 - 43 = 7$. As a result, $n(A)$ will be 22, the sum of the cardinalities of its two non-overlapping pieces $A - B$ and $A \cap B$.

In the above example, we have repeatedly used the cardinality formula $n(C) + n(C^c) = n(S)$ for any event C. Another cardinality formula that is often useful is: $n(A \cup B) = n(A) + n(B) - n(A \cap B)$. It is often called the *union rule* and is easy to verify once the Venn diagram is drawn.

Once the concepts of a sample space and compound events are understood, the next step is to assign a positive fraction to each of the listed outcomes in a mathematically coherent way that is consistent with reality. This positive fraction will be called the *probability* of the associated outcome. In fact, this is the step where believers in the frequentist or classical notion of probability differ philosophically from the believers in subjective probability. The former will define the probability of an outcome as the "long-term" ratio between the number of times that particular outcome has occurred and the number of trials (i.e., the number of times the experiment has been repeated). In more mathematical terms, the probability of an outcome x is $p(x) = \lim_{n \to \infty} \frac{\#x}{n}$, where n is the number of trials and $\#x$ is the frequency of occurrence of that outcome in n trials. The existence of this limit is ensured by the so-called strong law of large numbers (SLLN). On the other hand, a believer in subjective probability may assign a probability to an outcome that reflects his/her degree of confidence in its occurrence. For example, suppose somebody is playing a dice game where each time he/she rolls a "6" with a fair die, he/she wins \$ 100. If a "6" turned up in 5 of the last 10 rolls by chance, he/she may very well think "I am feeling lucky" and assume that the chances are 50% that the next roll will also produce a "6." This is based on perception and has nothing to do with the long-term proportion of occurrence of the outcome "6" for a fair die, which is $\frac{1}{6}$. In this book, we stick to the frequency-based definition of probability, although there are some problems with it in practice. We simply do not know how many times a random experiment has to be repeated in order to be able to determine the "true" probabilities of its outcomes accurately. For example, if one wants to determine the probability of the outcome "6" in a die-rolling experiment by repeated trials, usually the quantity $\frac{\#6}{n}$ (where $n=$ number of trials) fluctuates a lot for small or moderate values of n, thereby producing misleading estimates. This ratio is ultimately guaranteed to stabilize around the "true" probability, but does "ultimately" mean after 10,000 trials or 1,000,000? That is why it is more convenient to "reason out" these probabilities using mathematical reasoning, based on the precise description of the experiment. For example, in the experiment of tossing a *fair* or *unbiased* coin, the adjective "fair" or "unbiased" leads us to the logical conclusion that none of the two outcomes H and T should be more likely than the other. As a result, each of them should be assigned a probability of $\frac{1}{2}$, since the total probability of the entire sample space must always be 1 or 100%. Likewise, in the case of rolling a *perfect* or *balanced* die, those adjectives will lead to the logical conclusion that each of the six outcomes should receive the same probability, which must therefore be $\frac{1}{6}$ in order to keep the total 1. However, if a special die is used whose three sides are marked with the number "3," another two sides are marked with "2" and the remaining side with "1," our logical reasoning will not lead to assign equal probabilities to the three possible outcomes 1, 2 and 3. Instead we will conclude that "2" should be twice as likely and "3" should be three times as likely as "1," which automatically determines the probabilities to be $P(1) = \frac{1}{6}$, $P(2) = \frac{1}{3}$ and $P(3) = \frac{1}{2}$. In this case, the outcomes are *unequally likely*, whereas in the previous two examples, the "fairness" adjectives made the outcomes *equally likely*. Probabilities determined in this way are called *theoretical* or *model-based* probabilities. How different these are from the long-term frequency-based probabilities will depend on how realistic our models are and/or how correct our logical reasoning is. For instance, if our random experiment is picking a ball without looking from a box containing 50 balls of identical size and 5 different colors red, black, white, purple and yellow (10 balls of each color), our logical reasoning will lead us to assign a probability of $\frac{1}{5}$ to each of the outcomes $\{R, B, W, P, Y\}$. This will indeed coincide with the long-term frequency of each of the colors because the balls are otherwise identical. However, if our experiment was to catch a fish (without looking) using a line and baits from a tank containing 50 fish of 5 different colors R, B, W, P and Y (10 of each color) and releasing it back to the tank, our logical reasoning might once again lead us to assign the same probability (1/5) to each color. That is if we continued to think of the fish as colored balls in a box. But the reality may be different. Some fish may have a bigger appetite than others or be more aggressive in nature, which makes them more likely to bite the bait. In this case, the model-based probabilities may differ from the ones based on long-term frequencies since our model failed to take into account the full reality. Having said this, we should also point out that in many practical applications, modeling the fish as identical balls of different colors will be "good enough," in the sense that the resulting discrepancy in the probabilities will hardly matter. That is why we fit probability models to real-life phenomena, knowing that they are semi-realistic approximations at best. As George Box, a renowned statistician, once said: "No model is correct; but some are useful."

Having assigned nonnegative fractions (or probabilities) to the outcomes in a sample space in one way or another, now we go on to define probabilities of events. The probability $P(A)$ of an event A is formally defined as $P(A) = \sum_{x:x \in A} P(x)$. However, for experiments with equally likely outcomes, it reduces to a simple formula: $P(A) = \frac{n(A)}{n(S)}$. For example, in the experiment of rolling a balanced die, if A is the event that an odd number turns up and B is the event of a number ≤ 2 turning up, $P(A) = \frac{1}{6} + \frac{1}{6} + \frac{1}{6} = \frac{1}{2} = \frac{3}{6} = \frac{n(A)}{n(S)}$. Similarly, $P(B) = \frac{1}{3} = \frac{n(B)}{n(S)}$. But the above formula does not hold for unequally likely outcomes. Recall the special die with the number "3" on three sides, "2" on two sides and "1" on one side? For it, if B denotes the same event, the correct $P(B)$ is $\frac{1}{6} + \frac{1}{3} = \frac{1}{2}$, which is different from $\frac{n(B)}{n(S)} = \frac{2}{3}$. Some direct consequences of this computationally convenient formula in the equally likely case are (a) $P(A) = 1 - P(A^c)$ for any event A and (b) $S(A \bigcup B) == P(A) + P(B) - P(A \bigcap B)$ for any two events A and B, which come from the corresponding cardinality formulae mentioned earlier.

Probabilities defined in this way satisfy the following:
Probability axioms:
Axiom 1. $0 \leq P(A) \leq 1$ for any event A in S;

Axiom 2. $P(S) = 1$;

Axiom 3. For any countable collection $E_1, E_2, E3, \ldots$ of mutually exclusive events in S, $P(E_1 \bigcup E_2 \bigcup E_3 \bigcup \ldots) = P(E_1) + P(E_2) + P(E_3) + \ldots$. In other words, $P(\bigcup_{i=1}^{\infty} E_i) = \lim_{k \to \infty} \sum_{i=1}^{k} P(E_i)$.

Any way of defining probabilities of events (objective or subjective) should satisfy these axioms in order to be called consistent. The last of the three axioms is called *countable additivity*. Some prefer replacing it by the *finite additivity* axiom of De Finetti, but here we stick to the former (actually, countable additivity implies finite additivity but not vice versa). In any case, we are now in a position to compute probabilities of various compound events.

Example 11.2. For the experiment of drawing a card, without looking, from a (well-shuffled) deck of 52 cards, recall that we defined events $A = \{$ the card drawn is an ace $\}$, $B = \{$ the card drawn is black $\}$, $C = \{$ The card drawn is a face card $\}$ and $D = \{$ the card drawn is a diamond $\}$. Let us compute the probabilities (a) $P(A \bigcup B)$, (b) $P(A \bigcap D)$, (c) $P(B \bigcap C)$, (d) $P(B \bigcap D)$, (e) $P(A \bigcup C)$, (f) $P(A \Delta D)$, (g) $P(A \bigcup C \bigcup D)$. $P(A \bigcup B)$ is 28/52 or 7/13 by direct counting, since there are 28 cards altogether in the deck that are either ace or black (or both). It could also be found using the union rule, as $P(A) = 1/13$, $P(B) = 1/2$ and $P(A \bigcap B) = 1/26$. $P(A \bigcap D)$ is 1/52, since there is only one diamond ace. The answers to (c), (d) and (e) are 3/26, 0 and 4/13 respectively (not counting the aces as face cards). $P(A \Delta D)$ is 15/52, since $P(A - B) = 3/52$, $P(B - A) = 12/52 = 3/13$ and $A \Delta D$ is the disjoint union of these two (notice that for two disjoint events E and F, the union formula $P(E \bigcup F) = P(E) + P(F) - P(E \bigcap F)$ simply reduces to $P(E \bigcup F) = P(E) + P(F)$ since the intersection term vanishes). Finally, $P(P(A \bigcup C \bigcup D)$ is 25/52, by direct counting.

Remark 11.1. The union rule for a three-set Venn diagram is a little more complicated. Just as a Venn diagram with two overlapping sets A and B has $2^2 = 4$ disjoint components $A - B$, $B - A$, $A \bigcap B$ and $(A \bigcup B)^c$, a Venn diagram with three mutually overlapping sets has $2^3 = 8$ disjoint components that can be appropriately named using set-theoretic symbols. In view of this, it is not too difficult to see why $P(E_1 \bigcup E_2 \bigcup E_3) = P(E_1) + P(E_2) + P(E_3) - P(E_1 \bigcap E_2) - P(E_2 \bigcap E_3) - P(E_3 \bigcap E_1) + P(E_1 \bigcap E_2 \bigcap E_3)$.

Let us now focus our attention on event pairs such as A and B, A and D or B and C. Since $P(A) = 1/13$, $P(B) = 1/2$ and $P(A \bigcap B) = 1/26$, clearly an interesting relation holds, namely, $P(A)P(B) = P(A \bigcap B)$. Similar is the case for the other two pairs. One might get the impression that it happens all the time. But what about the event pairs B and D? If we define a new event E as the card drawn being a spade, what about the pair B and E or D and E? For B and D, $P(B \bigcap D) = 0 \neq P(B).P(D)$. Similar is the case for D and E. Regarding B and E, we have $P(B)P(E) = (1/2)(1/4) = 1/8$ which is different from $P(B \bigcap E) = 1/4$. So the equation that the pairs $\{A, B\}$, $\{A, D\}$ and $\{B, C\}$ satisfy is a special one and when it holds, the event-pair involved is called an *independent* pair of events. So the pairs $\{B, E\}$, $\{D, E\}$ and $\{B, D\}$ are *dependent* pairs.

Example 11.3. Suppose our experiment is to choose a whole number randomly from among $1, 2, \ldots, 20$. Now, the phrase "choosing randomly" means choosing in such a way that all possible outcomes are equally likely. How would one carry out the experiment to ensure this? One way could be to write the numbers 1 through 20 on twenty identical-looking paper chips or plastic tokens, put them in a bowl or a hat, mix them well and then draw one without looking. In any case, if A is the event that the chosen number is odd and B, the chosen number is a multiple of 3, then A and B are independent (verify it). If C is the event that the chosen number is ≤ 10 and D is the event that it is a multiple of 4, then C and D are dependent (verify it too).

Now we will probe into it a little bit farther and try to understand *why* some event-pairs are independent and others are not. If we examine the independent event-pairs in the above examples carefully, we will see that within each pair, the events do *not* carry any information regarding one another. For instance, if your instructor is standing at the lecture podium and drawing a card, without looking, from a (well-shuffled) deck of 52 cards, and you are sitting in the audience, trying to predict whether the card drawn will be an ace. The instructor looks at the card drawn but does not tell you what it is. To you, the chances of it being an ace are $1/13$. If, at this point, the instructor changes his/her mind and reveals that the card drawn is actually black, does it change the probability of it being an ace in your mind? Earlier, you were thinking that it would be 1 in 13 since there are 4 aces in the deck of 52 cards. Now, in view of this additional information about the color, your brain will automatically stop thinking about the 26 red cards that are irrelevant and focus only on the black cards. But the proportion of aces is still the same—2 out of 26 or $1/13$. In other words, the color of a card has *no* information about the "ace or non-ace" classification, and vice versa. When you are trying to predict the color of a card drawn by somebody else, any information about whether it is an ace does not affect your chances of making a correct prediction either. This *lack of mutual information* is the real explanation behind probabilistic independence. The color of a card does not carry any information about whether it is a face card or a number card, nor does the name of its suit about whether it is an ace.

If independence is another name for lack of information, a pair of dependent events must contain some information about each other. The question is: *how much?* One way of quantifying the information that an event carries about another is to measure the amount of change in the latter's probability, given the knowledge of the former's occurrence. In Example 11.3, the ordinary probability of D is $5/20$ or $1/4$. But as soon as it is revealed that the chosen number has turned out to be ≤ 10, our focus shifts to only the outcomes $1, \ldots, 10$ and since there are only two multiples of 4 in this range, the new probability of D is 2 out of 10 or $1/5$. So the additional piece of information that C has occurred has a "shrinking effect" on the denominator of the familiar formula $P(D) = \frac{n(D)}{n(S)}$ and changes it from the total count of the entire sample space to the count of the "relevant" part of the sample space (i.e., $n(C \bigcap S)$

or $n(C)$). At the same time, the numerator changes from $n(D)$ to the count of the "relevant" part of D (i.e., $n(D \cap C)$). The resulting formula $\frac{n(D \cap C)}{n(C)}$ is called the *conditional probability* of D given C and is denoted by $P(D \mid C)$. It is easy to see that $P(D \mid C) = \frac{P(D \cap C)}{P(C)}$, which yields the following important equation:

$$P(C \cap D) = P(C)P(D \mid C). \tag{11.1}$$

This is the so-called *long multiplication rule*, which in fact generalizes to any finite number of events. If $E_1, E_2, \ldots, E_n$ are n events, then

$$P\left(\bigcap_{i=1}^{n} E_i\right) = P(E_1)P(E_2 \mid E_1)P(E_3 \mid E_1 \cap E_2)\ldots P(E_n \mid E_1 \cap \ldots \cap E_{n-1}).$$
$$\tag{11.2}$$

Notice that if two events A and B are independent, $P(A \mid B)$ reduces to $P(A)$ and, similarly, $P(B \mid A)$ reduces to $P(B)$. In fact, this can be used as the definition of independence. Let us now see some more examples of conditional probabilities.

Example 11.4. In the context of tossing a fair coin twice, so that $S = \{HH, HT, TH, TT\}$ and all four outcomes are equally likely, what is the conditional probability that both tosses show "heads up," given that none of them produces a tail? Since the intersection of the two events has probability $= 1/4$ and that for the latter event is $3/4$, the desired conditional probability will be $1/3$.

Example 11.5. Suppose our random experiment is rolling a balanced die twice. The sample space S will have 36 outcomes, each being an ordered pair of whole numbers (x, y) with $1 \leq x, y \leq 6$. And they are all equally likely, each having probability $1/36$. What will be the conditional probability of exactly one of x and y being odd (call this event A), given that x and y are different (call it B)? Once again, since the probability of their intersection is $18/36$ or $1/2$ (can be verified by direct counting or reasoned out easily) and the denominator probability is $30/36$ or $5/6$ (all but the outcomes $(1,1), (2,2), \ldots, (6,6)$), the answer is $3/5$. Do you think that A and B are independent?

Example 11.6. Suppose you are drawing two cards one by one, without looking, from a (well-shuffled) deck of 52 cards *with* replacement. The sample space will have $52^2 = 2704$ ordered pairs of cards, all of which are equally likely. Now, what is the conditional probability of both being face cards, given that both are spades? Clearly, the intersection of these events has $3^2 = 9$ outcomes in it whereas the second event itself has $13^2 = 169$. So the answer is $9/169$.

Example 11.7. The file cabinet in your office is locked and you have a bunch of n keys in a key-ring, exactly one of which will open it. You start trying the keys one by one and once a key has been tried, it is not tried again. What are the chances that you succeed at the kth attempt ($k \leq n$)? If we define events $E_1, E_2, \ldots, E_{k-1}$ as $E_i = \{$failing at the ith attempt$\}$ and define E_k as "succeeding at the kth attempt," then the desired event $E = \bigcap_{i=1}^{k} E_i$. But by (11.2),

$$P(E) = P(E_1)P(E_2 \mid E_1)\ldots P(E_{k-1} \mid \bigcap_{i=1}^{k-2} E_i)P(E_k \mid \bigcap_{i=1}^{k-1} E_i),$$

which in this case is $(\frac{n-1}{n})(\frac{n-2}{n-1})\ldots(\frac{n-k+1}{n-k+2})(\frac{1}{n-k+1})$. But this is simply $1/n$. In other words, somewhat counter-intuitively, the chances of succeeding at any particular attempt are the same.

Example 11.8. Let us assume that in a certain population, 5% of the people carry a certain virus (i.e., the virus infection has a *prevalence* rate of 5%). A clinical diagnostic test that is used to detect the infection will pick up the infection correctly 99% of the times (i.e., has a *false negative* rate of 1%). The test raises a false alarm (i.e., gives a positive result even when the infection is not there) 2% of the times. If a person walks into the clinic, takes the test and gets a positive result, what are the chances that he/she actually has the infection? It is a legitimate question since the test appears to be slightly imperfect. Now, let us translate this story to symbols. Let $A =$ {a randomly chosen person from that population actually has the infection} and $B =$ {a randomly chosen person from that population tests positive}. Then, in terms of these, what do we know and what is the question? We know that $P(A) = 0.05$, $P(B \mid A) = 0.99$ and $P(B \mid A^c) = 0.02$. We are supposed to find $P(A \mid B)$. As we know, it is simply $\frac{P(A \cap B)}{P(B)}$. But, according to (11.1), the numerator is $P(A)P(B \mid A) = (0.05)(0.99) = 0.0495$. What about the denominator? Notice that $B = B \cap S = B \cap (A \cup A^c) = (B \cap A) \cup (B \cap A^c)$ due to the distributive property. So $P(B) = P(B \cap A) + P(B \cap A^c)$. We already know the first term. The second term can be found similarly as $P(A^c)P(B \mid A^c) = (1 - 0.05)(0.02) = 0.019$. So $P(B) = 0.0495 + 0.019 = 0.0685$. So the answer is 0.7226 or 72.26%.

There is something interesting in this last example, namely, the way we computed $P(B)$. Realizing that the sample space $S = A \cup A^c$, we decomposed B into two disjoint pieces ($B \cap A$ and $B \cap A^c$) and computed $P(B)$ as the sum of the probabilities of those pieces. This can be generalized to the following scenario. Suppose $A_1, A_2, \ldots, A_n$ are mutually disjoint events such that $\bigcup_{i=1}^{n} A_i = S$. Such a collection of events is called a *partition* of the sample space. So, in Example 11.7, A and A^c formed a partition of the sample space. Any event B can be written as $B = \bigcup_{i=1}^{n} B \cap A_i$, which is a disjoint union. So

$$P(B) = \sum_{i=1}^{n} P(B \cap A_i) = \sum_{i=1}^{n} P(B \mid A_i)P(A_i), \text{ due to (11.1).} \qquad (11.3)$$

This is known as the *law of total probability*. Using this, any of the "reverse" conditional probabilities $P(A_j \mid B)$ can be computed as

$$P(A_j \mid B) = \frac{P(A_j \cap B)}{P(B)} = \frac{P(B \mid A_j).P(A_j)}{\sum_{i=1}^{n} P(B \mid A_i)P(A_i)}. \qquad (11.4)$$

This is known as the *Bayes theorem*. In this context, the $P(A_i)$'s are called the *prior* probabilities of the partition-cells, the "forward" conditional probabilities $P(B \mid A_i)$'s are called the *likelihoods* and the "reverse" conditional probabilities $P(A_i \mid B)$'s are called the *posterior* probabilities. This simple result is so powerful that an entire subfield of statistics is based on this idea. Bayesian inference now dominates the cutting-edge applications of statistics in the highly challenging data-analysis problems posed by modern science. We will see more of it later. For the time being, let us see some examples of the Bayes theorem in action.

Example 11.9. In the experiment of drawing a card, without looking, from a (well-shuffled) deck of 52 cards, what are the chances that the card is a diamond, given that it is a face card? Let A_1, A_2, A_3 and A_4 be the events that the card drawn belongs to the diamond suit, the spade suit, the heart suit and the club suit respectively. Together, they form a partition of the sample space. Let B be the event that the card drawn is a face card. Then the prior probabilities are $P(A_i) = \frac{1}{4}$ for $i = 1, \ldots, 4$, the likelihoods are $P(B \mid A_i) = \frac{3}{13}$ and, therefore, the posterior probabilities are $\frac{(3/13)(1/4)}{(4)(3/13)(1/4)}$ or $\frac{1}{4}$. In other words, the prior and the posterior probabilities are identical. What does it tell you? That the event B is independent of the partition cells (we already knew that, right?).

Example 11.10. If you are trying to buy an air ticket the day before the 4th of July weekend to fly from Green Bay, Wisconsin, to Chicago, Illinois, there are several choices: United Airlines (UA), Northwest Airlines (NA), Midwest Airlines (MA), American Airlines (AA) and Delta ComAir (DCA). They respectively operate 15%, 20%, 25%, 30% and 10% of the flights that depart from the Green Bay airport daily. The chances of getting a seat on a day like that are small: 3% on UA, 5% on NA, 3% on MA, 2% on AA and 5% on DCA. Given that you ultimately managed to fly to Chicago, what are the chances that you chose American Airlines? Well, the five airlines form a five-cell partition with prior probabilities given by the percentages of the daily flights out of Green Bay that they operate. Using the likelihoods of getting a seat on various airlines, The Bayes theorem yields the following posterior probability of having chosen AA, given that you managed to fly:

$$\frac{(0.3)(0.02)}{(0.15)(0.03) + (0.2)(0.05) + (0.25)(0.03) + (0.3)(0.02) + (0.1)(0.05)} = \frac{0.006}{0.033},$$

which is about 0.182. Notice that in an example like this, the prior probabilities of the partition cells are more likely to be subjective, because often one's choice of a carrier is motivated by lower fares or the desire to earn frequent flier miles or the quality of in-flight service. The percentages of daily flights operated by the airlines would indeed be the correct priors if one chose a carrier randomly, which is seldom the case. In real-life applications of the Bayes theorem, there is often a lot of controversy regarding the appropriate choice of prior probabilities, but the theorem works irrespective of how they are chosen.

Since we compute the probability of an event A as $n(A)/n(S)$ in the "equally likely" scenario, efficient computation of probabilities depends on our ability to quickly enumerate sample spaces and events. In the examples discussed so far, counting the number of outcomes in a sample space or an event has been a piece of cake, but the degree of difficulty quickly increases with the size of the set. For example, any counting for experiments such as 5 rolls of a die or 10 tosses of a coin is not so trivial. So we must develop some clever counting tricks. Here is a motivating example.

Example 11.11. Suppose you are synthesizing proteins *in vitro* in a laboratory. Recall from Chapter 1 that the building blocks of proteins are amino acids (AA) and there are 20 of them. If you decide to synthesize a 10AA oligopeptide chain consisting of distinct AA's by randomly choosing from the full collection of 20, with the restriction that the first AA in the chain must be a methionine, what are the chances of ending up with the chain methionine-valine-threonine-glycine-alanine-arginine-leucine-ceistine-proline-isoleucine? To answer this question, the first order of business is to find out $n(S)$. What is a typical outcome here? It is an ordered arrangement of 10 distinct AA's chosen out of 20 so that the first one is a methionine. So the best way to find $n(S)$ is to first find the number of different subcollections of 9 AA's chosen out of 19 (i.e., all but methionine) and then multiply it with the number of different ways in which 9 distinct AA's can be rearranged. In what follows, we first address the "subcollections" question and then the "rearrangement" question.

If you have a bag containing just one item (let it be an apple or A), how many subcollections of 1 item can you form from it? Just one (i.e., $\{A\}$). If you have two items in your bag instead of one (say, an apple and a banana, or simply A and B), how many single-item subcollections can you form from it? Two (i.e., $\{A\}$ and $\{B\}$). How many double-item subcollections can be formed? Only one (namely, $\{A, B\}$). Next, if you have three items in your bag (say, an apple, a banana and a cucumber, or simply A, B and C), the numbers of single-item, double-item and triple-item subcollections that can be formed are 3, 3 and 1 respectively. List them all and verify. If you have one additional item in the bag so that it now contains four items (say, A, B, C and D), once again the different single-item, double-item, triple-item and quadruple-item subcollections that can be obtained from it are easy to list. Their numbers are 4, 6, 4 and 1 respectively. Now, in order to facilitate the recognition of a pattern in these counts, let us ask the "silly" question: "How many *empty* subcollections can be formed?" in each of the above cases. The answer is always 1, since an empty subcollection is an empty subcollection – there is no variety of it. With this, let us write those counts in the following way:

$$
\begin{array}{ccccccccc}
 & & & & 1 & & 1 & & \\
 & & & 1 & & 2 & & 1 & \\
 & & 1 & & 3 & & 3 & & 1 \\
 & 1 & & 4 & & 6 & & 4 & & 1
\end{array}
$$

The pattern that emerges in this triangular arrangement of numbers (widely known as a *Pascal's triangle*) is that each non-terminal number in it is the sum of its two nearest neighbors from the preceding row (called its "parents"). The terminal numbers in each row have just one "parent" apiece. So, how does it help answer our "subcollection" question in general? In order to find the number of different k-item subcollections that can be formed from a bag containing n distinct items, we just need to look up the $(k + 1)^{th}$ entry in the nth row of this table. That entry, usually denoted by $_nC_k$, is nothing but $\frac{n!}{k!(n-k)!}$.

Now, to rearrangements. If you just have an apple (or A), the number of different rearrangements of it is just 1. If, instead, you have two items (A and B), there are two possible rearrangements (AB and BA). For three items A, B and C, the number is 6 (ABC, ACB, BAC, BCA, CAB and CBA). For 4 items A,B,C and D, it is 24. If we still try to list all the 24 rearrangements (we will quit this habit shortly, once we know the formula), the following is the most efficient way of doing it. For the moment, ignore D and focus on the remaining three items A, B and C. From the previous step, we know that they have 6 distinct rearrangements. List them all and add the missing letter D at the end of each of them. You have generated 6 distinct rearrangements of A,B,C and D. Next, repeat this process with A,B and D only, ignoring C. You generate 6 more. Ultimately you will generate another 6+6=12 by ignoring B and A one at a time. That is why the total number is 24. In view of this "ignoring one at a time" algorithm, the number of different rearrangements of 5 distinct items will be 24+24+24+24+24=120. In any case, is there a pattern emerging here too? The numbers of distinct rearrangements of n items were 1 for $n = 1$, 2=(2)(1) for $n = 2$, 6=(3)(2)(1) for $n = 3$, 24=(4)(3)(2)(1) for $n = 4$, and so forth. So in general, the number of different rearrangements (also called *permutations*) of n distinct items is $n(n-1)(n-2)\ldots(3)(2)(1)$, usually denoted by $n!$. This simple formula, however, fails if some of the items concerned are indistinguishable. For example, two identical A's can be arranged in only one way, not two. Similarly, two identical A's and a B can be arranged in just three ways (AAB, ABA, BAA), not six. If you have n objects, k_1 of which are identical of type 1, ..., k_L of which are identical of type L ($\sum_{i=1}^{L} k_i = n$), then the number of different rearrangements drastically reduces from $n!$ to $\frac{n!}{(k_1!)\ldots(k_L!)}$.

Having said all these, let us go back to our original question of oligopeptide synthesis. 9 distinct amino acids can be chosen from a "bag" of 19 in $_{19}C_9 = \frac{19!}{(9!)(10!)}$ different ways. A set of 9 distinct amino acids can be permuted in 9! different ways. So, the total number of possible 10AA oligopeptide chains consisting of distinct AA's and starting with methionine is $(\frac{19!}{(9!)(10!)})(9!) = \frac{19!}{10!}$. In general, the number of different *ordered* permutations of k distinct objects chosen out of n distinct objects is $\frac{n!}{(n-k)!}$, which is denoted by $_nP_k$.

Example 11.12. In the experiment of drawing two cards, without looking, from a (well-shuffled) deck of 52 cards one by one without replacement, what are the chances that both are hearts? The answer is $n(A)/n(S)$ where $n(S) =_{52} C_2 = \frac{52!}{(2!)(50!)} = 1326$ and $n(A) =_{13} C_2 = \frac{13!}{(2!)(11!)} = 78$. So the chances that both of them are hearts are about 5.88%.

Example 11.13. If you are randomly permuting the four letters of our genetic alphabet (i.e., the four nucleotides adenine, guanine, thymine and cytosine or A,G,T and C), what are the chances that the purines and the pyrimidines are together? For this problem, $n(S)$ is of course $4! = 24$. To find $n(A)$, let us look at it this way. The "purine block" and the "pyrimidine block" can be permuted in $2! = 2$ ways and within each block, the two bases can be permuted in $2! = 2$ ways. So $n(A) = (2)(2)(2) = 8$ and the answer is $1/3$.

11.2 Random variables and probability distributions

For many real-life random experiments, the sample space is very large and the individual outcomes are not of our interest; instead we are interested in clusters of outcomes sharing a common feature. So it is enough to summarize the sample space as a list of those clusters and the corresponding probabilities. Usually, each such cluster corresponds to a unique value of a "quantity of interest" (called a *random variable*), so listing the clusters is equivalent to listing the distinct values of the associated random variable. The resulting table, with values of the random variable in one row and the corresponding probabilities in the other, is called a *probability mass function* table or p.m.f. table. Here are some examples.

Example 11.14. In the experiment of tossing a fair coin three times, which has $S = \{$ HHH, HHT, HTH, THH, HTT, THT, TTH, TTT$\}$ with equally likely outcomes, if the "quantity of interest" or random variable (call it X) is simply the number of heads among the three tosses, the p.m.f. table will be

Values	0	1	2	3
Probs	$\frac{1}{8}$	$\frac{3}{8}$	$\frac{3}{8}$	$\frac{1}{8}$

Example 11.15. In the experiment of rolling a balanced die twice, with S having 36 equally likely ordered pairs of whole numbers (x, y), $1 \le x, y \le 6$, if the random variable of interest (call it Y) is the sum of the two numbers that turned up, the p.m.f. table will be

Values	2	3	4	5	6	7	8	9	10	11	12
Probs	$\frac{1}{36}$	$\frac{2}{36}$	$\frac{3}{36}$	$\frac{4}{36}$	$\frac{5}{36}$	$\frac{6}{36}$	$\frac{5}{36}$	$\frac{4}{36}$	$\frac{3}{36}$	$\frac{2}{36}$	$\frac{1}{36}$

Example 11.16. Suppose ten million tickets of a state lottery have been sold for a dollar a piece. Only one of them carries a super bumper prize of

$ 5,000,000, 5 of them carry mega-prizes of $ 500,000 each, 25 of them carry second prizes of $ 50,000 each, 50 of them carry third prizes of $ 10,000 each and another 100 of them carry consolation prizes of $ 1000 each. If you go to a store and randomly buy a ticket for this lottery, what is the p.m.f. table of your earning (call it W)? Notice that irrespective of whether you win a prize or not, you always pay the ticket price. So the p.m.f. table of your earning will be as follows:

Amounts	$ 4999999	$ 499999	$ 49999	$ 9999	$ 999	$ -1
Probs	$\frac{1}{10000000}$	$\frac{1}{2000000}$	$\frac{1}{400000}$	$\frac{1}{200000}$	$\frac{1}{100000}$	$\frac{9999819}{10000000}$

In this last example, the important question is what your *average* (or *expected*) earning will be if you repeatedly buy tickets for this lottery. Will it be as high as the grand prizes this lottery advertises? The *expected value* or *mean* of a random variable W (denoted by $E(W)$ or μ_W) is defined as $\sum_{i=1}^{m} w_i p_i$ if W takes the values $w_1, \ldots, w_m$ with probabilities $p_1, \ldots, p_m$ respectively. It is simply a weighted average of its values, with the weights being the probabilities. For this lottery, $E(W) = -0.065$. That is, on an average you *lose* six and a half cents! This is why the lottery business is profitable to its host.

While the expected value is a useful device for comparing two random variables X and Y (after all, we cannot compare them value-by-value or probability-by-probability since their p.m.f. tables may not even be of the same length), it is by no means the whole story. Two completely different random variables may end up having the same expected value, so we need some other summary-measures to capture the other differences. Any p.m.f. table can be pictorially represented by a *stick plot* or a *probability histogram*. In a stick plot, the distinct values are marked along the horizontal axis and a stick is erected over each value whose height is the corresponding probability. In a probability histogram, the sticks are replaced by rectangular bars of equal width centered at the values. Various features of this plot such as how "fat" or "thin" its tails are, how asymmetric it is around its expected value and how peaked or flat-topped it is, are connected to its *moments*. The rth *raw moment* of X for any real number r is defined as $E(X^r)$, i.e., as the expected value of the random variable X^r which takes the values $x_1^r, x_2^r, \ldots$ with probabilities $p_1, p_2, \ldots$ ($x_1, x_2, \ldots$ being the values of X). The rth *central moment* of X is defined as $E(X - E(X))^r$, i.e., as the expected value of the random variable $(X - E(X))^r$ which takes the values $(x_1 - \mu_X)^r, (x_2 - \mu_X)^r, \ldots$ with probabilities $p_1, p_2, \ldots$. The 2nd central moment of a random variable is called its *variance* (denoted by σ_X^2), and it is related to the "fatness" or "thinness" of the histogram tail. Usually the heavier the tail, the greater the variance. The square root of the variance is known as the *standard deviation* (denoted by σ_X). The 3rd central moment is related to the degree of *skewness* (i.e., lack of symmetry) of the histogram and the 4th central moment is related to its degree of peakedness. Two important properties of expected values are (i) $E(X + Y) = E(X) + E(Y)$ for any two random variables X and Y and (ii) $E(cX) = cE(X)$ for any random variable X and any constant c. Simple algebraic expansions and repeated use of the above two properties will show that the raw moments and the central moments are related. For example, $\sigma_X^2 = E(X^2) - (E(X))^2$. Also,

$E(X - E(X))^3 = E(X^3) - 3E(X)E(X^2) + 2(E(X))^3$. All the raw moments of a random variable can be conveniently obtained from its *moment generating function* (MGF). Just as a thin popcorn-bag from the supermarket shelf, when microwaved, swells up and produces lots of delicious popcorns, the MGF is like a handy storage device that produces raw moments of many different orders when repeatedly differentiated. Formally, the MGF of X is defined as $M_X(t) = E(e^{tX})$ for t in some interval around 0 on the real line. So it is the expected value of the random variable e^{tX} that takes the values $e^{tx_1}, e^{tx_2}, \ldots$ with probabilities $p_1, p_2, \ldots$. Its rth derivative, evaluated at $t = 0$, is the rth raw moment of X. An important and useful fact about MGF's is that there is a one-to-one correspondence between the form of the MGF and the form of the associated p.m.f., so it is possible to identify the p.m.f. by looking at the MGF (call it "MGF fingerprinting" if you will!).

Example 11.17. For the random variable Y in Example 11.15, $\mu_Y = E(Y) = \frac{252}{36} = 7$, $\sigma_X^2 = E(Y^2) - (E(Y))^2 = \frac{1974}{36} - 7^2 = 5.833$, the 3rd raw moment $E(Y^3) = \frac{16758}{36} = 465.5$ and the 3rd central moment $E(Y - E(Y))^3 = 465.5 - 3(7)(\frac{1974}{36}) + 2(7^3) = 0$. The 3rd central moment is 0 for any p.m.f. that is *symmetric* with respect to its mean (the converse is not necessarily true, though). Finally, the MGF of Y is

$$M_Y(t) = \frac{e^{2t} + e^{12t} + 2(e^{3t} + e^{11t}) + 3(e^{4t} + e^{10t}) + 4(e^{5t} + e^{9t}) + 5(e^{6t} + e^{8t}) + 6e^{7t}}{36}.$$

Its the first derivative of it w.r.t. t, evaluated at $t = 0$, is $E(Y) = 7$, its second derivative evaluated at $t = 0$ is $E(Y^2) = 54.833$ and its third derivative evaluated at $t = 0$ is $E(Y^3) = 465.5$ (verify these).

Now consider the random experiment of closing your eyes and touching a 6-inch ruler with a needle-tip. What if we define our random variable X to be the distance between the point where the needle touched the ruler and the left end-point (or zero-point) of the ruler? A needle-tip being so sharp and pointed, X can actually take any value between 0 and 6. That is, the set of possible values of X is the entire interval $[0, 6]$ and is, therefore, uncountably infinite. This immediately implies that if we wanted to write down a p.m.f. table for X or draw its probability histogram, we would fail. Also, despite the fact that your eyes being closed gives this experiment an "equally likely" flavor, in the sense that intuitively each real number in the interval $[0, 6]$ should have the same probability, what *is* that probability? Just as each of the 6 outcomes in the roll of a balanced die has probability $\frac{1}{6}$, will it be $\frac{1}{\infty}$ or 0 in this case? Then we are in the strange situation that each single outcome has probability 0, yet one of those countless outcomes will definitely occur! This kind of random variables, which lands us in this bizarre situation, is called the class of *continuous* random variables. In contrast, the random variables with p.m.f. tables that we saw earlier will be called *discrete*. By the way, from our examples above, one should not get the wrong impression that *all* discrete random variables have finite value-sets (or *supports*). Examples of discrete random variables with countably infinite supports are forthcoming.

In any case, getting back to continuous random variables, how does one compute probabilities for them? For example, for the X in the "needle and ruler" experiment, what is $P(X \leq 2)$ or $P(3 \leq X \leq 5)$? Well, think about the natural discrete counterpart of this experiment. Since this experiment basically says: "Pick any real number randomly from within $[0,6]$," its natural discrete counterpart would be: "Pick any whole number randomly from among 1,2,3,4,5 and 6". It should look familiar since it is nothing but the balanced die-rolling experiment. If for this experiment, we defined X to be the whole number that turned up, how would we compute $P(X \leq 2)$ and $P(3 \leq X \leq 5)$? One way would be to compute them directly from the p.m.f. table of X or by adding the heights of appropriate sticks in its stick plot, but these are not relevant to our present situation where there is no p.m.f. table or stick plot. Another way of computing those probabilities would be by the *area method* from the probability histogram of X. The probability histogram of an "equally likely" experiment such as rolling a balanced die looks like a flat or rectangular box. To find $P(X \leq 2)$, we just need to find the total area of the bars in this histogram that correspond to this event (i.e., the two leftmost bars). Since each rectangular bar has a base-width of 1 unit and height $= \frac{1}{6}$, the total area of those two bars is $\frac{1}{3}$. Similarly, $P(3 \leq X \leq 5) = \frac{1}{2}$, the sum of the areas of three such bars. If we now imagine following the same area method to compute those probabilities in the continuous case, except at a much, much finer scale with "needle-thin" bars (because there are countless values crammed together in a small space), we begin to understand how one deals with continuous random variables. We are actually taking the limit of a histogram with n bars as $n \to \infty$, keeping the base-width of the histogram fixed. As we do this, the upper contour of the histogram (which would have a rugged broken-line structure in a discrete histogram) starts to appear like a smooth curve. This curve, which still encloses a total area of 1 underneath it like the discrete histograms, will be called the *probability density curve* of X. The function $f(x) : D \to \mathcal{R}$, whose graph is the probability density curve, will be called the *probability density function* (p.d.f.) of X (D being the value-range or support of X). In general, any nonnegative-valued integrable function defined on the real line, which encloses a total area of 1, could potentially be the p.d.f. of some continuous random variable coming from some underlying experiment.

So, in our "needle and ruler" experiment, the density curve is a flat line on the interval $[0,6]$ and the p.d.f. is $f(x) = \frac{1}{6}I(x \in [0,6])$, where $I(.)$ is the indicator function that takes the value 1 only if the condition in it is satisfied (otherwise 0). This is known as the *Uniform* p.d.f. $P(X \leq 2)$ is the area underneath this flat line between 0 and 2. Similarly, $P(3 \leq X \leq 5)$ is the area between 3 and 5. Formally speaking, for any two real numbers a and b,

$$P(a \leq X \leq b) = \int_a^b f(x)dx. \tag{11.5}$$

This gives us the formal reason why, for a continuous random variable X, $P(X = x) = 0$ for any particular value x, because it is simply the integral in (11.5) with $a = b$. Now, if we define a function $F(t) : \mathcal{R} \to [0, 1]$ as $F(t) = \int_{-\infty}^{t} f(x)dx$, then $F(t)$ is nothing but the area underneath the p.d.f. $f(x)$ up to the point t or, equivalently, $P(X \leq t)$. This F is called the *cumulative distribution function* (c.d.f.) corresponding to the p.d.f. f. In this case, the Uniform[0,6] c.d.f. is

$$F(t) = 0 \text{ if } t \leq 0; \quad \int_{0}^{t} \frac{1}{6}dx \text{ or } \frac{t}{6} \text{ if } t \in (0, 6); \ 1 \text{ if } t \geq 6. \quad (11.6)$$

By the fundamental theorem of integral calculus, such an $F(t)$ is a continuous (if fact differentiable) function with $F'(x) = f(x)$. Incidentally, we could also have defined a c.d.f. for a discrete random variable, but those c.d.f.'s would be *step functions*, i.e., their graphs could consist of flat line-segments with jumps in-between (explain to yourself why). But irrespective of discrete or continuous, all c.d.f.'s have the following properties:

Result 11.2. The c.d.f. $F(.)$ of any random variable X satisfies
(i) $\lim_{t \to -\infty} F(t) = 0$ and $\lim_{t \to \infty} F(t) = 1$;
(ii) $F(t)$ is *right-continuous*, i.e., $\lim_{t \to s+} F(t) = F(s)$, where "$t \to s+$" means that t approaches s from its right side;
(iii) $F(a) \leq F(b)$ for any $a, b \in \mathcal{R}$ with $a \leq b$.

The density curves of different continuous random variables can have a great variety of geometric shapes. Here is another example.

Example 11.18. Suppose you touch a 6-inch ruler with a needle with your eyes closed, and your friend does the same thing independently of you. Let Y_1 be the distance between the touching point and the zero-point on the ruler in your case, and Y_2 be that in your friend's case. If we define $Y = Y_1 + Y_2$, this Y will have a *triangular* density on the support [0,12] with p.d.f. $f(y)$ given by

$$f(y) = \frac{y}{36}I(0 \leq y \leq 6) + \frac{12 - y}{36}I(6 < y \leq 12). \quad (11.7)$$

How do we know? One way would be to go through the same limiting process that led us to the Uniform[0,6] density earlier. For this experiment, the natural discrete counterpart is two persons rolling a balanced die each, independently of one another. The discrete histogram corresponding to the sum of the two numbers is essentially triangle-shaped (ignoring the rugged upper contours of the bars).

Now that we know how continuous random variables behave, all the numerical summaries we defined for a discrete random variable (i.e., raw and central moments) can be easily generalized to them. For a continuous X with p.d.f. $f(x)$ and support D, the rth raw and central moments are defined as

$$E(X^r) = \int_{D} x^r f(x)dx, \ E(X - \mu_X)^r = \int_{D} (x - \mu_X)^r f(x)dx, \quad (11.8)$$

provided that the integrals exist. Like before, various features of the density curve (e.g., lightness or heaviness of tails, skewness, flatness, etc.) are connected to its moments and the central moments are expressible in terms of the raw moments. The MGF of X is also defined analogously as $M_X(t) = E(e^{tX}) = \int_D e^{tx} f(x)dx$ if this integral exists on some interval around 0, and the mechanism by which the raw moments are generated from this MGF is the same as before.

Example 11.19. The MGF of the Uniform$[0,6]$ random variable X mentioned earlier is

$$M_X(t) = \int_0^6 e^{tx} \frac{1}{6} dx = \frac{1}{6} \frac{e^{6t} - 1}{t} \text{ for } t \neq 0.$$

Remember the definition of two events being independent? It can be easily generalized to any finite number of events. The events $E_1, \ldots, E_n$ are independent if and only if $P(\bigcap_{i=1}^n E_i) = \prod_{i=1}^n P(E_i)$. The definition of the independence of a bunch of random variables is analogous. The random variables $X_1, \ldots, X_n$ with supports $D_1, \ldots, D_n$ respectively are said to be independent if for any subsets A_i of D_i $(i = 1, \ldots, n)$, $P(X_1 \in A_1, \ldots, X_n \in A_n) = \prod_{i=1}^n P(X_i \in A_i)$. An immediate consequence of this definition is that the expected value of the product of a bunch of independent random variables turns out to be the product of the individual expected values. We will now focus on some frequently encountered and useful random variables— both discrete and continuous.

A vast majority of real-life random experiments producing discrete random variables can be modeled by the following:

(i) the number of heads in n independent tosses of a (possibly biased) coin whose $P(\text{head}) = p$ in a single toss;
(ii) the number of independent tosses of a (possibly biased) coin with $P(\text{head}) = p$ that are needed to get the kth head;
(iii) the number of black sheep in a random sample of n sheep drawn without replacement from a population of $N(\geq n)$ sheep containing r black sheep;
(iv) the (limiting) number of heads in a very large number (n) of independent tosses of a highly biased coin with a very small $P(\text{head}) = p$ so that np is moderate.

We will briefly describe the resulting random variable and its properties in each of these cases.

If X is the number of heads in n independent tosses of a coin whose $P(\text{head})$ in each toss is $p \in (0, 1)$, the p.m.f. of X is given by

$$P(X = x) =_n C_x p^x (1 - p)^{n-x}, \text{ for } x = 0, 1, \ldots, n. \tag{11.9}$$

This can be easily derived using the independence between tosses and the fact that there are $_nC_x$ different permutations of x identical H's and $(n - x)$ identical T's that give rise to the same value of X. This is known as the *Binomial*(n, p) p.m.f., which is also called a *Bernoulli* trial for $n = 1$. For this X, $\mu_X = E(X) = np$, $\sigma_X^2 = np(1 - p)$ and $M_X(t) = (1 - p + pe^t)^n$.

If X is the number of independent tosses of a coin with $P(\text{head}) = p$ that are needed to see the *first* head, the p.m.f. of X is given by

$$P(X = x) = p(1 - p)^{x-1}, \text{ for } x = 1, 2, \ldots . \tag{11.10}$$

This is once again easy to derive using the independence between tosses. This is known as the *Geometric(p)* p.m.f., which has mean $\mu_X = \frac{1}{p}$, variance $\sigma_X^2 = \frac{1-p}{p^2}$ and MGF $M_X(t) = \frac{pe^t}{1-(1-p)e^t}$. An interesting and useful feature of this distribution is its *memoryless* property. Given that no head has appeared until the kth toss, the conditional probability that there will be no head until the $(k + m)^{th}$ toss is the *same* as the (unconditional) probability of seeing no head in the first m tosses, for any two positive integers k and m. That is, $P(X > k + m \mid X > k) = P(X > m) = (1 - p)^m$.

If X is the number of independent tosses of a coin with $P(\text{head}) = p$ that are needed to see the rth head $(r \geq 1)$, the p.m.f. of X is given by

$$P(X = x) = _{x-1}C_{r-1}p^r(1 - p)^{x-r}, \text{ for } x = r, r + 1, \ldots . \tag{11.11}$$

This is equally easy to derive using the independence between tosses and the fact that the first $r - 1$ heads could occur anywhere among the first $x - 1$ tosses. This is known as the *Negative Binomial(r, p)* p.m.f., with mean $\mu_X = \frac{r}{p}$, variance $\sigma_X^2 = \frac{r(1-p)}{p^2}$ and MGF $M_X(t) = \left(\frac{pe^t}{1-(1-p)e^t}\right)^r$. It is related to the Geometric(p) p.m.f. in the sense that if $Y_1, \ldots, Y_r$ are independent and identically distributed (i.i.d.) random variables each having a Geometric(p) p.m.f., $\sum_{i=1}^{r} Y_i$ will have a Negative Binomial(r, p) distribution.

Suppose a population of N sheep consists of r black sheep and $N - r$ white sheep. If a random sample of n $(\leq r)$ is drawn without replacement from this population and X is the number of black sheep in this sample, then X has the p.m.f.

$$P(X = x) = (_rC_x)(_{N-r}C_{n-x})/_NC_n, \text{ for } x = 0, 1, \ldots, n. \tag{11.12}$$

This is not difficult to see, in view of the Pascal's triangle and the associated formula for finding the number of different subcollections. This is known as the *Hypergeometric (N, r, n)* p.m.f., with mean $\mu_X = \frac{nr}{N}$ and variance $\sigma_X^2 = \frac{nr(N-r)(N-n)}{N^2(N-1)}$. The MGF does not simplify to a convenient expression, but can still be written down as a summation.

If X counts how many heads there are in a very large number n of independent tosses of a highly biased coin with $P(\text{head}) = p$ very small such that np is moderate (call it λ), then the binomial p.m.f. of X can be well-approximated by the following p.m.f.:

$$P(X = x) = \frac{e^{-\lambda}\lambda^x}{x!}, \text{ for } x = 0, 1, \ldots . \tag{11.13}$$

This can be shown formally by taking the double limit of a Binomial(n, p) p.m.f. as $n \to \infty$ and $p \to 0$ in such a way that np remains a constant λ.

This is called the *Poisson(λ)* p.m.f., with mean $\mu_X = \lambda$, variance $\sigma_X^2 = \lambda$ and MGF $M_X(t) = e^{\lambda(e^t-1)}$. This is a widely used probability model for random experiments producing count data, except that it is only suitable for datasets with the mean approximately equal to the variance. However, there are ways to modify this p.m.f. so that it can accommodate datasets with the variance higher or lower than the mean (see, for example, Shmueli et al., *Applied Statistics* (2005)). Also, the Poisson(λ) p.m.f. has the *reproductive* property that if X_1 and X_2 are independent random variables with $X_i \sim$ Poisson(λ_i), then $X_1 + X_2 \sim$ Poisson($\lambda_1 + \lambda_2$).

Now, the continuous case. A vast majority of real-life random experiments that produce continuous random variables can be modeled by the following probability density functions, in addition to the Uniform$[a, b]$ p.d.f. discussed a short while ago (with mean $\frac{a+b}{2}$ and variance $\frac{(b-a)^2}{12}$).

A random variable X is said to have a *Normal* (or *Gaussian*, after Carl Friedrich Gauss) p.d.f. with mean $\mu_X \in \mathcal{R}$ and variance σ_X^2 ($\sigma_X > 0$) if its p.d.f. is given by

$$f(x) = \frac{1}{(2\pi\sigma^2)^{0.5}} e^{-(x-\mu)^2/2\sigma^2}, \text{ for } x \in (-\infty, \infty), \qquad (11.14)$$

where we have omitted the subscript X to reduce clutter. The MGF of X is $M_X(t) = e^{t\mu_X + t^2\sigma_X^2/2}$. The special case with mean 0 and variance 1 is called a *standard normal* p.d.f. and the associated random variable is usually denoted by Z. The $N(\mu_X, \sigma_X^2)$ p.d.f. has the *linearity* property, that is, $X \sim N(\mu_X, \sigma_X^2) \implies aX + b \sim N(a\mu_X + b, a^2\sigma_X^2)$. This can be shown by first deriving the MGF of $aX + b$ from that of X and then using "MGF fingerprinting." As a result, $X \sim N(\mu_X, \sigma_X^2) \implies \frac{X-\mu_X}{\sigma_X} \sim N(0, 1)$ or standard normal. The ratio $\frac{X-\mu_X}{\sigma_X}$ is usually called the *z-score* of X. Another celebrated result that plays a fundamental role in probability and statistics is the *central limit theorem* (CLT). In its most simplified form, it basically says that if $\{X_1, \ldots, X_n\}$ is a random sample from some p.m.f. or p.d.f. having mean μ_X and standard deviation σ_X (i.e., if $X_1, \ldots, X_n$ are i.i.d. random variables from the same p.m.f. or p.d.f.), then the c.d.f. of $(\frac{1}{n}\sum_{i=1}^{n} X_i - \mu_X)/(\sigma_X/n^{\frac{1}{2}})$ approaches that of $N(0, 1)$ in the limit as $n \to \infty$. In other words, for n sufficiently large, the c.d.f. of the z-score of $\bar{X} = \frac{1}{n}\sum_{i=1}^{n} X_i$ can be well-approximated by a standard normal c.d.f. This has far-reaching consequences, one of which is the large-sample normal approximation to discrete p.m.f.'s. A normal p.d.f. also has the *reproductive* property that if $X_i \sim N(\mu_i, \sigma_i^2)$ for $i = 1, \ldots, n$ and they are independent, the sum $\sum_{i=1}^{n} X_i$ has a $N(\sum_{i=1}^{n} \mu_i, \sum_{i=1}^{n} \sigma_i^2)$. Since probability computation using a normal p.d.f. is not possible analytically, extensive tables are available for ready reference.

A random variable X is said to have an *Exponential*(β) p.d.f. if the p.d.f. is given by

$$f(x) = \frac{1}{\beta}e^{-\frac{x}{\beta}}, \text{ for } x \in (0, \infty) \text{ and } \beta > 0. \qquad (11.15)$$

This p.d.f. has mean β, variance β^2 and MGF $(1 - \beta t)^{-1}$ for $t < \frac{1}{\beta}$. The c.d.f. is $F(t) = 1 - e^{-\frac{t}{\beta}}$. Two interesting and useful facts about this p.d.f. are (a) its *memoryless* property and (b) its relation to the geometric p.d.f. mentioned earlier. The fact that $P(X > t + s \mid X > t) = P(X > s) = e^{-\frac{s}{\beta}}$ for any positive t and s is known as the memoryless property (verify it). If Y is defined as the largest integer $\leq X$ (also called the "floor" of X), then $Y + 1$ has a Geometric(p) p.m.f. with $p = 1 - e^{-\frac{1}{\beta}}$. This is because $P(Y + 1 = y + 1) = P(Y = y) = P(y \leq X < y + 1) = (1 - e^{-\frac{1}{\beta}})e^{-\frac{1}{\beta}y}$. Sometimes the exponential p.m.f. is reparametrized by setting $\lambda = \beta^{-1}$, so the p.d.f. looks like $f(x) = \lambda e^{-\lambda x} I(x > 0)$ with mean λ^{-1} and variance λ^{-2}. If $X_1, \ldots, X_n$ are i.i.d. random variables having the p.d.f. in (11.15), then the minimum of $\{X_i\}_{i=1}^n$ also has an exponential p.d.f. and the sum $\sum_{i=1}^n X_i$ has a *Gamma*$((n, \beta))$ p.d.f.

A random variable X is said to have a *Gamma*(α, β) p.d.f. if the p.d.f. is given by

$$\frac{1}{\Gamma(\alpha)\beta^\alpha}e^{-\frac{x}{\beta}}x^{\alpha-1} \text{ for } x > 0, \alpha > 0 \text{ and } \beta > 0, \qquad (11.16)$$

where $\Gamma(\alpha) = \int_0^\infty e^{-y}y^{\alpha-1}dy$ is the gamma function with the property that $\Gamma(\nu + 1) = \nu\Gamma(\nu)$. This, in particular, implies that $\Gamma(n + 1) = n!$ for any nonnegative integer n. The parameter α is called the *shape parameter* as the density curve has different shapes for its different values. Its mean is $\alpha\beta$, variance is $\alpha\beta^2$ and MGF is $(1 - \beta t)^{-\alpha}$ for $t < \frac{1}{\beta}$. Clearly, an exponential p.d.f. is a special case of (11.16) with $\beta = 1$. Also, for $\alpha = \frac{n}{2}$ and $\beta = 2$, it is called a *chi-squared* (χ^2) p.d.f. with n *degrees of freedom*, which is widely useful because of the connection that $Z \sim N(0, 1) \implies z^2 \sim \chi^2$ with 1 degree of freedom. The Gamma(α, β) p.d.f. has the *reproductive* property in the sense that if $X_i \sim$ Gamma(α_i, β) for $i = 1, \ldots, n$ and they are independent, then $\sum_{i=1}^n X_i \sim$ Gamma($\sum_{i=1}^n \alpha_i, \beta$).

A random variable X is said to have a *Beta*(a, b) p.d.f. if the p.d.f. looks like

$$f(x) = \frac{\Gamma(a + b)}{\Gamma(a)\Gamma(b)}x^{a-1}(1 - x)^{b-1}, \text{ for } x \in (0, 1), a > 0 \text{ and } b > 0. \qquad (11.17)$$

This p.d.f. has mean $\frac{a}{a+b}$ and variance $\frac{ab}{(a+b)^2(a+b+1)}$. Notice that for $a = b = 1$, this is nothing but the Uniform[0,1] p.d.f. It also has a connection with the Gamma(α, β) p.d.f. To be precise, if $X \sim$ Gamma(α_1, β), $Y \sim$ Gamma(α_2, β) and X and Y are independent, the ratio $\frac{X}{X+Y}$ will have a Beta(a, b) p.d.f. with $a = \alpha_1$ and $b = \alpha_2$. More generally, if $X_1, \ldots, X_k, X_{k+1}, \ldots, X_m$ are independent random variables with $X_i \sim$ Gamma(α_i, β) ($\beta > 0$ and $\alpha_i >$

0 for $i = 1, \ldots, n$), then the p.d.f. of $\sum_{i=1}^{k} X_i / \sum_{i=1}^{m} X_i$ is Beta(a, b) with $a = \sum_{i=1}^{k} \alpha_i$ and $b = \sum_{i=k+1}^{m} \alpha_i$.

So far we have dealt with individual random variables. We now move on to the case where we have a bunch of them at once. If X and Y are two discrete random variables, then the *joint p.m.f.* of the *random vector* (X, Y) is given by $P(\{X = x\} \cap \{Y = y\})$ for all value-pairs (x, y). It can be imagined as a two-dimensional table with the values of X and Y being listed along the left margin and top margin respectively and the joint probabilities being displayed in various cells. So each row of this table corresponds to a particular value of X and each column corresponds to a particular value of Y. If we *collapse* this table column-wise, i.e., add all the probabilities displayed in each row, thereby ending up with a single column of probabilities for the values of X, this single column of probabilities gives us the *marginal* p.m.f. of X. Similarly, by collapsing the table row-wise (i.e., summing all the probabilities in each column), we get the *marginal* p.m.f. of Y. Now, for each fixed value x_i of X, the *conditional* p.m.f. of Y given that $X = x_i$ is nothing but $\{\frac{P(Y=y_j \text{ and } X=x_i)}{P(X=x_i)}\}_{j=1}^{m}$, assuming that Y takes the values $y_1, \ldots, y_m$. Similarly, for each fixed value y_i of Y, the conditional p.m.f. of X given that $Y = y_i$ can be defined. The raw and central moments and the MGF of X computed using such a conditional p.m.f. will be called its conditional moments and conditional MGF. Similar is the terminology for Y.

All these concepts can be easily generalized to the case of an $n \times 1$ random vector $(X_1, \ldots, X_n)$. The joint p.m.f. of these can be imagined as an n-dimensional table, whose cells contain $P\{\bigcap_{i=1}^{n} (X_i = x_i)\}$ for various values x_i of X_i $(i = 1, \ldots, n)$. We can now talk about several types of marginal and conditional p.m.f.'s, such as the *joint marginal* p.m.f. of $X_{i_1}, \ldots, X_{i_k}$ which is obtained by collapsing the n-dimensional table w.r.t. all other indices except $i_1, \ldots, i_k$, or the *joint conditional* p.m.f. of $X_{i_1}, \ldots, X_{i_k}$ given $X_{j_1}, \ldots, X_{j_l}$ for two disjoint subsets of indices $\{i_1, \ldots, i_k\}$ and $\{j_1, \ldots, j_l\}$ which is obtained by dividing the joint marginal p.m.f. of $X_{i_1}, \ldots, X_{i_k}, X_{j_1}, \ldots, X_{j_l}$ with that of $X_{j_1}, \ldots, X_{j_l}$. We can also talk about joint moments (marginal or conditional). For example, the $(j_1, \ldots, j_k)^{th}$ order joint raw moment of $X_{i_1}, \ldots, X_{i_k}$ is defined as $E(\prod_{s=1}^{k} X_{i_s}^{j_s})$, where the expected value is based on the joint marginal p.m.f. of $X_{i_1}, \ldots, X_{i_k}$. In the case of a conditional joint moment, this expected value would be based on the conditional joint p.m.f. of $X_{i_1}, \ldots, X_{i_k}$ given another disjoint set of variables.

In a situation where we have a bivariate random vector (X, Y), the numerical summary that describes the nature (i.e., direction) of the *joint variability* of X and Y is called the *covariance* (denoted by COV(X, Y) or σ_{XY}). It is defined as

$$COV(X, Y) = E\{(X - \mu_X)(Y - \mu_Y)\} = E(XY) - \mu_X \mu_Y. \qquad (11.18)$$

A positive value of it indicates a *synergistic* or *direct* relation between X and Y, i.e., in general X increases if Y does so. A negative covariance, on the other hand, shows an *antagonistic* or *inverse* relation whereby an increase in X will be associated with a decrease in Y in general. In order to measure the strength of the linear association between X and Y (which is difficult to assess from the covariance since it is an absolute measure, not a relative one), we convert the covariance to *correlation* (denoted by ρ_{XY}). It is defined as

$$\rho_{XY} = \frac{\sigma_{XY}}{\sigma_X \sigma_Y}, \qquad (11.19)$$

where σ_X and σ_Y are the marginal standard deviations of X and Y. It is bounded below and above by -1 and 1 respectively (which is easy to see by the *Cauchy-Schwarz inequality* that says: $E \mid UV \mid \leq (E(U^2))^{\frac{1}{2}}(E(V^2))^{\frac{1}{2}}$ for any random variables U and V), so that any value close to its boundaries is considered evidence of a strong linear association between X and Y. Values close to the center of this range testify to a relatively weak linear association. In the case where we have an $n \times 1$ random vector, there are $_nC_2 = \frac{n(n-1)}{2}$ pairwise covariances to talk about. If we construct an $n \times n$ matrix Σ whose $(i,j)^{th}$ entry is $COV(X_i, X_j) = \sigma_{ij}$ (say), then it will be a symmetric matrix with the ith diagonal entry being the variance of X_i or σ_i^2 ($i = 1, \ldots, n$). This matrix is referred to as the *variance-covariance matrix* or the *dispersion matrix* of the random vector. Sometimes it is more convenient to work with the matrix obtained by dividing the $(i,j)^{th}$ entry of Σ with $\sigma_i \sigma_j$ (i.e., the product of the standard deviations of X_i and X_j). This symmetric $n \times n$ matrix with all diagonal entries $= 1$ is called the *correlation matrix* of $(X_1, \ldots, X_n)$ due to obvious reasons. In case the dispersion matrix Σ has an inverse (an $n \times n$ matrix W is said to be the *inverse* of another $n \times n$ matrix W^* if $WW^* = W^*W = I$, the $n \times n$ identity matrix with all diagonal entries $= 1$ and all off-diagonal entries $= 0$), the inverse is called the *precision matrix*. Here are some examples.

Example 11.20. Suppose the joint p.m.f. of (X,Y) is given by the table

	5	10	15	20
0	$\frac{1}{48}$	$\frac{1}{48}$	$\frac{1}{12}$	$\frac{1}{8}$
1	$\frac{1}{12}$	$\frac{1}{12}$	$\frac{1}{6}$	$\frac{1}{6}$
2	$\frac{1}{8}$	$\frac{1}{12}$	$\frac{1}{48}$	$\frac{1}{48}$

Here, the marginal p.m.f. of X is actually Binomial$(2,\frac{1}{2})$ and that of Y is

value	5	10	15	20
prob	$\frac{11}{48}$	$\frac{9}{48}$	$\frac{13}{48}$	$\frac{15}{48}$

So X has mean=$(2)(1/2)=1$ and variance=$(2)(1/2)(1-1/2)=1/2$, and those for Y are 13.33 and respectively. The conditional p.m.f. of X given that $Y = 15$ is $P(X = 0 \mid Y = 15) = \frac{4}{13}$, $P(X = 1 \mid Y = 15) = \frac{8}{13}$ and $P(X = 2 \mid Y = 15) = \frac{1}{13}$.

Example 11.21. An agency that conducts nationwide opinion polls has decided to take a random sample (with replacement) of 1000 people from the entire U.S. population and ask each sampled individual about the effectiveness of the Kyoto protocol to control global warming. Each individual can say "effective" or "not effective" or "no opinion" (let us classify answers such as "Don't even know what the Kyoto protocol is!" as "no opinion" for the sake of simplicity). If the true (but unknown) percentage of people in the U.S. who think that the Kyoto protocol would be effective is 25%, that of those who consider it ineffective is 35% and the rest are in the third category, what are the chances that in a sample of size 1000, half will have no opinion (including those who are unaware of the whole issue) and an equal number of people will consider the protocol effective or ineffective? Let X and Y be the number of sampled individuals in the "effective" and "ineffective" groups respectively. Then the answer will come from the joint p.m.f. of (X, Y). What is it? Since we are dealing with randomly sampled individuals from a huge population, it is reasonable to assume that each person's opinion is independent of everybody else's. Since there are three possible answers, asking each individual about the Kyoto protocol is like tossing an unbalanced "three-sided coin" for which, the probabilities associated with the three sides are 0.25, 0.35 and 0.4. Also, since it is sampling without replacement, these probabilities remain unchanged from "toss" to "toss". So, just as the probability of k heads in n tosses of a regular (i.e., two-sided) coin with $P(\text{head}) = p$ is $_nC_k p^k (1-p)^{n-k}$ (see 11.9), where the coefficient $_nC_k$ is nothing but the number of different rearrangements of k identical H's and $n - k$ identical T's, the answer in the present case will be

$$P(X = 250, Y = 250) = \frac{1000!}{(250!)(250!)(500!)} (0.25)^{250}(0.35)^{250}(1-0.25-0.35)^{500},$$

where the coefficient in front is the number of different rearrangements of 250 identical "yes" answers, 250 identical "no" answers and 500 identical "no opinion" answers. In general, when the sample-size is n, $P(X = k_1 \text{ and } Y = k_2)$ has the same expression with 250 and 250 replaced by k_1 and k_2 for any k_1 and k_2 that add up to something $\leq n$. This is known as the *Trinomial*(n, p_1, p_2) p.m.f. (here $n = 1000, p_1 = 0.25$ and $p_2 = 0.35$). It can be easily extended to the situation where the question has $m(> 3)$ possible answers and the true (but unknown) population-proportions associated with those answers are $p_1, \ldots, p_m$ respectively ($\sum_{i=1}^{m} p_i = 1$). If X_i counts the number of sampled individuals that gave the ith answer ($i = 1, \ldots, m - 1$), then the joint p.m.f. of the X_i's looks like

$$P(x_1, \ldots, x_{m-1}) = \frac{n!}{(x_1!) \ldots (x_{m-1}!)((n - \sum_{i=1}^{m-1} x_i)!)} \left(\prod_{i=1}^{m-1} p_i^{x_i} \right) p_m^{n - \sum_{i=1}^{m-1} x_i},$$

for any nonnegative integers $x_1, \ldots, x_{m-1}$ with $\sum_{i=1}^{m-1} x_i \leq n$. This is called the *Multinomial*$(n, p_1, \ldots, p_{m-1})$ p.m.f. Marginally, each X_i is Binomial(n, p_i), so that its mean is np_i and variance is $np_i(1 - p_i)$. The covariance between X_i and X_j ($i \neq j$) is $-np_i p_j$, which intuitively makes sense since a large value of X_i will force X_j to be small in order to keep the sum constant ($= n$).

For continuous random variables, the analogous concepts are defined as follows. The joint p.d.f. $f(x_1, \ldots, x_n)$ of the continuous random variables $X_1, \ldots, X_n$ is a nonnegative-valued function defined on a subset D of $\mathcal{R}^n$ such that its n-fold integral over D is 1. The subset D, which is the Cartesian product of the value-ranges of the individual X_i's, is called the support of this joint p.d.f. The marginal joint p.d.f. of the random variables $X_{i_1}, \ldots, X_{i_k}$ is obtained by integrating the joint p.d.f. with respect to all the other variables over their full ranges. The conditional joint p.d.f. of a subcollection of random variables $\{X_{i_1}, \ldots, X_{i_k}\}$ given another disjoint subcollection $\{X_{j_1}, \ldots, X_{j_l}\}$ is obtained by dividing the marginal joint p.d.f. of the combined collection $X_{i_1}, \ldots, X_{i_k}, X_{j_1}, \ldots, X_{j_l}$ by that of $\{X_{j_1}, \ldots, X_{j_l}\}$ only. The marginal and conditional joint moments are the expected values of products of random variables computed using their marginal and conditional joint p.m.f.'s respectively. Pairwise covariances and correlations are defined in the same way as in (11.18) and (11.19). Here is an example.

Example 11.22. A random vector $\mathbf{X} = (X_1, X_2)$ is said to have a *bivariate normal* joint p.d.f. with mean vector $\boldsymbol{\mu} = (\mu_1, \mu_2)$ and dispersion matrix $\Sigma = \begin{bmatrix} \sigma_1^2 & \rho\sigma_1\sigma_2 \\ \rho\sigma_1\sigma_2 & \sigma_2^2 \end{bmatrix}$ for some $\sigma_1 > 0$, $\sigma_2 > 0$ and $\rho \in (-1, 1)$ if its joint p.d.f. looks like

$$f(x_1, x_2) = \frac{1}{2\pi\sigma_1\sigma_2(1 - \rho^2)^{1/2}} \, e^{-\frac{1}{2}(\mathbf{X}-\boldsymbol{\mu})\Sigma^{-1}(\mathbf{X}-\boldsymbol{\mu})^t},$$

the superscript "t" meaning "transpose." The exponent $-\frac{1}{2}(\mathbf{X} - \boldsymbol{\mu})\Sigma^{-1}(\mathbf{X} - \boldsymbol{\mu})^t$

$$= -\frac{1}{2(1 - \rho^2)} \left[\left(\frac{x_1 - \mu_1}{\sigma_1}\right)^2 - 2\rho\frac{(x_1 - \mu_1)(x_2 - \mu_2)}{\sigma_1\sigma_2} + \left(\frac{x_2 - \mu_2}{\sigma_2}\right)^2 \right]$$

is a *quadratic form* in x_1 and x_2. It can be shown by integrating this joint p.d.f. with respect to x_2 over its full range $(-\infty, \infty)$ that the marginal p.d.f. of X_1 is Normal(μ_1, σ_1^2). Similarly, integrating out the variable x_1 will yield the marginal p.d.f. of X_2, which is Normal(μ_2, σ_2^2). From the structure of the dispersion matrix, it should be clear that the correlation between X_1 and X_2 is ρ. Also, the conditional p.d.f. of X_2 given $X_1 = x_1$ is normal with mean and variance

$$E(X_2 \mid X_1 = x_1) = \mu_2 + \rho\frac{\sigma_2}{\sigma_1}(x_1 - \mu_1); \; \mathrm{Var}(X_2 \mid X_1 = x_1) = \sigma_2^2(1 - \rho^2).$$

The above formula for $E(X_2 \mid X_1 = x_1)$ is called the *regression* of X_2 on X_1. Likewise, the conditional p.d.f. of X_1 given $X_2 = x_2$ will be normal with mean $E(X_1 \mid X_2 = x_2) = \mu_1 + \rho\frac{\sigma_1}{\sigma_2}(x_2 - \mu_2)$ and variance $\mathrm{Var}(X_1 \mid X_2 = x_2) = \sigma_1^2(1-\rho^2)$. All these can be generalized to an $n \times 1$ random vector $(X_1, \ldots, X_n)$ whose joint p.d.f. will be called *multivariate normal* (MVN) with mean vector $\boldsymbol{\mu} = (\mu_1, \ldots, \mu_n)$ and dispersion matrix $\Sigma_{n \times n}$. The $(i, j)^{th}$ and $(j, i)^{th}$ entries of Σ would both be $\rho_{ij}\sigma_i\sigma_j$, where ρ_{ij} is the correlation between X_i and X_j. Once again, the marginal p.d.f.'s and the conditional p.d.f.'s will all be normal. The MVN($\boldsymbol{\mu}, \Sigma$) joint p.d.f. looks like

$$f(x_1,\ldots,x_n) = \frac{1}{(2\pi)^{n/2}(|\ \Sigma\ |)^{1/2}}e^{-\frac{1}{2}(\mathbf{X}-\boldsymbol{\mu})\Sigma^{-1}(\mathbf{X}-\boldsymbol{\mu})^t},$$

where $|\ \Sigma\ |$ denotes the *determinant* of Σ (see, for example, Leon (1998) for the general definition of the determinant of a $n \times n$ matrix).

This MVN($\boldsymbol{\mu}, \Sigma$) p.d.f. and its univariate version (11.14) are members of a much more general class that includes many other familiar p.d.f.'s and p.m.f.'s such as (11.13), (11.15) and (11.16). It is called the *exponential family* of distributions. An $n \times 1$ random vector $\mathbf{X}$ is said to have a distribution in the exponential family if its p.d.f. (or p.m.f.) has the form

$$f(\mathbf{x}) = c(\boldsymbol{\theta})d(x)e^{\sum_{i=1}^{k}\pi_i(\boldsymbol{\theta})t_i(\mathbf{X})} \tag{11.20}$$

for some (possibly vector) parameter $\boldsymbol{\theta}$ and some functions $\pi_1,\ldots,\pi_k, t_1,\ldots,t_k$, c and d. The vector $(\pi_1(\boldsymbol{\theta}),\ldots,\pi_k(\boldsymbol{\theta}))$ is called the *natural parameters* for (11.20). Often (11.20) is written in a slightly different way as:

$$f(\mathbf{x}) = \exp\{\sum_{i=1}^{k}\pi_i(\boldsymbol{\theta})t_i(\mathbf{x}) + C(\boldsymbol{\theta}) + D(\mathbf{x})\}, \tag{11.21}$$

where $C(\boldsymbol{\theta}) = \ln c(\boldsymbol{\theta})$ and $D(\mathbf{x}) = \ln d(\mathbf{x})$. This is known as the *canonical form* of the exponential family. Can you identify the natural parameters $\boldsymbol{\theta}$ and the functions $C(.), D(.), \pi_i(.)$ and $t_i(.)$ ($i = 1,\ldots,n$) for the p.m.f.'s and p.d.f.'s in (11.13)–(11.16)?

Two quick notes before we close our discussion of multivariate random vectors. From the definition of covariance, it should be clear that two independent random variables have zero covariance (and hence, zero correlation). So, if the components of a random vector $\{X_1,\ldots,X_n\}$ are independent random variables, their dispersion matrix Σ is diagonal (i.e., has zeroes in all off-diagonal positions). However, the converse is not necessarily true. There exist random vectors with dependent components that have diagonal dispersion matrices (can you construct such an example?). Also, for a random vector with independent components, the joint p.m.f. (or p.d.f.) turns out to be the *product* of the individual p.m.f.'s (or p.d.f.'s). And in general, the joint p.d.f. $f(x_1,\ldots,x_n)$ of a continuous random vector $(X_1,\ldots,X_n)$ can be written as

$$f(x_1)f(x_2\mid X_1 = x_1)f(x_3\mid X_1 = x_1\&X_2 = x_2)\ldots f(x_n\mid X_1 = x_1,\ldots,X_{n-1}$$
$$\tag{11.22}$$

$= x_{n-1})$, which is analogous to (11.2).

Next we turn to the relation between the p.m.f.'s (or p.d.f.'s) of two random variables that are functionally related. We will explore some methods of deriving the p.m.f. (or p.d.f.) of a function of a random variable that has a familiar p.m.f. or p.d.f. If we were only interested in computing expected values, this would not be essential, since for a discrete random variable X with values $x_1, x_2, \ldots$ and corresponding probabilities $p_1, p_2, \ldots$, the expected value of $Y = g(X)$ is simply $E(g(X)) = \sum_i g(x_i)p_i$. Likewise, for a continuous random variable X with support D and p.d.f. $f(x)$, we have $E(g(X)) = \int_D g(x)f(x)dx$. But sometimes we do need to know the exact p.m.f. or p.d.f. of $g(X)$ and there are three main methods to derive it from that of X: (a) the *c.d.f.* method, (b) the *Jacobian* method and (c) the *MGF* method. We illustrate each of these by means of an example.

Example 11.23. Let X have an Exponential(β) p.d.f. Suppose we want to derive the p.d.f. of $Y = aX + b$ for some $a > 0$ and $b \in \mathcal{R}$. Let us start with its c.d.f. $G(t)$ (say). $G(t) = P(Y \leq t) = P(aX + b \leq t) = P(X \leq \frac{t-b}{a}) = 1 - e^{-\frac{(t-b)/a}{\beta}}$. So the p.d.f. of Y will be $G'(t) = \frac{1}{a\beta}e^{-\frac{(t-b)}{a\beta}}$. This is sometimes called the *negative exponential* p.d.f. with location parameter b and scale parameter $a\beta$.

Example 11.24. We will once again derive the p.d.f. of $Y = aX+b$ where $X \sim$ Exponential(β), but by a different method this time. If we call the mapping $X \longrightarrow aX + b = Y$ the *forward* transformation, the *inverse* transformation will be $Y \longrightarrow \frac{Y-b}{a} = X$. The *Jacobian* of this inverse transformation is the quantity $\left| \frac{d}{dy} \frac{y-b}{a} \right|$, which is simply $\frac{1}{a}$ here. Now, if we write the p.d.f. of X having replaced all the x's in it by $\frac{y-b}{a}$ and then multiply it with the Jacobian obtained above, we get $(\frac{1}{\beta}e^{-\frac{(y-b)/a}{\beta}})(\frac{1}{a})$. This is precisely the p.d.f. we got in Example 11.23.

Example 11.25. It was mentioned earlier that moment generating functions have a one-to-one correspondence with p.m.f.'s (or p.d.f.'s), so that "MGF fingerprinting" is possible. For instance, an MGF of the form $M(t) = e^{t\mu + t^2\sigma^2/2}$ corresponds only to a Normal(μ, σ^2) p.d.f. So, if $X \sim$ Normal(μ, σ^2), what is the p.d.f. of $Y = aX + b$ for some $a > 0$ and $b \in \mathcal{R}$? Let us first find its MGF. $M_Y(t) = E(e^{tY}) = E(e^{t(aX+b)}) = E(e^{taX}e^{tb}) = e^{tb}E(e^{taX})$, the last equality following from the linearity property of expected values mentioned earlier in this section. But $E(e^{taX})$ is nothing but the MGF of X evaluated at "ta," i.e., it is $M_X(ta) = e^{ta\mu + t^2a^2\sigma^2/2}$. So $M_Y(t) = e^{tb}e^{ta\mu + t^2a^2\sigma^2/2} = e^{t(a\mu+b)+t^2a^2\sigma^2/2}$. This corresponds only to a Normal($a\mu+b, a^2\sigma^2$) p.d.f., which must be the p.d.f. of Y.

Example 11.26. Let X be a continuous random variable with a p.d.f. $f(x)$ and a (strictly monotonically increasing) c.d.f. $F(t)$. What will be the p.d.f. of $Y = F(X)$? Let us derive it via the c.d.f. method. The c.d.f. of Y is $G(t) = P(Y \leq t) = P(F(X) \leq t)$. Take some $t \in (0, 1)$. If F^{-1} is the inverse function of F (two functions $h(x)$ and $k(x)$ are called inverses of one another if $h(k(x)) = x$ for all x in the domain of k and $k(h(x)) = x$ for all x in the domain of h), then $P(F(X) \leq t) = P(F^{-1}(F(X)) \leq F^{-1}(t)) = P(X \leq F^{-1}(t))$. But this is, by definition, $F(F^{-1}(t))$, which is nothing but t. In other words, $G(t) = t$ for all $t \in (0, 1)$. Also, obviously, $G(t) = 0$ for $t \leq 0$ and $G(t) = 1$ for $t \geq 1$. Can you identify this c.d.f.? It should be clear from (11.6) that $G(t)$ is the Uniform(0,1) c.d.f. (verify it through differentiation). In other words,

$F(X)$ has a Uniform(0,1) p.d.f. The transformation $X \longrightarrow F(X)$ is known as the *probability integral transform*. Turning it around, let us now start with an $X \sim$ Uniform(0,1) and let $Y = F^{-1}(X)$. What distribution will it have? Its c.d.f. is $P(Y \le t) = P(F^{-1}(X) \le t) = P(F(F^{-1}(X)) \le F(t)) = P(X \le F(t)) = F(t)$, the last equality following from the nature of the Uniform(0,1) c.d.f. Hence we conclude that Y has the c.d.f. $F(t)$.

This last example is particularly important in the context of statistical simulation studies, since it enables one to generate random samples from various probability distributions that have closed-form expressions for their c.d.f.'s. In order to generate n i.i.d. random variables $X_1, \ldots, X_n$ from the Exponential(β) p.d.f., simply generate n i.i.d. Uniform(0,1) random variables $U_1, \ldots, U_n$ (which computers can readily do) and set $X_i = F^{-1}(U_i)$ $(i = 1, \ldots, n)$, where $F(t) = 1 - e^{-x/\beta}$. Even if the c.d.f. is not strictly monotonically increasing (so that its inverse does not exist), as is the case for discrete random variables, the above technique can be slightly modified to serve the purpose.

We conclude this section by listing a number of useful probability inequalities that are relevant to our book. The second one is a generalization of the Cauchy-Schwarz inequality mentioned earlier.

Result 11.3. Let X and Y be random variables such that $E(|X|^r)$ and $E(|Y|^r)$ are both finite for some $r > 0$. Then $E(|X + Y|^r)$ is also finite and $E(|X + Y|^r) \le c_r(E(|X|^r) + E(|Y|^r))$, where $c_r = 1$ if $0 < r \le 1$ or $c_r = 2^{r-1}$ if $r > 1$. This is often called the c_r inequality.

Result 11.4. Let $p > 1$ and $q > 1$ be such that $\frac{1}{p} + \frac{1}{q} = 1$. Then $E(|XY|) \le (E(|X|^p))^{1/p}(E(|Y|^q))^{1/q}$. This is well known as the Holder inequality. Clearly, taking $p = q = 2$, we get the Cauchy-Schwarz inequality.

Result 11.5. For any number $p \ge 1$, $(E(|X+Y|^p))^{1/p} \le (E(|X|^p))^{1/p} + (E(|Y|^p))^{1/p}$. This is popularly known as the Minkowski inequality.

Result 11.6. $P(|X| \ge \varepsilon) \le \frac{E(|X|^r)}{\varepsilon^r}$ for any $\varepsilon > 0$ and $r > 0$ such that $E(|X|^r)$ is finite. This is known as Markov's inequality. If Y is a random variable with mean μ and standard deviation σ, then letting $Y - \mu$ and $c\sigma$ play the roles of X and ε respectively in the above inequality with $r = 2$, we get: $P(|Y - \mu| \ge c\sigma) \le \frac{1}{c^2}$. This is called the Chebyshev-Bienayme inequality (or simply Chebyshev's inequality).

11.3   Basics of stochastic processes

A sequence of random variables $\{X_1, X_2, X_3, \ldots\}$ with the same support or value-set (say, $V = \{v_1, v_2, v_3, \ldots\}$) is called a *stochastic process*. V is called its *state space*. The index $i = 1, 2, 3, \ldots$ usually refers to time. Many stochastic processes have the property that for any indices i and j, the conditional probability of $X_i = v_j$ given the entire past history (i.e., $X_1, \ldots, X_{i-1}$) is only a function of $X_{i-1}, \ldots, X_{i-m}$ for some fixed m. If $m = 1$, the stochastic process is called a *Markov chain* (after the famous Russian mathematician A.A. Markov). For a Markov chain, the conditional probabilities of state-to-state transitions, i.e., $P(X_i = v_j \mid X_{i-1} = v_k) = p_{kj}^{(i)}$ (say) are often assumed to be independent of the time-index i so we can drop the superscript "(i)." In that case, it is called a *time-homogeneous* Markov chain. For such a chain, these p_{kj}'s can be arranged in a matrix A whose $(k, j)^{th}$ entry is p_{kj}. This A is called the *one-step transition matrix*. Each row of A sums up to 1, since there will definitely be a transition to *some* state from the current state. Such a matrix is called a *row stochastic* matrix. For some Markov chains, the columns of A also add up to 1 (i.e., a transition is guaranteed to each state from some other state). In that case, it is called a *doubly stochastic* matrix. Notice that, although the current state of a Markov chain only has a direct influence on where the chain will be at the very next time-point, it is possible to compute probabilities of future transitions (e.g., 2-step or 3-step transitions) by multiplying the matrix A with itself repeatedly. For example, the $(k, j)^{th}$ entry of A^2 is $\sum_i p_{ki} p_{ij}$ and this is nothing but $P(X_{n+2} = v_j \mid X_n = v_k)$. To understand why, write it as

$$\frac{P(X_{n+2} - v_j \ \& \ X_n = v_k)}{P(X_n = v_k)} = \frac{\sum_i P\{(X_{n+2} = v_j) \cap (X_{n+1} = v_i) \cap (X_n = v_k)\}}{P(X_n = v_k)}$$

and notice that the latter quantity is nothing but

$$\sum_i P\{X_{n+2} = v_j \mid (X_{n+1} = v_i) \cap (X_n = v_k)\} P(X_{n+1} = v_i \mid X_n = v_k),$$

which, by the Markov property, is $\sum_i P(X_{n+2} = v_j \mid X_{n+1} = v_i) P(X_{n+1} = v_i \mid X_n = v_k) = \sum_i p_{ij} p_{ki}$. Similarly, it can be shown that $P(X_{n+m} = v_j \mid X_n = v_k)$ is the $(j, k)^{th}$ entry of A^m. Now we will see some examples.

Example 11.29. In a laboratory, a certain kind of bacteria are grown in a nutrient-rich medium and monitored every half hour around the clock. The bacteria can either be in a dormant state (call it state 1) or a vegetative growth state (state 2) or a rapid cell division state (state 3), or it can be dead (state 4). Suppose it is reasonable to assume that the observed state of the bacteria colony at any particular inspection time is determined solely by its state at the previous inspection time. If the colony is in state 1 now, there is a 60% chance that it will move to state 2 in the next half hour, a 30% chance that it will move to state 3 and a 10% chance that it will move to state 4. If it is in state 2 at the moment, the chances of it moving to state 1, state 3 and state 4 in the next half hour are 70%, 25% and 5% respectively. If it is in state 3 at present, the chances of it moving to state 1, state 2 and state 4 are 20%, 78% and 2% respectively. However, once it gets to state 4, there is no moving back to any other state (such a state is known as an *absorbing barrier*). So the one-step transition probability matrix for this Markov chain is

$$
A = \begin{bmatrix}
0 & 0.6 & 0.3 & 0.1 \\
0.7 & 0 & 0.25 & 0.05 \\
0.2 & 0.78 & 0 & 0.02 \\
0 & 0 & 0 & 1
\end{bmatrix}
$$

Notice that the above transition scheme does not allow any "self-loop" for the states 1, 2 and 3, but it would if the first three diagonal entries were positive. In any case, the two-step transition probabilities will be given by A^2 which is

$$
\begin{bmatrix}
0 & 0.6 & 0.3 & 0.1 \\
0.7 & 0 & 0.25 & 0.05 \\
0.2 & 0.78 & 0 & 0.02 \\
0 & 0 & 0 & 1
\end{bmatrix}
\begin{bmatrix}
0 & 0.6 & 0.3 & 0.1 \\
0.7 & 0 & 0.25 & 0.05 \\
0.2 & 0.78 & 0 & 0.02 \\
0 & 0 & 0 & 1
\end{bmatrix}
=
\begin{bmatrix}
0.48 & 0.234 & 0.15 & 0.136 \\
0.05 & 0.615 & 0.21 & 0.125 \\
0.546 & 0.12 & 0.255 & 0.079 \\
0 & 0 & 0 & 1
\end{bmatrix}
$$

Example 11.30. Here is a Markov chain with a countably infinite state space. Suppose an ant is crawling on an infinite sheet of graphing paper that has an one-inch grid (i.e., has horizontal and vertical lines printed on it with each two consecutive horizontal or vertical lines being 1 inch apart). Starting from the origin at time 0, the ant crawls at a speed of an inch a minute and at each grid-point (i.e., the intersection of a vertical and a horizontal line), it randomly chooses a direction from the set { left, right, vertically up, vertically down }. If we observe the ant at one-minute intervals and denote its position after i minutes by X_i, then $\{X_1, X_2, \ldots\}$ is a Markov chain with state space $\{(j, k) : j \in \mathcal{Z}, k \in \mathcal{Z}\}$, where $\mathcal{Z}$ is the set of all integers. Given that it is currently at the grid-point (j^*, k^*), the one-step transition probability to any of its four nearest neighbor grid-points is $\frac{1}{4}$ and the rest are all zeroes. This is known as a *two-dimensional random walk*.

The k-step transition probability matrix $(k \geq 1)$ of a Markov chain is filled with the *conditional* probabilities of moving from one state to another in k steps, but what about the *unconditional* probabilities of being in various states after k steps? To get those, we need an *initial distribution* for the chain which specifies the probabilities of it being in various states at time 0. In our bacteria colony example, suppose the probability that the colony is in a dormant state at time 0 is $\pi_0(1) = 0.7$, that it is in a vegetative growth state is $\pi_0(2) = 0.2$ and that it is in a rapid cell-division state is $\pi_0(3) = 0.1$ ($\pi_0(4)$ being zero). Let us write it as $\pi = (0.7, 0.2, 0.1, 0)$. Then the (unconditional) probability that the chain will be in a dormant state (state 1) at time 1 is $\sum_{i=1}^{4} P(\text{state 1 at time 1 \& state i at time 0})$

$$
= \sum_{i=1}^{4} P(\text{state 1 at time 1} \mid \text{state i at time 0}) P(\text{state i at time 0}), \quad (11.23)
$$

which is nothing but $\sum_{i=1}^{4} a_{i1} \pi_0(i)$ or the first entry of the 4×1 vector πA. In this case, it is 0.16. Similarly, the (unconditional) probability that the chain will be in state j ($j = 2, 3, 4$) is the jth entry of πA. It should be clear from

(11.23) that the (unconditional) probabilities of being in various states at time k are given by πA^k.

A Markov chain with a finite state-space is said to be *regular* if the matrix A^k has all nonzero entries for some $k \geq 1$. One can prove the following important result about regular Markov chains (see, for example, Bhat (1984)):

Result 11.7. Let $\{X_1, X_2, \ldots\}$ be a regular Markov chain with an $m \times m$ one-step transition probability matrix A. Then there exists an $m \times m$ matrix A^* with identical rows and nonzero entries such that $\lim_{k\to\infty} A^k = A^*$.

The entries in any row of this limiting matrix (with identical rows) are called the *steady-state* transition probabilities of the chain. Let us denote the $(i, j)^{th}$ entry a_{ij}^* of this matrix simply by a_j^*, since it does not vary with the row-index i. The vector $(a_1^*, \ldots, a_m^*)$ represents a p.m.f. on the state space, since $\sum_{j=1}^m a_j^* = 1$. It is called the *steady state distribution* (let us denote it by Π). If k is large enough so that A^k is equal to or very close to A^*, multiplying A^k with A will yield A^* again because $A^k A = A^{k+1}$ will also be exactly or approximately equal to A^*. Since the rows of A^* are nothing but Π, this explains the following result:

Result 11.8. Let $\{X_1, X_2, \ldots\}$ be a regular Markov chain with one-step transition matrix A. Its steady-state distribution Π can be obtained by solving the system of equations: $\Pi A = \Pi$, $\Pi \mathbf{1}^t = 1$ (where $\mathbf{1}^t$ is the $m \times 1$ column vector of all ones).

Example 11.30. Consider a two-state Markov chain with $A = \begin{bmatrix} 0.6 & 0.4 \\ 0.2 & 0.8 \end{bmatrix}$.

In order to find its steady-state distribution, we need to solve the equations $0.6a_1^* + 0.2a_2^* = a_1^*$, $0.4a_1^* + 0.8a_2^* = a_2^*$ and $a_1^* + a_2^* = 1$. The solutions are $a_1^* = \frac{1}{3}$ and $a_2^* = \frac{2}{3}$.

We conclude our discussion of Markov chains by stating a few more results and introducing a few more concepts.

Result 11.9. Let $\{X_1, X_2, \ldots\}$ be a regular Markov chain with steady-state distribution $\Pi = (a_1^*, \ldots, a_m^*)$. If τ_i denotes the time taken by the chain to return to state i, given that it is in state i right now, then $E(\tau_i) = \frac{1}{a_i^*}$ for $i = 1, \ldots, m$.

Result 11.10. Recall from Example 11.29 that a state i is called an *absorbing barrier* if the $(i, i)^{th}$ entry in the one-step transition probability matrix A is 1. Let B be the set of all absorbing barriers in the (finite) state space of a Markov chain. Assume that there is a *path* from every state outside B to at least one state in B (a path being a sequence of states $i_1 i_2 \ldots i_L$ such that $p_{i_j i_{j+1}} > 0$ for $1 \leq j \leq L-1$). Then, letting ν_s denote the time needed by the chain to go from a state s outside B to some state in B, we have: $P(\nu_s \leq k) = \sum_{r \in B} p_{ir}^{(k)}$, where $p_{ir}^{(k)}$ is the $(i, r)^{th}$ element of A^k. This ν_s is often called the *time to absorption* of a non-absorbing state s.

For any state s in the state space of a Markov chain $\{X_1, X_2, \ldots\}$, define $T_s^{(0)} = 0$ and $T_s^{(k)} = \inf\{n > T_s^{(k-1)} : X_n = s\}$ for $k \geq 1$. This $T_s^{(k)}$ is usually known as the time of the kth return to s. Let $\eta_{rs} = P(T_s^{(1)} < \infty \mid X_0 = r)$.

Then intuitively it should be clear that $P(T_s^{(k)} < \infty \mid X_0 = r) = \eta_{rs}\eta_{ss}^{(k-1)}$. In other words, if we start at r and want to make k visits to s, we first need to go to s from r and then return $k - 1$ times to s. A state s is said to be *recurrent* if $\eta_{ss} = 1$. If $\eta_{ss} < 1$, it is called *transient*. A subset Δ of the state space is called *irreducible* if $r \in \Delta$ and $s \in \Delta \Longrightarrow \eta_{rs} > 0$. A subset Δ^* of the state space is called *closed* if $r \in \Delta^*$ and $\eta_{rs} > 0 \Longrightarrow s \in \Delta^*$.

Result 11.11. (Contagious nature of recurrence). If r is a recurrent state and $\eta_{rs} > 0$, then s is recurrent as well and $\eta_{sr} = 1$.

Result 11.12. (Decomposition of the recurrent states). If $R = \{r : \eta_{rr} = 1\}$ is the collection of recurrent states in a state space, then R can be written as a disjoint union $\cup_i R_i$ where each R_i is irreducible and closed.

A stochastic process $\{X_0, X_1, X_2, \ldots\}$ is called *stationary* if for any two non-negative integers k and n, the joint distributions of the random vectors $(X_0, \ldots, X_n)$ and $(X_k, \ldots, X_{k+n})$ are the same. Let $\{X_0, X_1, X_2, \ldots\}$ be a Markov chain with steady-state distribution $\Pi = (a_1^*, a_2^*, \ldots)$. If the initial distribution (i.e., the distribution of X_0) is also Π, then the chain will be stationary. This, incidentally, is the reason why the steady-state distribution is also often called the *stationary distribution*.

Suppose a Markov chain $\{X_1, X_2, \ldots\}$ has a stationary distribution $\Pi = (a_1^*, a_2^*, \ldots)$ such that $a_i^* > 0$ for all i. It can be shown that all its states will be recurrent and we will have a decomposition of the state space as suggested in Result 11.11. If, in that decomposition, there is only one component (i.e., if the entire sample space is irreducible), the chain is called *ergodic*. Actually, the definition of ergodicity is much more general and under the above conditions, the irreducibility and the ergodicity of a Markov chain are equivalent. Since the full generality is beyond the scope of this book, we use this equivalence to define ergodicity.

11.4   Hidden Markov models

Let us now stretch our imagination a little bit more and think of a scenario where $\{X_1, X_2, \ldots\}$ is a Markov chain but is no longer directly observable. Instead, when the event $\{X_i = s\}$ occurs for any state s, we only get to see a "manifestation" of it. The question is how accurately one can guess (or "estimate") the true state of the underlying Markov chain. More formally, when the underlying chain visits the state s, an observable manifestation M_i is chosen according to a p.m.f. on the set of possible manifestations $\mathbf{M} = \{M_1, \ldots, M_L\}$. This is actually a conditional p.m.f. given s. So there are five items to keep in mind here: the state space $\{s_1, s_2, \ldots, s_m\}$ of the underlying Markov chain, its one-step transition probability matrix A, the set of manifestations $\mathbf{M}$ (also sometimes called an *alphabet* of *emitted letters*), the conditional p.m.f. $\mathbf{P}_i = (p_i(M_1), \ldots, p_i(M_L))$ on $\mathbf{M}$ given the true underlying state s_i and the initial distribution $\boldsymbol{\pi} = (\pi_1, \ldots, \pi_m)$.

Example 11.31. Suppose you are playing a dice game with a partner who occasionally cheats. The rule of the game is that you earn $ 100 from your partner if you roll a 4, 5 or 6 and you pay him/her $ 100 if any of the other three numbers turn up. If played with a "perfect" or "balanced" die, it is a fair game, which is easy to see once you write down the p.m.f. table for your earning and compute the expected value. But your partner, who supplies the die, is occasionally dishonest and secretly switches to an unbalanced die from time to time. The unbalanced die has $P(1) = P(2) = P(3) = \frac{1}{4}$ and $P(4) = P(5) = P(6) = \frac{1}{12}$. If the latest roll has been with the balanced die, he/she will switch to the unbalanced one for the next roll with probability 0.05 (i.e., stay with the balanced one with probability 0.95). On the other hand, the chances of his/her switching back from the unbalanced one to the balanced one is 0.9. Here, if Y_i denotes the outcome of your i^{th} roll, then $\{Y_1, Y_2, \ldots\}$ follows a hidden Markov model (HMM) with an underlying Markov chain $\{X_1, X_2, \ldots\}$ whose state space is { balanced, unbalanced } and one-step transition probability matrix is $A = \begin{bmatrix} 0.95 & 0.05 \\ 0.9 & 0.1 \end{bmatrix}$. The alphabet of emitted letters here is $\{1, 2, 3, 4, 5, 6\}$ and the conditional p.m.f. on it switches between $\{\frac{1}{6}, \frac{1}{6}, \frac{1}{6}, \frac{1}{6}, \frac{1}{6}, \frac{1}{6}\}$ and $\{\frac{1}{4}, \frac{1}{4}, \frac{1}{4}, \frac{1}{12}, \frac{1}{12}, \frac{1}{12}\}$, depending on the true nature of the die used. If the chances are very high (say, 99%) that your partner will not begin the game by cheating, the initial distribution for the underlying chain will be $(\pi_1 = 0.99, \pi_2 = 0.01)$.

For an HMM like this, there are three main questions: (a) Given A, $\mathbf{P}$ and $\boldsymbol{\pi} = (\pi_1, \ldots, \pi_m)$, how to efficiently compute the likelihood (or joint probability) of the observed manifestations $\{Y_1, Y_2, \ldots, Y_K\}$? (b) Given $\{Y_1, Y_2, \ldots\}$, how to efficiently estimate the true state-sequence $\{x_1, x_2, \ldots\}$ of the underlying Markov chain with reasonable accuracy? (c) Given the "connectivity" or "network structure" of the state space of the underlying chain (i.e., which entries of A will be positive and which ones will be 0), how to efficiently find the values of A, $\mathbf{P}$ and $\boldsymbol{\pi}$ that maximize the observed likelihood mentioned in (a)? Here we will briefly outline the algorithms designed to do all these. More details can be found, for example, in Ewens and Grant (2001).

The forward and backward algorithms: The goal is to efficiently compute the likelihood of the observed Y_i's given A, $\mathbf{P}$ and $\boldsymbol{\pi}$. We begin by computing $\gamma_i^{(k)} = P(Y_1 = y_1, \ldots, Y_k = y_k \text{ and } X_k = s_i)$ for $1 \leq i \leq m$ and $1 \leq k \leq K$, because the desired likelihood is nothing but $\sum_{i=1}^{m} \gamma_i^{(K)}$. The first one, $\gamma_i^{(1)}$, is clearly $\pi_i p_i(y_1)$. Then, realizing that

$$\gamma_i^{(k+1)} = \sum_{j=1}^{m} P(Y_1 = y_1, \ldots, Y_{k+1} = y_{k+1}, X_k = s_j \text{ and } X_{k+1} = s_i),$$

we immediately get the *induction* relation: $\gamma_i^{(k+1)} = \sum_{j=1}^{m} \gamma_j^{(k)} p_{ji} p_i(y_{k+1})$ where p_{ji} is the $(j, i)^{th}$ entry of A. This is an expression for $\gamma_i^{(k+1)}$ in terms of $\{\gamma_j^{(k)}; j = 1, \ldots, m\}$. So, first we compute $\gamma_i^{(1)}$ for $i = 1, \ldots, m$ and then, using the induction formula, compute $\gamma_i^{(2)}$ for all i's which will subsequently produce the values of $\gamma_i^{(3)}$'s through the induction formula again. Continuing in this manner, we will get the $\gamma_i^{(K)}$'s for all i's and then we are done. This is called the *forward* algorithm because of the forward induction it uses.

For the *backward* algorithm, our aim is to compute $\delta_i^{(k)} = P(Y_k = y_k, \ldots Y_K = y_K \mid X_k = s_i)$ for $i = 1, \ldots, m$ and $1 \leq k \leq K - 1$. We initialize the process by setting $\delta_j^{(K)} = 1$ for all j's. Then we notice that $\delta_i^{(k-1)} = \sum_{j=1}^{m} p_{ij} p_j(y_{k-1}) \delta_j^{(k)}$, due to the conditional independence of the Y_j's given the state of the underlying chain. Using this backward induction formula, we gradually compute $\delta_i^{(K-1)}$ for $1 \leq i \leq m$, $\delta_i^{(K-2)}$ for $1 \leq i \leq m$, and so forth. Once we get $\delta_i^{(0)}$ for all i's, we multiply each of them by the corresponding initial probability π_i to obtain $P(Y_1 = y_1, \ldots, Y_K = y_K$ and $X_1 = s_i)$. The desired likelihood is nothing but the sum of this quantity over i from 1 to m.

The Viterbi algorithm. Given an observed sequence of manifestations $y_1, \ldots, y_K$, our goal is to efficiently estimate the state-sequence $x_1, \ldots, x_K$ of the underlying chain that has the highest conditional probability of occurring. The Viterbi algorithm first computes

$$\max_{x_1, \ldots, x_K} P(X_1 = x_1, \ldots, X_K = x_k \mid Y_1 = y_1, \ldots, Y_K = y_K) \qquad (11.24)$$

and then traces down a vector $(x_1, \ldots, x_K)$ that gives rise to this maximum. We begin by defining

$$\xi_i^{(k)} = \max_{x_j : 1 \leq j \leq k-1} P(X_1 = x_1, \ldots, X_{k-1} = x_{k-1}, X_k = s_i, Y_1 = y_1, \ldots, Y_k = y_k)$$

for some k $(1 \leq k \leq K)$ and some i $(1 \leq i \leq m)$. For $k = 1$, it is simply $P(X_1 = s_i$ and $Y_1 = y_1)$. Then it is not difficult to see that (11.24) is simply $\max_i \xi_i^{(K)}$ divided by $P(Y_1 = y_1, \ldots, Y_K = y_K)$. So the sequence of states $(x_1, \ldots, x_K)$ that maximizes the conditional probability in (11.24) will also correspond to $\max_i \xi_i^{(K)}$. So we focus on the $\xi_i^{(k)}$'s and compute them using forward induction. Clearly, $\xi_i^{(1)} = \pi_i p_i(y_1)$ for all i. Then, $\xi_i^{(k)} = \max_{1 \leq j \leq m} \xi_j^{(k-1)} p_{ji} p_i(y_k)$ for $k \in \{2, \ldots, K\}$ and $i \in \{1, \ldots, m\}$. Now suppose that $\xi_{i_1}^{(K)} = \max_{1 \leq i \leq m} \xi_i^{(K)}$. Then we put $X_K = s_{i_1}$. Next, if i_2 is the index for which $\xi_i^{(K-1)} p_{i i_1}$ is the largest, we put $X_{K-1} = s_{i_2}$. Proceeding in this manner, we get all the remaining states. This algorithm does not produce *all* the state sequences of the underlying chain that give rise to the maximum in (11.24) – just one of them.

The Baum-Welch algorithm. Here the goal is to efficiently estimate the unknown parameters π_i's, p_{ij}'s and $p_i(M_j)$'s using the observed manifestations $y_1, \ldots, y_K$. Actually, the data we need will be a collection of observed sequences $\{y_1^{(t)}, \ldots, y_K^{(t)}\}_{t=1,2,\ldots}$. We will denote the corresponding true-state sequences for the underlying chain by $\{x_1^{(t)}, \ldots, x_K^{(t)}\}_{t=1,2,\ldots}$. The first step will be to initialize all the parameters at some values chosen arbitrarily from the respective parameter spaces or chosen to reflect any *a priori* knowledge about them. Using these initial values, we compute $\hat{\pi}_i$ = the expected proportion of times in the state s_i at the first time-point, given the observed data,

$$\hat{p}_{ij} = \frac{E(U_{ij}|y_1^{(t)},...,y_K^{(t)})}{E(U_i|y_1^{(t)},...,y_K^{(t)})} \text{ and } \hat{p}_i(M_j) = \frac{E(U_i(M_j)|y_1^{(t)},...,y_K^{(t)})}{E(U_i|y_1^{(t)},...,y_K^{(t)})}, \text{ where } U_{ij} \text{ is the ran-}$$

dom variable counting how often $X_r^{(t)} = s_i$ and $X_{r+1}^{(t)} = s_j$ for some t and r, U_i is the random variable counting how often $X_r^{(t)} = s_i$ for some t and r, and $U_i(M_j)$ is the random variable counting how often the events $\{X_r^{(t)} = s_i\}$ and $\{Y_r^{(t)} = M_j\}$ jointly occur for some t and r. Once these are computed, we set these to be the updated values of the parameters. It can be shown that by doing so, we increase the likelihood function of the observed data $\{Y_1,\ldots,Y_K\}$. As a result, if we continue this process until the likelihood function reaches a local maximum or the increment between successive iterations is very small, we will get the maximum likelihood estimates of the parameters. So we now focus on the computations of the above quantities. Let us define $\zeta_r^{(t)}(i,j)$ as

$$P(X_r^{(t)} = s_i, X_{r+1}^{(t)} = s_j \mid y_1^{(t)},\ldots,y_K^{(t)}) = \frac{P(X_r^{(t)} = s_i, X_{r+1}^{(t)} = s_j, y_1^{(t)},\ldots,y_K^{(t)})}{P(y_1^{(t)},\ldots,y_K^{(t)})}.$$

$$(11.25)$$

Notice that both the numerator and the denominator of (11.25) can be efficiently computed using the forward and backward algorithms. For example, one can write the numerator as

$$P(X_r^{(t)} = s_i, X_{r+1}^{(t)} = s_j \& y_1^{(t)},\ldots,y_K^{(t)}) = \gamma_i^{(r)} p_{ij} p_j(y_{r+1}^{(t)}) \delta_j^{(r+1)}.$$

We then define indicators $I_r^{(t)}(i)$ as $I_r^{(t)}(i) = 1$ only if $X_r^{(t)} = s_i$ (and zero otherwise). It is easy to see that $\sum_t \sum_r I_r^{(t)}(i)$ is the number of times the underlying Markov chain visits the state s_i, whereas $\sum_t \sum_r E(I_r^{(t)}(i) \mid y_1^{(t)},\ldots,y_K^{(t)})$ is the conditional expectation of the number of times the state s_i is visited by the underlying chain given $\{y_1^{(t)},\ldots,y_K^{(t)}\}$. This conditional expectation is actually $= \sum_t \sum_r \sum_{j=1}^m \zeta_r^{(t)}(i,j)$, since

$$E(I_r^{(t)}(i) \mid y_1^{(t)},\ldots,y_K^{(t)}) = P(X_r^{(t)} = s_i \mid y_1^{(t)},\ldots,y_K^{(t)}) = \sum_{j=1}^m \zeta_r^{(t)}(i,j).$$

Likewise, $\sum_t \sum_r \zeta_r^{(t)}(i,j)$ can be shown to be the expected number of transitions from s_i to s_j given the observed manifestations. Finally, defining indicators $J_r^{(t)}(i,M_s)$ as $J_r^{(t)}(i,M_s) = 1$ only if the events $\{X_r^{(t)} = s_i\}$ and $\{Y_r^{(t)} = M_s\}$ jointly occur (and zero otherwise), we realize that $E(J_r^{(t)}(i,M_s) \mid y_1^{(t)},\ldots,y_K^{(t)})$ is the (conditional) expected number of times the tth underlying process $\{X_1^{(t)}, X_2^{(t)},\ldots\}$ is in the state s_i at time r and the corresponding manifestation is M_s, given the observed data $\{y_1^{(t)},\ldots,y_K^{(t)}\}$. So, the $E(U_i(M_s) \mid y_1^{(t)}, \ldots, y_K^{(t)})$ from the previous page is nothing but

$$\sum_t \sum_r E(J_r^{(t)}(i, M_s) \mid y_1^{(t)}, \ldots, y_K^{(t)}) = \sum_t \sum_r \sum_{(t,r):y_r^{(t)}=M_s} \sum_{j=1}^m \zeta_r^{(t)}(i,j).$$

This concludes our discussion of hidden Markov models.

11.5   Frequentist statistical inference

We now turn to various methods of learning about, or drawing inferences regarding, unknown parameters from data *without* assuming any prior knowledge about them. Usually this is done by either *estimating* the parameters or *testing* the plausibility of statements/assertions made about them (called *hypotheses*). Sometimes our goal is just to identify the dataset (among a bunch of datasets) associated with the largest or smallest parameter-value, which is known as a *selection* problem. But here we focus primarily on estimation and hypothesis-testing. As indicated in Section 11.1, the link between the unknown quantities (i.e., the parameters) and the observed data is a probability model which is assumed to have generated the data. Under the simplest setup, the data $\{X_1, X_2, \ldots, X_n\}$ are considered to be an i.i.d. sample from the assumed probability model. In other words, the X_i's are considered independent and identically distributed, each following that probability model. In case the data are discrete (i.e., the values are on a countable grid such as whole numbers), the X_i's will have a common p.m.f.. In the continuous case, they will have a common p.d.f. This p.m.f. or p.d.f. will depend on the unknown parameter(s). We will denote both a p.m.f. and a p.d.f. by $f(x; \boldsymbol{\theta})$, where $\boldsymbol{\theta}$ is a vector of parameters. As defined in an earlier section, the *likelihood function* of the observed data is the joint p.m.f. or p.d.f. of $(X_1, \ldots, X_n)$, which boils down to $\prod_{i=1}^n f(x_i; \boldsymbol{\theta})$ due to independence. The $n \times n$ dispersion matrix of $(X_1, \ldots, X_n)$ is $I_{n \times n}$. Under this setup, we first talk about estimation.

The raw and central moments of $f(x; \boldsymbol{\theta})$ will involve one or more of the parameters. If we have k parameters, sometimes equating the formula for the rth raw moment of $f(x; \boldsymbol{\theta})$ to the corresponding raw moment of the sample $(r = 1, \ldots, k)$ gives us a system of k equations and solving this system, we get the *method of moments* (MM) estimates of the parameters. Notice that, for an i.i.d. sample from a univariate p.m.f. or p.d.f. $f(x; \boldsymbol{\theta})$, if we define a function $\hat{F}_n(t) : \mathcal{R} \to [0, 1]$ as $\hat{F}_n(t) = \frac{\#X_i \text{ that are } \leq t}{n}$, this function satisfies all the properties of a c.d.f. in Result 11.2. This is known as the *empirical c.d.f.* of the data and the moments corresponding to it are called *sample moments*. For instance, the first raw sample moment is the sample mean $\bar{X} = \frac{1}{n} \sum_{i=1}^n X_i$. In general, the rth raw sample moment is $\frac{1}{n} \sum_{i=1}^n X_i^r$. The second central sample moment is the sample variance $s^2 = \frac{1}{n} \sum_{i=1}^n (X_i - \bar{X})^2$. For reasons that will be clear shortly, some people use $n - 1$ instead of n in the sample variance

denominator, which then becomes algebraically the same as $\frac{n\sum_i^n X_i^2 - (\sum_{i=1}^n X_i)^2}{n(n-1)}$.
Here are some examples of MM estimates.

Example 11.32. Let $X_1, \ldots, X_n$ be i.i.d. from a Poisson(λ) p.m.f. Then $E(X_i) = \lambda$ and equating it to the first raw sample moment, we get $\hat{\lambda}_{MM} = \frac{1}{n}\sum_{i=1}^n X_i$.

Example 11.33. Let $X_1, \ldots, X_n$ be i.i.d. from a Gamma(α, β) p.d.f. Then the mean $E(X_i) = \alpha\beta$ and the variance $E(X_i - E(X_i))^2 = \alpha\beta^2$. So, equating these two to the corresponding sample moments, we get:

$$\alpha\beta = \frac{1}{n}\sum_{i=1}^n X_i = \bar{X}; \; \alpha\beta^2 = \frac{1}{n-1}\sum_{i=1}^n (X_i - E(X_i))^2 = s^2.$$

Solving these, we get $\hat{\beta}_{MM} = s^2/\bar{X}$ and $\hat{\alpha}_{MM} = \bar{X}^2/s^2$.

In our discussion of the Viterbi algorithm for hidden Markov models, we saw how to find the true underlying state-sequence that maximizes the probability of joint occurrence of the observed manifestations. If we apply the same principle here and try to find those values of the parameter(s) $\boldsymbol{\theta}$ that maximize the joint probability of occurrence (or the likelihood function) of the observed data $x_1, \ldots, x_n$, they will be called the *maximum likelihood* (ML) estimate(s) of the parameter(s). Usually, once the likelihood is written down as a function of the parameter(s) (treating the observed data as fixed numbers), setting its partial derivative w.r.t. each parameter equal to zero gives us a system of equations. Solving it, we get the critical points of the likelihood surface. Some of these will correspond to local maxima, others to local minima and still others will be saddle points. Using standard detection techniques for local maxima and finding the one for which, the likelihood surface is the highest, we can get ML estimates of the parameter(s). Often the above procedure is applied to the natural logarithm of the likelihood function, which still produces the correct answers because the logarithmic transformation is monotone (i.e., preserves "increasingness" or "decreasingness"). Clearly, the ML estimate of a parameter is not necessarily unique. It will be so if the likelihood surface is, for example, unimodal or log-concave. Here are some examples of ML estimation.

Example 11.34. Let $X_1, \ldots, X_n$ be i.i.d. Poisson(λ). Then the likelihood function is $L(\lambda; x_1, \ldots, x_n) = e^{-n\lambda}\lambda^{\sum_1^n x_i}/\prod_1^n x_i!$. So,

$$\frac{d}{d\lambda}\log L(\lambda; x_1, \ldots, x_n) = -n + \frac{\sum_{i=1}^n x_i}{\lambda} = 0 \Rightarrow \lambda = \frac{1}{n}\sum_{i=1}^n X_i,$$

which indeed corresponds to the global maximum of the log-likelihood function (and hence, of the likelihood function) since $\frac{d^2}{d\lambda^2}\log L(\lambda; x_1, \ldots, x_n) = -\lambda^{-2}\sum_{i=1}^n X_i$ is negative for all $\lambda \in (0, \infty)$. So, at least in this case, the MM estimator of λ coincided with its ML estimator. But this is not true in general.

Example 11.35. Let X_i's be i.i.d. Gamma($1, \beta$), that is, Exponential(β). Then the log-likelihood function is $\log L(\beta; x_1, \ldots, x_n) = -n\log\beta - \frac{\sum_{i=1}^n x_i}{\beta}$, so that $\frac{d}{d\beta}\log L(\beta; x_1, \ldots, x_n) = 0 \Rightarrow \hat{\beta} = \frac{1}{n}\sum_{i=1}^n x_i$.

Example 11.36. Let X_i's be i.i.d. Normal(μ, σ^2). Then the log-likelihood function is $\log L(\mu, \sigma^2; x_1, \ldots, x_n) = -\frac{n}{2}\log(2\pi) - \frac{n}{2}\log\sigma^2 - \frac{\sum_{i=1}^{n}(x_i - \mu)^2}{2\sigma^2}$. We can certainly take its partial derivatives w.r.t. μ and σ^2 and equate them to zero, thereby getting a system of two equations. But in this case, we can play a different trick. For each fixed value of σ^2, we can maximize the log-likelihood w.r.t. μ and then, having plugged in whatever value of μ maximizes it, we can treat it as a function of σ^2 alone and find the maximizing value of σ^2. The first step is easy, once we observe that

$$\sum_{i=1}^{n}(x_i - \mu)^2 = \sum_{i=1}^{n}(x_i - \bar{x} + \bar{x} - \mu)^2 = \sum_{i=1}^{n}(x_i - \bar{x})^2 + \sum_{i=1}^{n}(\bar{x} - \mu)^2,$$

the last equality following from the fact that $\sum_{i=1}^{n}(x_i - \bar{x})(\bar{x} - \mu) = 0$. So it is clear that $\hat{\mu}_{ML} = \bar{x}$. Now, having replaced μ with $\bar{x}$ in the original log-likelihood function, we maximize it via differentiation w.r.t. σ^2 and get $\sigma^2_{ML} = \frac{1}{n}\sum_{i=1}^{n}(x_i - \bar{x})^2$.

Two quick points about ML estimators before we move on. The ML estimator of a one-to-one function of a parameter θ is the same one-to-one function of $\hat{\theta}_{ML}$. This is an advantage over method-of-moments estimators, since the latter do not have this property. Also, as we see in the case of $\hat{\sigma}^2_{ML}$ in Example 11.36, the ML estimator of a parameter is not necessarily *unbiased* (an unbiased estimator being defined as one whose expected value equals the parameter it is estimating). It can be shown that $E(\sigma^2_{ML}) = \frac{n-1}{n}\sigma^2$, so that an unbiased estimator of σ^2 is $\hat{\sigma}^2_U = \frac{n}{n-1}\hat{\sigma}^2_{ML} = \frac{1}{n-1}\sum_{i=1}^{n}(x_i - \bar{x})^2$. This is the reason why the commonly used sample variance formula has $n-1$ instead of n in the denominator.

Another frequently used estimation technique is called the *least squares* method. In real life, often we have multivariate data where each observation is a $k \times 1$ vector of measurements on k variables. Some of these variables are our primary focus, in the sense that we want to study how they vary (or whether they significantly vary at all) in response to changes in the underlying experimental conditions and want to identify all the sources contributing to their variability. These are usually called *response* variables. The other variables in the observation-vector may represent the experimental conditions themselves or the factors contributing to the variability in the responses. These are usually called the *explanatory* variables or *design* variables. Sometimes there is no underlying experiment and all the variables are measurements on different characteristics of an individual or an object. In that case, our primary concern is to determine the nature of association (if any) between the responses and the explanatory variables. For example, we may have bivariate measurements (X_i, Y_i) where the response Y_i is the yield per acre of a certain crop and X_i is the amount of rainfall or the amount of fertilizer used. Or X_i could simply be a *categorical* variable having values 1,2 and 3 (1=heavily irrigated, 2=moderately irrigated, 3=no irrigation). In all these cases, we are primarily interested in the "conditional" behavior of the Y's given the X's. In particular, we want to find out what kind of a functional relationship exists between X and $E(Y \mid X)$. In case X is a continuous measurement (i.e., not an indicator of categories), we may want to investigate whether a linear function such as $E(Y \mid X) = \gamma + \beta X$ is adequate to describe the relation between X and Y (if not, we will think of nonlinear equations such as polynomial or exponential). Such an equation is called the *regression* of Y on X and the adequacy of a linear equation is measured by the *correlation* between them. In case X is a categorical variable with values $1, \ldots, m$, we may investigate if a linear equation such as $E(Y \mid X = i) = \mu + \alpha_i$ is adequate to explain most of the variability in Y (if not, we will bring in the indicator(s) of some other relevant factor(s) or perhaps some additional continuous variables). Such an equation is called a *general linear model*. In case all the X's are categorical, we carry out an *analysis of variance* (ANOVA) for this model. If some X's are categorical and others are continuous (called *covariates*), we perform an *analysis of covariance* (ANCOVA) on this model. If such a linear model seems inadequate even after including all the relevant factors and covariates, the conditional expectation of Y may be related to them in a nonlinear fashion. Perhaps a suitable nonlinear transformation (such as log, arcsine or square-root) of $E(Y \mid \mathbf{X})$ will be related to $\mathbf{X}$ in a linear fashion. This latter modeling approach is known as a *generalized linear model* and the nonlinear transformation is called a *link function*.

In all these cases, the coefficients β, γ, μ and α_i are to be estimated from the data. In the regression scenario with response Y and explanatory variables $X_1, \ldots, X_k$, we write the conditional model for Y given $\mathbf{X}$ as

$$Y_i = \gamma + \beta_1 X_1^{(i)} + \ldots + \beta_k X_k^{(i)} + \epsilon_i \text{ for } i = 1, \ldots, n, \tag{11.26}$$

where the ϵ_i's are zero-mean random variables representing random fluctuations of the individual Y_i's around $E(Y \mid X_1, \ldots, X_k)$. In the ANOVA scenario with one categorical factor (having k levels or categories), we write the conditional model

$$Y_{ij} = \mu + \alpha_i + \epsilon_{ij} \text{ for } 1 \leq i \leq k \text{ and } 1 \leq j \leq n_i \left(\sum_{i=1}^{k} n_i = n \right) \tag{11.27}$$

and in the ANCOVA scenario with a categorical factor X and a continuous covariate Z, we write the conditional model

$$Y_{ij} = \mu + \alpha_i + \beta Z_{ij} + \epsilon_{ij} \text{ for } 1 \leq i \leq k \text{ and } 1 \leq j \leq n_i \left(\sum_{i=1}^{k} n_i = n \right). \tag{11.28}$$

In the simplest case, the ϵ_i's are assumed to be i.i.d. having a Normal$(0, \sigma^2)$ p.d.f. But more complicated models may assume correlated ϵ_i's with unequal variances and/or a non-normal p.d.f. (which may even be unknown). In these latter cases, maximum likelihood estimation of the coefficients is analytically intractable and numerically a daunting task as well. So we resort to the following approach. In the regression case, we look at the *sum of squared errors*

$$\sum_{i=1}^{n}(Y_i - \gamma - \beta_1 X_1^{(i)} - \ldots - \beta_k X_k^{(i)})^2 = \sum_{i=1}^{n} \epsilon_i^2 \qquad (11.29)$$

and try to minimize it w.r.t. γ and the β_i's. It can be done in the usual way, by equating the partial derivatives of (11.29) w.r.t. each of those parameters to zero and solving the resulting system of equations. The solution turns out to be $(\mathbf{X}^t \mathbf{V}^{-1} \mathbf{X})^{-1} \mathbf{X}^t \mathbf{V}^{-1} \mathbf{Y}$, where $\sigma^2 V$ is the variance-covariance matrix of the ϵ_i's. This is known as the *least-squares* estimator. In case the ϵ_i's are assumed to be normal, the MLE's of the parameters $\gamma, \beta_1, \ldots, \beta_k$ actually coincide with their least-squares estimators. In almost all real-life scenarios, the matrix V will be unknown and will, therefore, have to be estimated. In the special case where V is simply $I_{n \times n}$, the least-squares estimator of $(\gamma, \beta_1, \ldots, \beta_k)$ takes the more familiar form $(\mathbf{X}^t \mathbf{X})^{-1} \mathbf{X}^t \mathbf{Y}$.

Example 11.37. In order to verify if a heavily advertised brand of oatmeal indeed affects the blood cholesterol levels of its consumers, it was given to a randomly selected group of 30 volunteers twice a day for a few days. Different volunteers tried it for different numbers of days. Once each volunteer stopped eating the oatmeal, the change in his/her blood cholesterol level (compared to when he/she started this routine) was recorded. If X_i denotes the number of days for which the ith volunteer ate the oatmeal and Y_i denotes the change in his/her blood cholesterol level, the following are the (X_i, Y_i) pairs for the 20 volunteers: (7,1.6), (15,3.1), (5,0.4), (18,3.1), (6,1.2), (10,3.9), (12,3.0), (21,3.9), (8,2.0), (25, 5.2), (14,4.1), (4,0.0), (20,4.5), (11,3.5), (16,5.0), (3,0.6), (21,4.4), (2,0.0), (23,5.5), (13,2.8). First of all, the correlation between X and Y is 0.908, which is an indication of the strong linear association between them. If we write the regression model of Y on X in the matrix notation $\mathbf{Y} = \mathbf{X}\boldsymbol{\beta} + \boldsymbol{\epsilon}$, the response vector $\mathbf{Y}$ will consist of the 20 cholesterol levels, the design matrix $\mathbf{X}$ will look like

$$\begin{bmatrix} 1 & 1 & 1 & 1 & 1 & 1 & 1 & 1 & 1 & 1 & 1 & 1 & 1 & 1 & 1 & 1 & 1 & 1 & 1 & 1 \\ 7 & 15 & 5 & 18 & 6 & 10 & 12 & 21 & 8 & 25 & 14 & 4 & 20 & 11 & 16 & 3 & 21 & 2 & 23 & 13 \end{bmatrix}$$

and the least-squares estimate of $\boldsymbol{\beta}$ will be $(\mathbf{X}^t \mathbf{X})^{-1} \mathbf{X}^t \mathbf{Y} = (0.0165, 0.22626)$. So the regression equation of Y on X is $Y = 0.0165 + 0.22626X$, which indicates that for a one-unit change in X, our best estimate for the change in Y (in the "least squares" sense) is 0.22626. This equation also gives us our best prediction for Y corresponding to a hitherto-unseen value of X. Notice that all of the variability in Y cannot be explained by X, since Y will even vary a little bit for individuals with the same value of X. For example, look at the 8th and the 17th volunteers in the dataset above. The proportion of the overall variability in Y that can be explained by X alone is called the *coefficient of determination* and, in this simple model, it equals the square of the correlation (i.e., 82.45%).

We now move on to another important aspect of these estimators – their own random behavior. After all, an estimator is nothing but a formula involving the random sample $\{X_1, \ldots, X_n\}$ and so it is a random variable itself. Its value will vary from one random sample to another. What p.d.f. will it have? One way of getting an idea about this p.d.f. is to look at a histogram of many, many different values of it (based on many, many different random samples) and try to capture the "limiting" shape of this histogram as the number of bars in it goes to infinity. The p.d.f. of an estimator obtained in this way will be called its *sampling distribution*. For instance, the sampling distribution of the MM estimator in Example 11.32 approaches the shape of a normal (or Gaussian) distribution. Similar is the case for the MLE's in Examples 11.34, 11.35 and 11.36 (for the parameter μ). Is it a mere coincidence that the sampling distributions of so many estimators look Gaussian in the limit? Also, if the limiting distribution is Gaussian, what are its mean and variance? The answer to this latter question is easier, since in all the examples mentioned above, the estimator concerned is the sample mean $\bar{X} = \frac{1}{n}\sum_{i=1}^{n} X_i$. Due to the linearity property of expected values, $E(\bar{X}) = \frac{1}{n}E(\sum_{i=1}^{n} X_i) = \frac{1}{n}\sum_{i=1}^{n} E(X_i)$, so that the mean of the sampling distribution of $\bar{X}$ is the *same* as the mean of each X_i. If Σ denotes the variance-covariance matrix of the random vector $\mathbf{X} = (X_1, \ldots, X_n)$, it can be shown that the variance of any linear combination $\mathbf{a}^t\mathbf{X} = a_1 X_1 + \ldots + a_n X_n$ is nothing but $\mathbf{a}^t\Sigma\mathbf{a}$. In particular, the variance of $\sum_{i=1}^{n} X_i$ is $\sum_{i=1}^{n} \text{VAR}(X_i) + \sum_{i=1}^{n}\sum_{j\neq i} \text{COV}(X_i, X_j)$, which reduces to just $\sum_{i=1}^{n} \text{VAR}(X_i)$ if the X_i's are i.i.d. (i.e., the covariances are zero). Since $\text{VAR}(cY) = c^2\text{VAR}(Y)$ for any random variable Y and any constant c, it easily follows that $\text{VAR}(\frac{1}{n}\sum_{i=1}^{n} X_i) = \frac{1}{n}\text{VAR}(X_i)$ for i.i.d. X_i's. Now, to the question of the sampling distribution being Gaussian.

It turns out that the sampling distributions of the estimators in Examples 11.32, 11.34, 11.35 and 11.36 (for μ) were not Gaussian by coincidence. If X_i's are i.i.d. from a p.m.f. or p.d.f. with finite mean μ and finite variance σ^2, the sampling distribution of $n^{1/2}(\bar{X} - \mu)/\sigma$ will always be standard normal (i.e., $N(0,1)$) in the limit as $n \to \infty$. This is the gist of the so-called *central limit theorem* (CLT) proved by Abraham DeMoivre and Pierre-Simon Laplace. It is one of the most celebrated results in probability theory, with far-reaching consequences in classical statistical inference. Along with its multivariate counterpart for i.i.d. random vectors, this result opens up the possibility of another kind of estimation, called *interval estimation* in the case of a scalar parameter or *confidence-set estimation* in the case of a parameter-vector. The idea is the following. If $\boldsymbol{\theta}$ is the mean of a p.d.f. $f(x; \boldsymbol{\theta}, \sigma^2)$ having variance σ^2, and $X_1, \ldots, X_n$ are i.i.d. observations from this p.d.f., then according to the CLT, $P(-z_{\alpha/2} \leq n^{1/2}(\bar{X} - \boldsymbol{\theta})/\sigma \leq z_{\alpha/2})$ will be approximately $1 - \alpha$ for a "sufficiently large" sample size n, where $z_{\alpha/2}$ is the $100(1 - \alpha/2)^{th}$ percentile of the $N(0,1)$ p.d.f. (i.e., the point beyond which the tail-area of a $N(0,1)$ p.d.f. is $\alpha/2$). In other words,

$$P(\bar{X} - \frac{z_{\alpha/2}\,\sigma}{n^{1/2}} \leq \boldsymbol{\theta} \leq \bar{X} + \frac{z_{\alpha/2}\,\sigma}{n^{1/2}}) = 1 - \alpha,$$

and the interval $(\bar{X} - \frac{z_{\alpha/2}\,\sigma}{n^{1/2}}\ ,\ \bar{X} + \frac{z_{\alpha/2}\,\sigma}{n^{1/2}})$ is called a $100(1-\alpha)\ \%$ *confidence interval* for $\boldsymbol{\theta}$. However, the variance σ^2 will typically be unknown, rendering the computation of the two end-points of this confidence interval impossible. Fortunately, it can be shown that the distribution of $n^{1/2}(\bar{X} - \boldsymbol{\theta})/s$ is also approximately $N(0,1)$ for large values of n, where $s^2 = \frac{1}{n-1}\sum_{i=1}^{n}(X_i - \bar{X})^2$ is the sample variance. As a result, we get the approximate $100(1-\alpha)\ \%$ confidence interval

$$(\bar{X} - \frac{z_{\alpha/2}\,s}{n^{1/2}}\ ,\ \bar{X} + \frac{z_{\alpha/2}\,s}{n^{1/2}}) \tag{11.30}$$

for $\boldsymbol{\theta}$.

Actually, it can be shown using the multivariate version of the "Jacobian of inverse transformation" technique (see Example 11.24) that the p.d.f. of $\frac{U}{(V/\nu)^{1/2}}$, where U and V are two independent random variables following $N(0,1)$ and $\chi^2(\nu)$ respectively, is *Student's t* with ν degrees of freedom. It is a p.d.f. for which extensive tables of percentiles are available. So, in case the sample size n is not large, we should replace $z_{\alpha/2}$ in the above confidence interval formula by $t_{\alpha/2}(\nu)$, provided that the data came from a normal distribution. This is because if $\{X_1, \ldots, X_n\}$ is a random sample from a $N(\boldsymbol{\theta}, \sigma^2)$ distribution, it can be shown using a technique called Helmert's orthogonal transformation (see Rohatgi and Saleh (2001), p 342) that $\bar{X}$ and $(n-1)S^2/\sigma^2$ are independent and, of course, $n^{1/2}(\bar{X} - \boldsymbol{\theta})/\sigma \sim N(0,1)$ and $(n-1)S^2/\sigma^2 \sim \chi^2(n-1)$.

11.6   Bayesian inference

While "learning" to some of us means estimating or testing hypotheses about unknown quantities or parameters (assumed to be fixed) from the data (a random sample from the population concerned), other people interpret the word as updating our existing knowledge about unknown parameters (assumed to be random variables) from the data (considered fixed in the sense that we are only interested in the conditional behavior of the parameters given the data). This is the Bayesian approach. Recall the Bayes theorem from Section 11.2. There were two main ingredients: a likelihood and a bunch of prior probabilities, ultimately yielding a bunch of posterior probabilities. Following the same principle, we will now have a dataset $\{X_1, \ldots, X_n\}$ which, given the parameter $\boldsymbol{\theta} = (\theta_1, \ldots, \theta_p)$, will be conditionally i.i.d. having the common p.d.f. or p.m.f. $f(x \mid \boldsymbol{\theta})$. So their conditional likelihood given $\boldsymbol{\theta}$ is $\prod_{i=1}^{n} f(x_i \mid \boldsymbol{\theta})$. The prior knowledge about $\boldsymbol{\theta}$ will be in the form of a prior p.d.f. or p.m.f. $\pi(\boldsymbol{\theta})$. Depending on the situation, the θ_i's can be assumed to be independent *a priori*, so that $\pi(\boldsymbol{\theta}) = \prod_{i=1}^{p} \pi_i(\theta_i)$ where $\theta_i \sim \pi_i$. Or the θ_i's can be dependent, in which case, $\pi(\boldsymbol{\theta}) = \pi_1(\theta_1)\pi_2(\theta_2 \mid \theta_1) \ldots \pi_p(\theta_p \mid \theta_j, j < p)$. In what follows, we will assume this latter form with the understanding that $\pi_i(\theta_i \mid \theta_j, j < i) = \pi_i(\theta_i)$ in the case of independence. In any case, by (11.22), the joint distribution of

$\{X_1, \ldots, X_n, \theta_1, \ldots, \theta_p\}$ is

$$L(x_1, \ldots, x_n \mid \boldsymbol{\theta})\pi(\boldsymbol{\theta}) = \left(\prod_{i=1}^{n} f(x_i \mid \boldsymbol{\theta})\right) \pi_1(\theta_1)\pi_2(\theta_2 \mid \theta_1) \ldots \pi_p(\theta_p \mid \theta_j, j < p).$$
(11.31)

For the denominator of the Bayes formula, we need the marginal joint p.d.f. or p.m.f. of the X_i's. To get it from (11.31), we need to integrate it (or sum it) w.r.t. the θ_i's over their full ranges. Finally, the posterior joint p.d.f. or p.m.f. of $\theta_1, \ldots, \theta_p$ will be

$$\frac{(\prod_{i=1}^{n} f(x_i \mid \boldsymbol{\theta}))\, \pi_1(\theta_1)\pi_2(\theta_2 \mid \theta_1) \ldots \pi_p(\theta_p \mid \theta_j, j < p)}{\int_{\theta_1} \cdots \int_{\theta_p} (\prod_{i=1}^{n} f(x_i \mid \boldsymbol{\theta}))\, \pi_1(\theta_1)\pi_2(\theta_2 \mid \theta_1) \ldots \pi_p(\theta_p \mid \theta_j, j < p)d\theta_p \ldots d\theta_1}.$$
(11.32)

Let us denote it by $\pi_{\boldsymbol{\theta}\mid\mathbf{X}}(\boldsymbol{\theta})$. Once this is obtained, we could estimate the θ_i's by the values that correspond to one of the posterior modes (i.e., highest points of the posterior surface). Or we could use the set $\{(\theta_1, \ldots, \theta_p) : \pi_{\boldsymbol{\theta}\mid\mathbf{X}}(\boldsymbol{\theta}) \geq 0.95\}$ as a 95% confidence region for $\boldsymbol{\theta}$ (called a *highest-posterior-density credible region* or *HPDCR*). However, due to the high dimensions of the parameter-vectors in real-life problems, computing (11.32) is usually an uphill task – especially the high-dimensional integral in the denominator. Over the years, a number of remedies have been suggested for this. One of them involves being "clever" with your choice of the prior p.d.f. (or p.m.f.) so that its functional form "matches" with that of the data-likelihood and as a result, the product of these two (the joint distribution (11.31)) has a familiar form. Then the denominator of (11.32) does not actually have to be computed, because it will simply be the "normalizing constant" of a multivariate p.d.f. or p.m.f. This is the so-called *conjugacy* trick. When this cannot be done, there are some useful analytical or numerical approximations to high-dimensional integrals (such as the LaPlace approximation) which enables us to approximate the posterior. When none of these works, we can resort to *Markov chain Monte Carlo* (MCMC) techniques. See Section 11.3 for the basic idea behind a Markov chain. There are many variants of it, such as the *Gibbs sampler*, the *Metropolis* (or *Metropolis-Hastings*) algorithm, etc.

Another fundamental principle in the context of our search for a "good" estimator is the *Bayes principle*. If δ is a decision rule (e.g., an estimator, a test function, a classification rule, etc.) and $\mathcal{R}_\delta(\boldsymbol{\theta})$ is the associated risk function with respect to some loss function $\mathcal{L}(\boldsymbol{\theta}, \delta)$, the *Bayes risk* of δ under the prior distribution $\pi(\boldsymbol{\theta})$ is defined as $r(\pi, \delta) = E_\pi[\mathcal{R}_\delta(\boldsymbol{\theta})]$. According to the Bayes principle, we should prefer the decision rule δ_1 to a competing decision rule δ_2 if $r(\pi, \delta_1) < r(\pi, \delta_2)$. If we can find a decision rule δ_π that minimizes $r(\pi, \delta)$ over all decision rules, we call it the *Bayes rule*. The associated Bayes risk $r(\pi, \delta_\pi)$ is often simply called the Bayes risk of π.

For a decision rule δ, a measure of *initial precision* is a quantity computed by averaging over all possible datasets that *might be* observed. Such a measure can be computed even before observing any data. Examples are the risk $\mathcal{R}_\delta(\boldsymbol{\theta})$, the Bayes risk $r(\pi, \delta)$, the MSE of an estimator, the type I and type II error-probabilities of a test, etc. On the other hand, a measure of *final precision* is something that can be computed only after observing the data. Therefore, final precision is a measure *conditional* on the data. In Bayesian learning, final precision is considered the most appropriate way to assess the optimality of a decision rule. To a Bayesian, datasets that might have been (but were not) observed are not relevant to the inference-drawing activity. The only relevant dataset is the one actually observed. Does it mean that the underlying philosophy of Bayesian learning is incompatible with the Bayes principle itself? To resolve this apparent contradiction, we need to distinguish between the *normal* and the *extensive* forms of a Bayes rule. The normal method of finding a Bayes rule is the one discussed above. The extensive method is based on the fact that the Bayes risk $r(\pi, \delta)$ for a continuous parameter $\boldsymbol{\theta} \in \boldsymbol{\Theta}$ is $\int_{\boldsymbol{\Theta}} \mathcal{R}_\delta(\boldsymbol{\theta}) \pi(\boldsymbol{\theta}) d\boldsymbol{\theta}$

$$= \int_{\boldsymbol{\Theta}} [\int_{\mathcal{X}} \mathcal{L}(\boldsymbol{\theta}, \delta(\mathbf{x})) f(\mathbf{x} \mid \boldsymbol{\theta}) d\mathbf{x}] \pi(\boldsymbol{\theta}) d\boldsymbol{\theta} = \int_{\mathcal{X}} [\int_{\boldsymbol{\theta}} \mathcal{L}(\boldsymbol{\theta}, \delta(\mathbf{x})) f(\mathbf{x} \mid \boldsymbol{\theta}) \pi(\boldsymbol{\theta}) d\boldsymbol{\theta}] d\mathbf{x},$$

where the conditions for switching the order of integration are assumed to be satisfied and $\mathcal{X}$ denotes the sample space. So, in order to minimize $r(\pi, \delta)$, all we need to do is find a decision rule that minimizes the integrand (inside square brackets) in the last expression above. If we now denote the denominator of (11.32) by $m(\mathbf{x})$, where $\mathbf{x} = \{x_1, \dots, x_n\}$, it is easy to see that a decision rule will minimize $r(\pi, \delta)$ if it minimizes $\int_{\boldsymbol{\Theta}} \mathcal{L}(\boldsymbol{\theta}, \delta(\mathbf{x})) \left[\frac{f(\mathbf{x}|\boldsymbol{\theta})}{m(\mathbf{X})}\right] \pi(\boldsymbol{\theta}) d\boldsymbol{\theta}$. But this last quantity is nothing but $\int_{\boldsymbol{\Theta}} \mathcal{L}(\boldsymbol{\theta}, \delta(\mathbf{x})) \pi_{\boldsymbol{\theta}|\mathbf{x}}(\boldsymbol{\theta}) d\boldsymbol{\theta}$, which is usually called the *Bayes posterior risk* of the decision rule δ. In other words, a Bayesian always seeks a decision rule that minimizes the Bayes posterior risk (a measure of final precision) and this does not violate the Bayes principle. Next we see some examples.

Example 11.38 Let X be a single observation from a Binomial(n, θ) population, so its p.m.f. is $f(x \mid \theta) = {}_nC_x \theta^x (1-\theta)^{n-x}$. If $\pi(\theta)$ is the prior p.d.f. of θ, its posterior p.d.f. will be

$$\pi_{\theta|\mathbf{x}}(\theta) = \frac{\theta^x (1-\theta)^{n-x} \pi(\theta)}{\int_0^1 s^x (1-s)^{n-x} \pi(s) ds}.$$

As was mentioned earlier, often it will be good enough to know that $\pi_{\theta|\mathbf{x}}(\theta) \propto \theta^x (1-\theta)^{n-x} \pi(\theta)$. In any case, the integral in the denominator of $\pi_{\theta|\mathbf{x}}(\theta)$ turns out to be difficult to calculate analytically even for some simple priors $\pi(\theta)$. But one particular choice for $\pi(\theta)$ renders analytical calculation unnecessary and it is $\pi(\theta) = (1/B(a, b))\theta^{a-1}(1-\theta)^{b-1}$, the Beta$(a, b)$ p.d.f. Then $\pi_{\theta|\mathbf{x}}(\theta) \propto \theta^{x+a-1}(1-\theta)^{n-x+b-1}$ and clearly it is a Beta$(x+a, n-x+b)$ p.d.f. So, based on the observed data, Bayesian learning in this case would simply mean updating the prior p.d.f. parameters from (a, b) to $(x + a, n - x + b)$. This is the so-called "conjugacy trick." The Beta family of prior p.d.f.s is a *conjugate family* for binomial data. But are we restricting our choice for a prior p.d.f. heavily for the sake of convenience? The answer is "no," because the Beta family of densities provides quite a bit of flexibility in modeling the prior knowledge about θ as the parameters $a > 0$ and $b > 0$ vary.

Once the posterior p.d.f. is obtained, how do we use it to draw inference on the parameter θ? Intuitively, a reasonable estimate of θ seems to be the posterior mode. Since the mode of a Beta(s_1, s_2) density is $\frac{s_1-1}{s_1+s_2-2}$ (provided that $s_1 > 1$ and $s_2 > 1$), the posterior mode in the beta-binomial scenario is $\frac{x+a-1}{n+a+b-2}$ as long as $n > 1$ and $0 < x < n$. If in addition, both the prior parameters a and b are larger than 1, we can talk about the prior mode $\frac{a-1}{a+b-2}$. Recall from Section 11.5 that, based on a single observation from Binomial(n, θ), the MLE (as well as UMVUE) of θ is $\frac{X}{n}$. It may be interesting to observe that the posterior estimate of θ (i.e., the posterior mode) is a *weighted average* of the prior mode and the MLE. In other words,

$$\text{posterior mode} = \frac{n}{n+a+b-2} \left(\frac{X}{n}\right) + \left(1 - \frac{n}{n+a+b-2}\right)\left(\frac{a-1}{a+b-2}\right). \quad (11.33)$$

This is not the end of the story. If, instead of just deciding to use the posterior mode as the estimator, we take a decision-theoretic approach and try to find the "best" estimator $\hat{\theta}$ under the squared error loss (in the sense of minimizing the Bayes posterior risk $\int_0^1 (\hat{\theta}-\theta)^2 \pi_{\theta|x}(\theta)d\theta$), the answer is the posterior mean $\int_0^1 x\pi_{\theta|x}(\theta)d\theta$. In the beta-binomial scenario, the posterior mean is $\frac{x+a}{n+a+b}$. Observe once again that

$$\frac{x+a}{n+a+b} = \frac{n}{n+a+b} \left(\frac{X}{n}\right) + \left(1 - \frac{n}{n+a+b}\right)\left(\frac{a}{a+b}\right), \quad (11.34)$$

where X/n is the data mean (also the MLE and the UMVUE) and $a/(a + b)$ is the prior mean. These are examples of what we often see in a Bayesian inference problem: (1) A Bayesian point estimate is a weighted average of a commonly used frequentist estimate and an estimate based only on the prior distribution and (2) the weight allocated to the common frequentist estimate increases to 1 as the sample-size $n \longrightarrow \infty$. It is often said that the Bayesian point estimate *shrinks* the common frequentist estimate towards the exclusively prior-based estimate.

Another method of inference-drawing based on the posterior distribution is to construct a *credible region* for θ. A $100(1 - \alpha)$ % credible region for θ is a subset C of the parameter space Θ such that $P(\theta \in C \mid X = x) \geq 1 - \alpha$. As is the case for frequentist confidence intervals, usually there will be many candidates for C. An "optimal" $100(1 - \alpha)$ % credible region should have the smallest volume among them (or, equivalently, we should have $\pi_{\theta|x}(\theta) \geq \pi_{\theta|x}(\theta')$ for every $\theta \in C$ and every $\theta' \notin C$). This leads us to the concept of a *highest posterior density credible region* (HPDCR). The $100(1-\alpha)$ % HPDCR for θ is a subset C^* of Θ such that $C^* = \{\theta \in \Theta : \pi_{\theta|x}(\theta) \geq k_\alpha\}$, where k_α is the largest real number satisfying $P(\theta \in C^* \mid X = x) \geq 1 - \alpha$.

If there are Bayesian analogs of classical (or frequentist) point estimation and confidence-set estimation, you would expect a Bayesian analog of classical hypothesis testing too. Suppose, that we want to test $H_0 : \theta \in \Theta_0$ against $H_1 : \theta \in \Theta_1$, where Θ_0 and Θ_1 constitute a partition of the parameter space Θ. If we denote $P(\theta \in \Theta_0 \mid X = x)$ by γ, the *posterior odds ratio* is defined as $\frac{\gamma}{1-\gamma}$. Notice that this is a measure of final precision. If π_0 and $1 - \pi_0$ denote respectively the prior probabilities of θ being in Θ_0 and Θ_1, the *prior odds ratio* would be $\frac{\pi_0}{1-\pi_0}$. The *Bayes factor* is defined as the ratio of these two odds ratios. In other words, it is $\frac{\gamma(1-\pi_0)}{(1-\gamma)\pi_0}$. If the Bayes factor is smaller than 1, our degree of posterior belief in H_0 is smaller than that in H_1. Before we move to the next example, here is an interesting observation about the Bayes factor. If both Θ_0 and Θ_1 are singleton (i.e., we are under the Neyman-Pearson theorem setup), the Bayes factor indeed reduces to the likelihood-ratio used for finding the MP test in the Neyman-Pearson theorem.

Example 11.39. Let $\mathbf{X} = \{X_1, \ldots, X_n\}$ be a random sample from a $N(\mu, \sigma^2)$ p.d.f., where μ is unknown but σ^2 is known. Suppose that we choose a $N(\mu_0, \sigma_0^2)$ prior density for μ. Then the posterior density $\pi_{\mu|\mathbf{X}}(\mu)$ will be proportional to

$$\left[\exp\left(-\frac{1}{2\sigma_0^2}(\mu - \mu_0)^2 \right) \right] \left[\exp\left(-\frac{1}{2\sigma^2} \sum_1^n (x_i - \mu)^2 \right) \right],$$

which, in turn, is proportional to

$$\left[\exp\left(-\frac{1}{2\sigma_0^2}(\mu^2 - 2\mu\mu_0) \right) \right] \left[\exp\left(-\frac{n}{2\sigma^2}(\mu^2 - 2\mu\bar{x}) \right) \right].$$

After completing the square in the exponent, one can see that the posterior density of μ is proportional to

$$\exp\left(-\frac{1}{2\sigma_n^2}(\mu - \mu_n)^2 \right), \text{ where } \mu_n = \frac{\bar{x} + \mu_0(\sigma^2/n\sigma_0^2)}{1 + \sigma^2/n\sigma_0^2} \text{ and } \frac{1}{\sigma_n^2} = \left(\frac{1}{\sigma_0^2} + \frac{n}{\sigma^2} \right).$$

$$(11.35)$$

It should now be clear that the posterior density is $N(\mu_n, \sigma_n^2)$. In other words, the normal family of priors is a conjugate family for the mean of normal data. As before, the mode (which coincides with the mean) of this density can be used as a point estimate of μ. Once again, notice that this point estimate is a weighted average of the sample mean $\bar{x}$ and the prior mean μ_0.

Next we shed a little bit of light on the controversial issue of choosing a prior. The primary concerns that guide our choice of a prior p.d.f. (or p.m.f.) are (a) its ability to adequately represent the extent and nature of the prior information available and (b) computational convenience. There are three main ways of choosing a prior p.d.f. or p.m.f.: (a) subjective, (b) objective (informative) and (c) noninformative. A subjective choice is the most controversial, since it exclusively reflects the degree of one's personal belief about $\boldsymbol{\theta}$. For example, often in a real-life scientific experiment, expert's opinion may be available regarding the unknown parameters involved from people who are highly trained or experienced in that field. The problem with eliciting a prior from this kind of opinion is that often experts don't agree and put forward conflicting or contradictory opinions. Objective and informative priors are less controversial since they are based on either historical records about the parameter-values themselves or data from previous experiments that contain information about the parameters. In the latter case, the posterior densities or p.m.f.s obtained from such older datasets can be used as priors for the current problem at hand. Or one could possibly combine the older datasets with that for the current problem and enjoy a much bigger sample-size. The question that naturally comes to mind is when (if at all) these two approaches will yield the same posterior distributions for the parameters in the current study. It will happen only if the older datasets and the current dataset can be considered statistically independent (to be more precise, conditionally independent given the parameters). A noninformative prior is so called because it is supposed to reflect the extent of ignorance about the parameter(s). Such a prior is also sometimes called a *diffuse* prior or a *vague* prior or a *reference* prior. This concept may be confusing at times, since, for example, a *flat* prior (i.e., one that is constant over the parameter space) is not necessarily non-informative just because it is flat. In general, a noninformative prior is one that is "dominated" by the likelihood function in the sense that it does not change much over the region in which the likelihood is reasonably large, and also does not take large values outside that region. Such a prior has also been given the name *locally uniform*. Often we feel that the best reflection of our ignorance about $\boldsymbol{\theta}$ is a constant prior density over an infinite parameter space or a nonconstant one that is so heavy-tailed that it does not integrate to 1. A prior density of this sort is called an *improper prior*. Such a prior density is not automatically disallowed as invalid, because it may still lead to a proper posterior density that integrates to 1. In many complicated real-life problems, Bayesians resort to *hierarchical modeling*. Such models often provide better insights into the dependence structure of the observed data (that may consist of response variables, covariates, etc.) and the unknown parameters and also help break down the overall variability into different layers. For instance, in Example 11.39, assuming both μ and σ^2 to be unknown, we could choose a $N(\mu_0, \sigma_0^2)$ prior for μ and a $Gamma(\alpha, \beta)$ prior (independently of μ) for $1/\sigma^2$. If there is uncertainty in our mind about the parameters $\mu_0, \sigma_0^2, \alpha$ and β, we can capture this uncertainty by imposing prior densities on them (e.g., a normal prior on μ_0, uniform priors on σ_0^2, α and β). These will be called *hyperpriors* and their parameters, *hyperparameters*. We may be reasonably certain about the hyperparameters, or sometimes they are estimated from the observed data (an approach known as *empirical Bayes*). We conclude this discussion with an example of a special noninformative prior that is widely used.

Example 11.40. Suppose that $\boldsymbol{\theta} = (\theta_1, \ldots, \theta_m)$. Recall the definition of a Fisher information matrix (FIM). Let us denote it by $I(\boldsymbol{\theta})$. It can be shown that the $(i,j)^{th}$ entry of the $m \times m$ FIM is $-E\left[\frac{\partial^2 \log f(\mathbf{X}|\boldsymbol{\theta})}{\partial \theta_i \partial \theta_j}\right]$. The prior density $\pi(\boldsymbol{\theta}) \propto \det(I(\boldsymbol{\theta}))^{0.5}$ is known as the *Jeffreys' prior* (where "det" stands for the determinant of a matrix). When our data come from a Binomial(n, θ) distribution, the FIM is just a scalar $(= \frac{n}{\theta(1-\theta)})$. So the Jeffreys' prior in this case will be $\propto [\theta(1-\theta)]^{-0.5}$, which is nothing but a Beta$(\frac{1}{2}, \frac{1}{2})$ density. When our data $\{X_1, \ldots, X_n\}$ come from a Normal(θ_1, θ_2^2) p.d.f., the FIM is a diagonal matrix with diagonal entries n/θ_2^2 and $2n/\theta_2^2$. So the Jeffreys' prior will be $\pi(\theta_1, \theta_2) \propto \frac{1}{\theta_2^2}$ on the upper half of the two-dimensional plane.

We conclude with a brief discussion of Bayesian computation. As indicated earlier, MCMC algorithms play a central role in Bayesian inference. Before we take a closer look at them, let us try to understand the Monte Carlo principle. The idea of Monte Carlo (MC) simulation is to draw a set of i.i.d. observations $\{x_i\}_{i=1}^n$ from a target p.d.f. $f(x)$ on a high-dimensional space $\mathcal{X}$. This sample of size n can be used to approximate the target density with the empirical point-mass function $f_n(x) = \frac{1}{n}\sum_1^n \delta_{x_i}(x)$, where $\delta_{x_i}(x)$ denotes Dirac's delta function that takes the value 1 if $x = x_i$ and 0 otherwise. As a result, one can approximate integrals such as $\int_{\mathcal{X}} g(x)f(x)dx$ with sums like $\frac{1}{n}\sum_1^n g(x_i)$, because the latter is an unbiased estimator of the former and is strongly consistent for it as well (by the *strong law of large numbers*). In addition, the central limit theorem (CLT) gives us asymptotic normality of the MC approximation error. For example, if $f(x)$ and $g(x)$ are univariate functions (i.e., $\mathcal{X}$ is some subset of $\mathcal{R}$) such that the variance of g with respect to f is finite [i.e., $\sigma_g^2 = \int_{\mathcal{X}} g^2(x)f(x)dx - (\int_{\mathcal{X}} g(x)f(x)dx)^2 < \infty$], then according to the CLT,

$$\sqrt{n}\left\{\frac{1}{n}\sum_1^n g(x_i) - \int_{\mathcal{X}} g(x)f(x)dx\right\} \implies N(0, \sigma_g^2)$$

as $n \longrightarrow \infty$, where $\implies$ means convergence in distribution. So it should be clear that in order to take advantage of the MC principle, we must be able to *sample* from a p.d.f. If it is a standard p.d.f. (e.g., normal, exponential or something else whose CDF has a closed-form expression), there are straightforward procedures for sampling from it. However, if it is non-standard (e.g., one with an ugly, irregular shape and no closed-form CDF) or is known only upto a proportionality-constant, sampling from it will be tricky and we need some special technique. One of them is *rejection sampling*, where we sample from a *proposal density* $f^*(x)$ that is easy to sample from and satisfies: $f(x) \leq Kf^*(x)$ for some $K < \infty$. Having sampled an observation x_i^* from f^*, we use the following acceptance-rejection scheme: Generate a random variable $U \sim$ Uniform$[0, 1]$ and accept x_i^* if $U < f(x_i^*)/[Kf^*(x_i^*)]$; otherwise reject it and sample another observation from f^* to repeat this procedure. If we continue this process until we have n acceptances, it can be easily shown that the resulting sample of size n is from $f(x)$. Although it is an easy-to-implement scheme, it has serious drawbacks in the sense that it is often impossible to bound $f(x)/f^*(x)$ from above by a finite constant K uniformly over the entire space $\mathcal{X}$. Even if such a K can be found, often it is so large that the acceptance probability at each step is very low and the process therefore becomes inefficient. So an alternative procedure has been tried. It is known as *importance sampling*. Here we once again introduce a proposal density $f^*(x)$ and realize that the integral $\int_{\mathcal{X}} g(x)f(x)dx$ can be rewritten as $\int_{\mathcal{X}} g(x)w(x)f^*(x)dx$, where $w(x) = f(x)/f^*(x)$ is usually called the *importance weight*. As a result, all we need to do is to draw n i.i.d. observations $x_1^*, \ldots, x_n^*$ from f^* and evaluate $w(x_i^*)$ for each of them. Then, following the MC principle, we will approximate the integral $\int_{\mathcal{X}} g(x)f(x)dx$ by $\sum_1^n g(x_i^*)w(x_i^*)$. This is once again an unbiased estimator and, under fairly general conditions on f and f^*, is strongly consistent as well. Like rejection sampling, this procedure is easy to implement, but choosing an appropriate proposal density may be tricky. Often the criterion that is used for this choice is the minimization of the variance of the resulting estimator $\sum_1^n g(x_i^*)w(x_i^*)$. That variance, computed with respect to the proposal density f^*, is given by

$$\mathrm{var}_{f^*}(g(x)w(x)) = \int_{\mathcal{X}} g^2(x)w^2(x)f^*(x)dx - \left(\int_{\mathcal{X}} g(x)w(x)f^*(x)dx\right)^2.$$

Clearly, minimizing this with respect to f^* is equivalent to minimizing just the first term on the right-hand side. We can apply *Jensen's inequality* (which says that $E(h(Y)) \geq h(E(Y))$ for any nonnegative random variable Y and any convex function h such that $E(Y)$ and $E(h(Y))$ are finite) to get the lower bound $(\int_{\mathcal{X}} \mid g(x) \mid w(x)f^*(x)dx)^2$, which is nothing but $(\int_{\mathcal{X}} \mid g(x) \mid f(x)dx)^2$. So in order to achieve this lower bound, we must use the proposal density $f^*(x) = \mid g(x) \mid f(x)/[\int_{\mathcal{X}} \mid g(x) \mid f(x)dx]$. Sometimes it is easy to sample from this proposal density, but more often it is difficult. In those cases, one may have to resort to a Markov chain Monte Carlo (MCMC) sampling scheme such as the *Metropolis-Hastings* (MH) algorithm or the *Gibbs sampler* (GS). We first describe the MH algorithm, since GS can be viewed as a special case of it.

The MH algorithm is named after N. Metropolis who first published it in 1953 in the context of the Boltzmann distribution, and W. K. Hastings who generalized it in 1970. Suppose we want to sample from the density $f(x)$. This algorithm generates a Markov chain $\{x_1, x_2, \ldots\}$ in which each x_i depends only on the immediately preceding one (i.e., x_{i-1}). Assume that the current state of the chain is x_t. To move to the next state, the algorithm uses a *proposal density* $f^*(x; x_t)$ which depends only on the current state and is easy to sample from. Once a new observation x^* is drawn from the proposal density and a random variable U is generated from the Uniform[0,1] density, x^* is accepted to be the next state of the chain (i.e., we declare $x_{t+1} = x^*$) if $U < \min\{1, f(x^*)f^*(x_t; x^*)/[f(x_t)f^*(x^*; x_t)]\}$. Otherwise we declare $x_{t+1} = x_t$. One commonly used proposal density is the normal density centered at x_t and having some known variance σ^2 (or known variance-covariance matrix $\sigma^2 I$ in the multivariate case). This proposal density will generate new sample-observations that are centered around the current state x_t with a variance σ^2. Incidentally, this choice for f^* would be allowed under the original Metropolis algorithm too, which required the proposal density to be symmetric (i.e., $f^*(x_t; x^*) = f^*(x^*; x_t)$). The generalization by Hastings removed this "symmetry" constraint and even allowed proposal densities to be just a function of x_t (i.e., to be free from x^*). In this latter case, the algorithm is called *independence chain Metropolis-Hastings* (as opposed to *random walk Metropolis-Hastings* when the proposal density is a function of both x_t and x^*). While the "independence chain" version can potentially offer higher accuracy than the "random walk" version with suitably chosen proposal densities, it requires some *a priori* knowledge of the target density. In any case, once the Markov chain is initialized by a starting value x_0, it is left running for a long time (the "burn-in" period) to ensure that it gets "sufficiently close" to the target density. During the burn-in period, often some parameters of the proposal density (e.g., the variance(s) in the case of a normal proposal density) have to be "fine-tuned" in order to keep the acceptance rate moderate (i.e., slightly higher than 50%). This is because the acceptance rate is intimately related to the size of the proposal-steps. Large proposal-steps would result in very low acceptance rates and the chain will not move much. Small proposal-steps would lead to very high acceptance rates and the chain will move around too much and converge slowly to the target density (in which case, it is said to be *slowly mixing*).

Now suppose that we have a p-dimensional density $f(x_1, \ldots, x_p) = f(\mathbf{x})$ as our target. Also suppose that for each i ($1 \leq i \leq p$), the univariate conditional density of x_i given all the other variables (call it $f_i(x_i \mid x_1, \ldots, x_{i-1}, x_{i+1}, \ldots, x_p)$) is easy to sample from. In this case, it is a good idea to choose the proposal density: $f^*(\mathbf{x}^*; \mathbf{x}_t) = f_j(x_j^* \mid \mathbf{x}_{-j,t})$ if $\mathbf{x}_{-j}^* = \mathbf{x}_{-j,t}$ and $= 0$ otherwise. Here $\mathbf{x}_{-j,t}$ means the current state $\mathbf{x}_l$ with its jth coordinate removed and $\mathbf{x}_{-j}^*$ means the newly drawn sample-observation $\mathbf{x}^*$ with its jth coordinate removed. A simple calculation shows that, under this choice, the acceptance probability will be 1. In other words, after initializing the Markov chain by setting $x_i = x_{i,0}$ for $1 \leq i \leq p$, this is how we move from the current state $\mathbf{x}_t$ to the next state $\mathbf{x}_{t+1}$: First sample $x_{1,t+1}$ from $f_1(x_1 \mid x_{2,t}, \ldots, x_{p,t})$; next sample $x_{2,t+1}$ from $f_2(x_2 \mid x_{1,t+1}, x_{3,t}, \ldots, x_{p,t})$; …; finally sample $x_{p,t+1}$ from $f_p(x_p \mid x_{1,t+1}, \ldots, x_{p-1,t+1})$. This is known as the *deterministic scan Gibbs sampler*. Since it is a special case of the MH algorithm described above, we can actually insert MH steps within a Gibbs sampler without disrupting the properties of the underlying Markov chain. As long as the univariate conditional densities (often collectively called the *full conditionals*) are familiar or easy to sample from, we will follow the Gibbs sampling scheme, but if one of them is a bit problematic, we will deal with it using the MH technique and then get back to the Gibbs sampling scheme for the other ones.

Bibliography

[1] K.M. Abel. Fetal origins of schizophrenia: testable hypotheses of genetic and environmental influences. *Br J Psychiatry*, 184:383–385, 2004.

[2] R.B.M. Aggio, A. Mayor, S. Reade, C.S.J. Probert, and K. Ruggiero. Identifying and quantifying metabolites by scoring peaks of gc-ms data. *BMC Bioinformatics*, 15:374, 2014.

[3] A. Akalin, M. Kormaksson, S. Li, F.G. Bakelman, M. Figueroa, A. Melnick, and C. Mason. methylKit: a comprehensive R package for the analysis of genome-wide DNA methylation profiles. *Genome Biology*, 13(10):R87+, October 2012.

[4] A. Alonso, S. Marsal, and A. Julià. Analytical methods in untargeted metabolomics: state of the art in 2015. *Frontiers in Bioengineering and Biotechnology*, 3:Article 23, 2015.

[5] A. Alonso, M.A. Rodríguez, M. Vinaixa, R. Tortosa, X. Correig, A. Julià, and S. Marsal. Focus: a robust workflow for one-dimensional nmr spectral analysis. *Analytical Chemistry*, 86(2):1160–1169, 2014.

[6] D. Alonso-López, F.J. Campos-Laborie, M.A. Gutiérrez, L. Lambourne, M.A. Calderwood, M. Vidal, and J. De Las Rivas. Apid database: redefining protein–protein interaction experimental evidences and binary interactomes. *Database*, 2019:baz005, 2019.

[7] D. Alvarez-Ponce. Recording negative results of protein–protein interaction assays: an easy way to deal with the biases and errors of interactomic data sets. *Briefings in Bioinformatics*, 16(6):1017–1020, 2017.

[8] S. Anders and W. Huber. Differential expression analysis for sequence count data. *Genome Biology*, 11(10):R106, 2010.

[9] C. Angermueller, S.J. Clark, H.J. Lee, I.C. Macaulay, M.J. Teng, T.X. Hu, F. Krueger, S.A. Smallwood, C.P. Ponting, T. Voet, G. Kelsey, O. Stegle, and W. Reik. Parallel single-cell sequencing links transcriptional and epigenetic heterogeneity. *Nature Methods*, 13(3):229–232, 2016.

[10] E.G. Armitage and C. Barbas. Metabolomics in cancer biomarker discovery: Current trends and future perspectives. *Journal of Pharmaceutical and Biomedical Analysis*, 87:1–11, 2014.

[11] M.J. Aryee, A. Jaffe, H.C. Bravo, C. Ladd-Acosta, A.P. Feinberg, K.D. Hansen, and R.A. Irizarry. Minfi: a flexible and comprehensive bioconductor package for the analysis of infinium dna methylation microarrays. *Bioinformatics*, 30 (10):1363–9, 2014.

[12] M. Ashburner, C.A. Ball, et al. Gene ontology: Tool for the unification of biology. *Nature Genetics*, 25(1):25–29, 2000.

[13] M. Askenazi, J.R. Parikh, and J.A. Marto. mzapi: a new strategy for efficiently sharing mass spectrometry data. *Nature Methods*, 6(4):240–241, 2009.

[14] W. Astle, M. De Iorio, S. Richardson, D. Stephens, and T. M. Ebbels. A bayesian model of nmr spectra for the deconvolution and quantification of metabolites in complex biological mixtures. *Journal of the American Statistical Association*, 107:1259–1271, 2012.

[15] P.L. Auer and R.W. Doerge. A two-stage poisson model for testing rna-seq data. *Statistical Applications in Genetics and Molecular Biology*, 10(1):26, 2011.

[16] D.N. Ayyala, D.E. Frankhouser, J.O. Ganbat, G. Marcucci, R. Bundschuh, P. Yan, and S. Lin. istatistical methods for detecting differentially methylated regions based on methylcap-seq data. *Briefings in Bioinformatics*, 2015.

[17] G.D. Bader and C.W. Hogue. Analyzing yeast protein-protein interaction data obtained from different sources. *Nature Biotechnology*, 20(10):991–997, 2002.

[18] A.-L. Barabasi and R. Albert. Emergence of scaling in random networks. *Science*, 206:509–512, 1999.

[19] A.-L. Barabasi and Z.N. Oltvai. Network biology: understanding the cell's functional organization. *Nature Reviews Genetics*, 5:101–113, 2004.

[20] J. Bartel, J. Krumsiek, and F. Theis. Statistical methods for the analysis of high-throughput metabolomics data. *Computational and Structural Biotechnology Journal*, 4(5):1–9, 2013.

[21] M.S. Bartolomei, S. Zemel, and S.M. Tilghman. Parental imprinting of the mouse h19 gene. *Nature*, 351:153–155, 1991.

[22] D. Baù, A. Sanyal, B.R. Lajoie, E. Capriotti, M. Byron, J.B. Lawrence, J. Dekker, and M.A. Marti-Renom. The three-dimensional folding of the a-globin gene domain reveals formation of chromatin globules. *Nature Structural & Molecular Biology*, 18(1):107–114, 2011.

[23] B.E. Metzger, L.P. Lowe, A.R. Dyer, E.R. Trimble, U. Chaovarindr, D.R. Coustan, D.R. Hadden, D.R. McCance, M. Hod, H.D. McIntyre, J.J. Oats, B. Persson, M.S. Rogers, and D.A. Sacks. Hyperglycemia and adverse pregnancy outcomes. *New England Journal of Medicine*, 358:1991–2002, 2008.

[24] J.A. Beagan and J.E. Phillips-Cremins. On the existence and functionality of topologically associating domains. *Nature Genetics*, 52:8–16, 2020.

[25] M. Bellew, M. Coram, M. Fitzgibbon, M. Igra, T. Randolph, P. Wang, D. May, J. Eng, R. Fang, C. Lin, J. Chen, D. Goodlett, J. Whiteaker, A. Paulovich, and M. McIntosh. A suite of algorithms for the comprehensive analysis of complex protein mixtures using high-resolution lc-ms. *Bioinformatics*, 22(15):1902–1909, 2006.

[26] M. Berdasco and M. Esteller. Aberrant Epigenetic Landscape in Cancer: How Cellular Identity Goes Awry. *Developmental Cell*, 19(5):698–711, 2010.

[27] N. Bhardwaj and H. Lu. Correlation between gene expression profiles and protein–protein interactions within and across genomes. *Bioinformatics*, 21(11):2730–2738, 2005.

[28] M. Bibikova, B. Barnes, C. Tsan, V. Ho, B. Klotzle, J.M. Le, D. Delano, L. Zhang, G.P. Schroth, K.L. Gunderson, J.B. Fan, and R. Shen. High density DNA methylation array with single CpG site resolution. *Genomics*, 98(4):288–295, 2011.

[29] A.P. Bird and A.P. Wolffe. Methylation-induced repression - belts, braces, and chromatin. *Cell*, 99(5):451–454, 2014.

[30] A.M. Bolger, M. Lohse, and B. Usadel. Trimmomatic: a flexible trimmer for illumina sequence data. *Bioinformatics*, 30(15):2114–2120, 2014.

[31] B.M. Bolstad, R.A. Irizarry, M. Astrand, and T.P. Speed. A comparison of normalization methods for high density oligonucleotide array data based on variance and bias. *Bioinformatics*, 19(2):185–193, 2003.

[32] K.M. Boycott, J.S. Parboosingh, B.N. Chodirker, R.B. Lowry, D.R. McLeod, J. Morris, C.R. Greenberg, A.E. Chudley, F.P. Bernier, J. Midgley, L.B. Moller, and A.M Innes. Clinical genetics and the hutterite population: a review of mendelian disorders. *American Journal of Medical Genetics*, 146A:1088–1098, 2008.

[33] H.C. Bravo and R.A. Irizarry. Model-based quality assessment and base-calling for second generation sequencing data. *Biometrics*, 66(3):665–674, 2010.

[34] A.B. Brinkman, F. Simmer, K. Ma, A. Kaan, J. Zhu, and H.G. Stunnenberg. Whole-genome dna methylation profiling using methylcap-seq. *Methods*, 52:232–236, 2010.

[35] R. Bro and A.K. Smilde. Principle component analysis. *Analytical Methods*, 6:2812–2831, 2014.

[36] D.I. Broadhurst and D.B. Kell. Statistical strategies for avoiding false discoveries in metabolomics and related experiments. *Metabolomics*, 2:171–196, 2008.

[37] J.H. Bullard, E. Purdom, K.D. Hansen, and S. Dudoit. Evaluation of statistical methods for normalization and differential expression in mrna-seq experiments. *BMC Bioinformatics*, 11:94, 2010.

[38] G. Cagney, P. Uetz, and S. Fields. Two-hybrid analysis of the saccharomyces cerevisiae 26s proteasome. *Physiological Genomics*, 7(1):27–34, 2001.

[39] V. Carey, L. Li, and R. Gentleman. *RBGL: An Interface to the BOOST Graph Library*, 2016.

[40] J.S. Carroll, X. S. Liu, A.S. Brodsky, W. Li, C.A. Meyer, A.J. Szary, J. Eeckhoute, W. Shao, E.V. Hestermann, T.R. Geistlinger, E.A. Fox, P.A. Silver, and M. Brown. Chromosome-wide mapping of estrogen receptor binding reveals long-range regulation requiring the forkhead protein FoxA1. *Cell*, 122(1):33–43, 2005.

[41] R. Caspi, R. Billington, L. Ferrer, H. Foerster, C.A. Fulcher, I.M. Keseler, A. Kothari, M. Krummenacker, M. Latendresse, L.A. Mueller, Q. Ong, S. Paley, P. Subhraveti, D.S. Weaver, and P.D. Karp. The metacyc database of metabolic pathways and enzymes and the biocyc collection of pathway/genome databases. *Nucleic Acids Research*, 44(1):D471–D480, 2015.

[42] R. Caspi, R. Billington, I.M. Keseler, A. Kothari, M. Krummenacker, P.E. Midford, W.K. Ong, S. Paley, P. Subhraveti, and P.D. Karp. The metacyc database of metabolic pathways and enzymes - a 2019 update. *Nucleic Acids Research*, 48(D1):D445–D453, 2020.

[43] R. Cavill, D. Jennen, J. Kleinjans, and J.J. Briedé. Transcriptomic and metabolomic data integration. *Briefings in Bioinformatics*, 17(5):1–11, 2016.

[44] R. Cavill, J.C.S. Kleinjans, and J.J. Briedé. Dynamic time warping for omics. *PLoS ONE*, 8:e71823, 2013.

[45] M.J. Chaisson and G. Tesler. Mapping single molecule sequencing reads using basic local alignment with successive refinement (blasr): application and theory. *BMC Bioinformatics*, 13:238, 2012.

[46] A. Chatr-aryamontri, A. Ceol, L.M. Palazzi, G. Nardelli, L. Schneider, M.V. Castagnoli, and G. Cesareni. Mint: the molecular interaction database. *Nucleic Acids Research*, 35(Database Issue):D572–D574, 2007.

[47] A. Chatr-Aryamontri, R. Oughtred, L. Boucher, J. Rust, C. Chang, N.K. Kolas, L. O'Donnell, S. Oster, C. Theesfeld, A. Sella, C. Stark, B.J. Breitkreutz, K. Dolinski, and M. Tyers. The biogrid interaction database: 2017 update. *Nucleic Acids Research*, 45(Database Issue):D369–D379, 2017.

[48] Chong Chen, C. Wu, L. Wu, Y. Wang, M. Deng, and R. Xi. scrmd: Imputation for single cell rna-seq data via robust matrix decomposition. *bioRxiv*, page 459404, 2018.

[49] J. Chen, H. Zheng, and M.L. Wilson. Likelihood ratio tests for maternal and fetal genetic effects on obstetric complications. *Genetic Epidemiology*, 33:526–538, 2009.

[50] S.X. Chen and Y. Qin. A two-sample test for high-dimensional data with applications to gene-set testing. *Annals of Statistics*, 38:808–835, 2010.

[51] T. Chen, Y. Cao, Y. Zhang, J. Liu, Y. Bao, C. Wang, W. Jia, and A. Zhao. Random forest in clinical metabolomics for phenotypic discrimination and biomarker selection. *Evidence-Based Complementary and Alternative Medicine*, 2013:298183, 2013.

[52] Y. Chen, A.T.L. Lun, and G.K. Smyth. Differential expression analysis of complex rna-seq experiments using edger. In S. Datta and D. Nettleton, editors, *Statistical Analysis of Next Generation Sequencing Data*. Springer, New York, 2014.

[53] T. Chiang and D. Scholtens. A general pipeline for quality and statistical assessment of protein interaction data using r and bioconductor. *Nature Protocols*, 4:535–546, 2009.

[54] T. Chiang, D. Scholtens, D. Sarkar, R. Gentleman, and W. Huber. Coverage and error models of protein-protein interaction data by directed graph analysis. *Genome Biology*, 8:R186, 2007.

[55] E.J. Childs, C.G.S. Palmer, K. Lange, and J.S. Sinsheimer. Modeling maternal-offspring gene-gene interactions: The extended-MFG test. *Genetic Epidemiology*, 34(5):512–521, 2010.

[56] Y.-R. Cho, W. Hwang, M. Ramanathan, and A. Zhang. Semantic integration to identify overlapping functional modules in protein interaction networks. *BMC Bioinformatics*, 8:265, 2006.

[57] S.J. Clark, R. Argelaguet, C.A. Kapourani, T.M. Stubbs, H.J. Lee, C. Alda-Catalinas, F. Krueger, G. Sanguinetti, G. Kelsey, J.C. Marioni, O. Stegle, and W. Reik. scnmt-seq enables joint profiling of chromatin accessibility dna methylation and transcription in single cells. *Nature Communications*, 9(1):781, 2018.

[58] A. Clauset, M.E.J. Newman, and C. Moore. Finding community structure in very large networks. *Physical Review E*, 70:066111, 2004.

[59] S.J. Cokus, S. Feng, X. Zhang, Z. Chen, B. Merriman, C.D. Haudenschild, S. Pradhan, S.F. Nelson, M. Pellegrini, and S.E. Jacobsen. Shotgun bisulphite sequencing of the Arabidopsis genome reveals DNA methylation patterning. *Nature*, 452(7184):215–219, 2008.

[60] F. Court, M. Baniol, Hélène Hagege, J.S.S. Petit, M.N.N. Lelay-Taha, F. Carbonell, M. Weber, G. Cathala, and T. Forne. Long-range chromatin interactions at the mouse Igf2/H19 locus reveal a novel paternally expressed long non-coding RNA. *Nucleic Acids Research*, 39(14):5893–5906, 2011.

[61] M.J. Cowley, M. Pinese, K.S. Kassahn, N. Waddell, J.V. Pearson, S.M. Grimmond, A.V. Biankin, S. Hautaniemi, and J. Wu. Pina v2.0: mining interactome modules. *Nucleic Acids Research*, 40:D862–865, 2012.

[62] G. Csardi and T. Nequsz. The igraph software package for complex network research. *InterJournal*, Complex Systems:1695, 2006.

[63] R.B. D'Agostino, W. Chase, and A. Albert Belanger. The appropriateness of some common procedures for testing the equality of two independent binomial populations. *The American Statistician*, 42:198–202, 2016.

[64] P. Damaschke. Finding hidden hubs and dominating sets in sparse graphs by randomized neighborhood queries. *Networks*, 57(4):344–350, 2011.

[65] B. Daniel, G. Nagy, and L. Nagy. The intriguing complexities of mammalian gene regulation: How to link enhancers to regulated genes. are we there yet? *FEBS Letters*, 2014.

[66] P.M. Das and R. Singal. DNA Methylation and Cancer. *Journal of Clinical Oncology*, 22(22):4632–4642, 2004.

[67] M. David, M. Dzamba, D. Lister, L. Ilie, and M. Brudno. Shrimp2: Sensitive yet practical short read mapping. *Bioinformatics*, 27(7):1011–1012, 2011.

[68] J. Davies, J. Telenius, S. Mcgowan, N. Roberts, S. Taylor, D. Higgs, and J. Hughes. Multiplexed analysis of chromosome conformation at vastly improved sensitivity. *Nature Methods*, 13:74–80, 2016.

[69] S. De Bodt, D. Proost, K. Vandepoele, P. Rouzé, and Y. Van de Peer. Predicting protein-protein interactions in arabidopsis thaliana through integration of orthology, gene ontology and co-expression. *BMC Genomics*, 10:288, 2009.

[70] A.M. De Livera, D.A. Dias, D. De Souza, T. Rupasinghe, J. Pyke, D. Tull, U. Roessner, M. McConville, and T.P. Speed. Normalizing and integrating metabolomics data. *Analytical Chemistry*, 84:10768–10776, 2012.

[71] E. de Silva, T. Thorne, P. Ingram, I. Agrafioti, J. Swire, C. Wiuf, and M.P.H. Stumpf. The effects of incomplete protein interaction data on structural and evolutionary inferences. *BMC Biology*, 4:39, 2006.

[72] C.P.E. de Souza, M. Andronescu, T. Masud, F. Kabeer, J. Biele, E. Laks, D.I. Lai, P.T. Ye, J. Brimhall, B.X. Wang et al. Epiclomal: Probabilistic clustering of sparse single-cell dna methylation data. *PLOS Computational Biology*, 16(9):578–592, 2020.

[73] E. de Wit. Capturing heterogeneity: single-cell structures of the 3d genome. *Nature Structural & Molecular Biology*, 24(5):437–438, 2017.

[74] T.M. DeChiara, E.J. Robertson, and Argiris Efstratiadis. Parental imprinting of the mouse insulin-like growth factor II gene. *Cell*, 64(4):849–859, 1991.

[75] J. Dekker. Gene regulation in the third dimension. *Science (New York, N.Y.)*, 319(5871):1793–1794, 2008.

[76] J. Dekker, K. Rippe, M. Dekker, and N. Kleckner. Capturing Chromosome Conformation. *Science*, 295(5558):1306–1311, 2002.

[77] A.P. Dempster, N.M. Laird, and D.B. Rubin. Maximum likelihood from incomplete data via the em algorithm. *Journal of the Royal Statistical Society: Series B (Methodological)*, 39(1):1–22, 1977.

[78] A. Denker and W. De Laat. The second decade of 3C technologies: Detailed insights into nuclear organization. *Genes and Development*, 30(12):1357–1382, 2016.

[79] K. Dettmer, P.A. Aronov, and B.D. Hammock. Mass spectrometry-based metabolomics. *Mass Spectrometry Reviews*, 26(1):51–78, 2007.

[80] E.W. Deutsch. mzml: A single, unifying data format for mass spectrometer output. *Proteomics*, 8(14):2776–2777, 2008.

[81] B. Dey, S. Thukral, S. Krishnan, M. Chakrobarty, S. Gupta, C. Manghani, and V. Rani. DNA-protein interactions: methods for detection and analysis. *Molecular and Cellular Biochemistry*, 365:279, 2012.

[82] R. Di Guida, J. Engel, J.W. Allwood, R.J.M. Weber, M.R. Jones, U. Sommer, M.R. Viant, and W.B. Dunn. Non-targeted uhplc-ms metabolomic data processing methods: a comparative investigation of normalisation, missing value imputation, transformation and scaling. *Metabolomics*, 12:93, 2016.

[83] F. Le Dily, F. Serra, and M.A. Marti-renom. Structure: is there a way to integrate and reconcile single cell and population experimental data? *WIREs Computational Molecular Science*, 7(5):e1308, 2017.

[84] J. Ding, S. Lin, and Y. Liu. Monte Carlo pedigree disequilibrium test for markers on the X chromosome. *American Journal of Human Genetics*, 79(3):567–573, 2006.

[85] M.T. Dittrich, G.W. Klau, A. Rosenwalk, T. Dandekar, and T. Müller. Identifying functional modules in protein–protein interaction networks: an integrated exact approach. *Bioinformatics*, 24(13):i223–i231, 2008.

[86] E. Dolzhenko and A. Smith. Using beta-binomial regression for high-precision differential methylation analysis in multifactor whole-genome bisulfite sequencing experiments. *BMC Bioinformatics*, 15(1):215+, 2014.

[87] C.H. Dong, W.D. Li, F. Geller, L. Lei, D. Li, O.Y. Gorlova, J. Hebebrand, C.I. Amos, R.D. Nicholls, and R.A. Price. Possible genomic imprinting 15 of three human obesity-related genetic loci. *American Journal of Human Genetics*, 76:421–437, 2005.

[88] J. Dostie, T.A. Richmond, R.A. Arnaout, R.R. Selzer, W.L. Lee, T.A. Honan, E.D. Rubio, A. Krumm, J. Lamb, C. Nusbaum, R.D. Green, and J. Dekker. Chromosome Conformation Capture Carbon Copy (5C): A massively parallel solution for mapping interactions between genomic elements. *Genome Research*, 16(10):1299–1309, 2006.

[89] T.A. Down, V.K. Rakyan, D.J. Turner, P. Flicek, H. Li, E. Kulesha, S. Graf, N. Johnson, J. Herrero, E.M. Tomazou et al. A bayesian deconvolution strategy for immunoprecipitation-based dna methylome analysis. *Nature Biotechnology*, 26:779–785, 2008.

[90] P. Du, W.A. Wibbe, and S.M. Lin. Improved peak detection inmass spectrum by incorporating continuous wavelet transform-based pattern matching. *Bioinformatics*, 22(17):2059–2065, 2006.

[91] X. Du and S.H. Zeisel. Spectral deconvolution for gas chromatography mass spectrometry-based metabolomics: current status and future perspectives. *Computational and Structural Biotechnology Journal*, 4:e201301013, 2013.

[92] W.B. Dunn, D.I. Broadhurst, H.J. Atherton, R. Goodacre, and J.L. Griffin. Systems level studies of mammalian metabolomes: The roles of mass spectrometry and nuclear magnetic resonance spectroscopy. *Chemical Society Reviews*, 40:387–426, 2011.

[93] W.R. Dunn, I.D. Wilson, A.W. Nicholls, and D. Broadhurst. The importance of experimental design and qc samples in large-scale and ms-driven untargeted metabolomic studies of humans. *Bioanalysis*, 4(18):2249–2264, 2012.

[94] C.A. Eads, K.D. Danenberg, K. Kawakami, L.B. Saltz, C. Blake, D. Shibata, P.V. Danenberg, and P.W. Laird. MethyLight: a high-throughput assay to measure DNA methylation. *Nucleic Acids Research*, 28(8):E32, 2000.

[95] F. Eckhardt, J. Lewin, R. Cortese, V.K. Rakyan, J. Attwood, M. Burger, J. Burton, T.V. Cox, R. Davies, T.A. Down, C. Haefliger, R. Horton, K. Howe, D.K. Jackson, J. Kunde, C. Koenig, J. Liddle, D. Niblett, T. Otto, R. Pettett, S. Seemann, C. Thompson, T. West, J. Rogers, A. Olek, K. Berlin, and S. Beck. Dna methylation profiling of human chromosomes 6, 20 and 22. *Nature Genetics*, 38:1378–1385, 2006.

[96] D.R.V. Edwards, J.R. Gilbert, L. Jiang, P.J. Gallins, L. Caywood, M. Creason, D. Fuzzell, C. Knebusch, C.E. Jackson, M.A. Pericak-Vance, J.L. Haines, and W.K. Scott. Successful aging shows linkage to chromosomes 6, 7, and 14 in the amish. *Annals of Human Genetics*, 75:516–528, 2011.

[97] E.E. Eichler, J. Flint, G. Gibson, A. Kong, S.M. Leal, J.H. Moore, and J.H. Nadeau. Missing heritability and strategies for finding the underlying causes of complex disease. *Nature Reviews Genetics*, 11(6):446–450, 2010.

[98] R. Elyanow, B. Dumitrascu, B.E. Engelhardt, and B.J. Raphael. netnmf-sc: leveraging gene-gene interactions for imputation and dimensionality reduction in single-cell expression analysis. *Genome Research*, 30(2):195–204, 2020.

[99] M. Epstein, C. Veal, R. Trembath, J. Barker, C. Li, and G. Satten. Genetic association analysis using data from triads and unrelated subjects. *American Journal of Human Genetics*, 76:592–608, 2005.

[100] Y. Erlich, P.P. Mitra, M. delaBastide, W.R. McCombie, and G.J. Hannon. Alta-cyclic: a self-optimizing base caller for next generation sequencing. *Nature Methods*, 5(8):679–682, 2008.

[101] A. Fabregat, K. Sidiropoulos, P. Garapati, M. Gillespie, K. Hausmann, R. Haw, B. Jassal, S. Jupe, F. Korninger, S. McKay, L. Matthews, B. May, M. Milacic, K. Rothfels, V. Shamovsky, M. Webber, J. Weiser, M. Williams, G. Wu, L. Stein, H. Hermjakob, and P. D'Eustachio. The reactome pathway knowledgebase. *Nucleic Acids Research*, 44(D1):D481–D487, 2016.

[102] M.E. Fahey, M.J. Bennett, C. Mahon, S. Jager, K. Pache, D. Kumar, A. Shapiro, K. Rao, S.K. Chanda, C.S. Craik, A.D. Frankel, and N.J. Krogan. Gps-prot: A web-based visualization platform for integrating host-pathogen interaction data. *BMC Bioinformatics*, 12:298, 2011.

[103] R. Fang, M. Yu, G. Li, S. Chee, T. Liu, A.D. Schmitt, and B. Ren. Mapping of long-range chromatin interactions by proximity ligation-assisted chip-seq. *Cell Research*, 26:1345–1348, 2016.

[104] M. Farlik, N.C. Sheffield, A. Nuzzo, P. Datlinger, A. Schönegger, J. Klughammer, and C. Bock. Single-cell dna methylome sequencing and bioinformatic inference of epigenomic cell-state dynamics. *Cell Reports*, 10(8):1386–1397, 2015.

[105] A.P. Feinberg and B. Tycko. The history of cancer epigenetics. *Nature Reviews Cancer*, 4(2):143–153, 2004.

[106] S.M. Fendt, J.M. Buescher, F. Rudroff, P. Picotti, N. Zamboni, and U. Sauer. Tradeoff between enzyme and metabolite efficiency maintains metabolic homeostasis upon perturbations in enzyme capacity. *Molecular Systems Biology*, 6:356, 2010.

[107] H. Feng, K.N. Conneely, and H. Wu. A Bayesian hierarchical model to detect differentially methylated loci from single nucleotide resolution sequencing data. *Nucleic Acids Research*, 42(8):gku154–e69, 2014.

[108] J. Feng, W. Li, and T. Jiang. Inference of isoforms from short sequence reads. *Journal of Computational Biology*, 18(3):305–321, 2011.

[109] A.C. Ferguson-Smith. Genomic imprinting: the emergence of an epigenetic paradigm. *Nature Reviews Genetics*, 12(8):565–575, 2011.

[110] R.J. Ferland, T.J. Cherry, P.O. Preware, E.E. Morrisey, and C.A. Walsh. Characterization of foxp2 and foxp1 mrna and protein in the developing and mature brain. *Journal of Comparative Neurology*, 460(2):266–279, 2003.

[111] A.F. Fernandez, Y. Assenov, J.I. Martin-Subero, B. Balint, R. Siebert, H. Taniguchi, H. Yamamoto, M. Hidalgo, Aik-Choon Tan, Oliver Galm, Isidre Ferrer, Montse Sanchez-Cespedes, Alberto Villanueva, Javier Carmona, Jose V. Sanchez-Mut, Maria Berdasco, Victor Moreno, Gabriel Capella, David Monk, Esteban Ballestar, Santiago Ropero, Ramon Martinez, Marta Sanchez-Carbayo, Felipe Prosper, Xabier Agirre, Mario F. Fraga, Osvaldo Graña, Luis Perez-Jurado, Jaume Mora, Susana Puig, Jaime Prat, Lina Badimon, Annibale A. Puca, Stephen J. Meltzer, Thomas Lengauer, John Bridgewater, Christoph Bock, and Manel Esteller. A DNA methylation fingerprint of 1628 human samples. *Genome Research*, 22(2):407–419, 2011.

[112] O. Fiehn. Combining genomics, metabolome analysis, and biochemical modelling to understand metabolic networks. *Comparative and Functional Genomics*, 2:155–168, 2001.

[113] S.C. Fields and O. Song. A novel genetic system to detect protein-protein interactions. *Nature*, 340(6230):245–246, 1989.

[114] M.F. Figueroa, O. Abdel-Wahab, C. Lu, P.S. Ward, J. Patel, A. Shih, Y. Li, N. Bhagwat, A. Vasanthakumar, H.F. Fernandez, M.S. Tallman, Z. Sun, K. Wolniak, J.K. Peeters, W. Liu, S.E. Choe, V.R. Fantin, E. Paietta, B. Löwenberg, J.D. Licht, L.A. Godley, R. Delwel, P.J. Valk, C.B. Thompson, R.L. Levine, and A. Melnick. Leukemic IDH1 and IDH2 mutations result in a hypermethylation phenotype, disrupt TET2 function, and impair hematopoietic differentiation. *Cancer Cell*, 18(6):553–567, 2010.

[115] E.H. Finn, G. Pegoraro, H.B. Brandão, A.L. Valton, M.E. Oomen, J. Dekker, L. Mirny, and T. Misteli. Extensive heterogeneity and intrinsic variation in spatial genome organization. *Cell*, 176(6):1502–1515.e10, 2020/06/09 2019.

[116] I.M. Flyamer, J. Gassler, M. Imakaev, H.B. Brandão, S.V. Ulianov, N. Abdennur, S.V. Razin, L.A. Mirny, and K. Tachibana-konwalski. Re-organization at oocyte-to-zygote transition. *Nature Publishing Group*, 544(7648):110–114, 2017.

[117] J.M. Fonville, S.E. Richards, R.H. Barton, C.L. Boulange, T.M.D. Ebbels, J.K. Nicholson, E. Holmes, and M.-E. Dumas. The evolution of partial least squares models and related chemometric approaches in metabonomics and metabolic phenotyping. *Journal of Chemometrics*, 24(11-12):639–649, 2010.

[118] A. Forcina and L. Franconi. Regression analysis with the beta-binomial distribution. *Rivista di Statistica Applicata*, 21:7–12, 1988.

[119] J. Forster, I. Famili, P. Fu, B.O. Palsson, and J. Neilsen. Genome-scale resonstruction of the *Saccharomyces cerevisiae* metabolic network. *Genome Research*, 13:244–253, 2003.

[120] D.E. Frankhouser, M. Murphy, J.S. Blachly, J. Park, M.W. Zoller, J.-O. Ganbat, J. Curfman, J.C. Byrd, S. Lin, G. Marcucci, P. Yan, and R. Bundschuh. PrEMeR-CG: inferring nucleotide level DNA methylation values from MethylCap-seq data. *Bioinformatics*, 30(24):3567–3574, 2014.

[121] D.E. Frankhouser, M. Murphy, J.S. Blachly, J. Park, M.W. Zoller, J.-O. Ganbat, J. Curfman, J.C. Byrd, S. Lin, G. Marcucci, P. Yan, and R. Bundschuh. Supplement to PrEMeR-CG: inferring nucleotide level DNA methylation values from MaethylCap-seq data. *Bioinformatics*, 30(24):3567–3574, 2014.

[122] J. Fraser, M. Rousseau, S. Shenker, M.A. Ferraiuolo, Y. Hayashizaki, M. Blanchette, and J. Dostie. Chromatin conformation signatures of cellular differentiation. *Genome Biology*, 10(4):R37, 2009.

[123] C.C. Friedel and R. Zimmer. Inferring topology from clustering coefficients in protein-protein interaction networks. *BMC Bioinformatics*, 7:519, 2006.

[124] M. Frommer, L.E. McDonald, D.S. Millar, C.M. Collis, F. Watt, G.W. Grigg, P.L. Molloy, and C.L. Paul. A genomic sequencing protocol that yields a positive display of 5-methylcytosine residues in individual DNA strands. *Proceedings of the National Academy of Sciences of the United States of America*, 89(5):1827–1831, 1992.

[125] M.J. Fullwood, M.H. Liu, Y.F. Pan, J. Liu, II. Xu, Y.B. Mohamed, Y.L. Orlov, S. Velkov, A. Ho, P.H. Mei, E.G.Y. Chew, P.Y. Huang, W.-J. Welboren, Y. Han, H.S. Ooi, P.N. Ariyaratne, V.B. Vega, Y. Luo, P.Y. Tan, P.Y. Choy, W., B. Zhao, K.S. Lim, S.C. Leow, J.S. Yow, R. Joseph, H. Li, K. V. Desai, J.S. Thomsen, Y.K. Lee, Karuturi, T. Herve, G. Bourque, H.G. Stunnenberg, X. Ruan, V.C.-Rataboul, W.-K. Sung, E.T. Liu, C.-LinWei, E. Cheung, and Y. Ruan. An oestrogen-receptor-[agr]-bound human chromatin interactome. *Nature*, 462(7269):58–64, 2009.

[126] E. Galeota, C. Gravila, F. Castiglione, M. Bernaschi, and G. Cesareni. The hierarchical organization of natural protein interaction networks confers self-organization properties on pseudocells. *BMC Systems Biology*, 9(Suppl 3):S3, 2015.

[127] A.-C. Gavin, M. Bösche, R. Krause, P. Paola Grandi, M. Marzioch, A. Bauer, J. Schultz, J.M. Rick, A.-M. Michon, C.-M. Cruciat, M. Remor, C. Höfert, M. Schelder, M. Brajenovic, H. Ruffner, A. Merino, K. Klein, M. Hudak, D. Dickson, T. Rudi, V. Gnau, A. Bauch, S. Bastuck, B. Huhse, C. Leutwein, M.-A. Marie-Anne Heurtier, R.R. Copley, A. Edelmann, E. Querfurth, V. Rybin, G. Drewes, M. Raida, T. Bouwmeester, P. Bork, B. Seraphin, B. Kuster, G. Neubauer, and G. Superti-Furga. Functional organization of the yeast proteome by systematic analysis of protein complexes. *Nature*, 415:141–147, 2002.

[128] A.C. Gavin, P. Aloy, P. Grandi, R. Krause, M. Boesche, M. Marzioch, C. Rau, L.J. Jensen, S. Bastuck, B. Dümpelfeld, A. Edelmann, M.A. Heurtier, V. Hoffman, C. Hoefert, K. Klein, M. Hudak, A.M. Michon, M. Schelder, M. Schirle, M. Remor, T. Rudi, S. Hooper, A. Bauer, T. Bouwmeester, G. Casari, G. Drewes, G. Neubauer, J.M. Rick, B. Kuster, P. Bork, R.B. Russell, and G. Superti-Furga. Proteome survey reveals modularity of the yeast cell machinery. *Nature*, 440(7084):631–636, 2006.

[129] R. Gentleman and W. Huber. Making the most of high-throughput protein-interaction data. *Genome Biology*, 8:112, 2007.

[130] G. Geva and R. Sharan. Identification of protein complexes from co-immunoprecipitation data. *Bioinformatics*, 27(1):111–117, 2011.

[131] F. Giacomoni, G. Le Corguillé, M. Monsoor, M. Landi, P. Pericard, M. Pétéra, C. Duperier, M. Tremblay-Franco, J.F. Martin, D. Jacob, S. Goulitquer, E.A. Thévenot, and C. Caron. Workflow4metabolomics: a collaborative research infrastructure for computational metabolomics. *Bioinformatics*, 31(9):1493–1495, 2015.

[132] N. Giannoukakis, C. Deal, J. Paquette, C.G. Goodyer, and C. Polychron. Parental genomic imprinting of the human igf2 gene. *Nature Genetics*, 4(1):98–101, 1993.

[133] C. Gieger, L. Geistlinger, E. Altmaier, M. Hrabe de Angelis, F. Kronenberg, T. Meitinger, H.-W. Mewes, H.-E. Wichmann, K.M. Weinberger, J. Adamski, T. Illig, and K. Suhre. Genetics meets metabolomics: a genome-wide association study of metabolite profiles in human serum. *PLoS Genetics*, 4(11):e1000282, 2008.

[134] J.J. Goeman and P. Bühlmann. Analyzing gene expression data in terms of gene sets: methodological issues. *Bioinformatics*, 23(8):980–987, 2007.

[135] J.J. Goeman, S.A. van de Geer, F. de Kort, and H.C. van Houwelingen. A global test for groups of genes: testing association with a clinical outcome. *Bioinformatics*, 20(1):93–99, 2004.

[136] D.S. Goldberg and F.P. Roth. Assessing experimentally derived interactions in a small world. *Proceedings of the National Academy of Sciences*, 100:4372–4376, 2003.

[137] R. Goodacre, S. Vaidyanathan, W.B. Dunn, G.G. Harrigan, and D.B. Kell. Metabolomics by numbers: acquiring and understanding global metabolite data. *Trends in Biotechnology*, 22:245–252, 2004.

[138] R. Gosden, J. Trasler, D. Lucifero, and M. Faddy. Rare congenital disorders, imprinted genes, and assisted reproductive technology. *Lancet*, 361:1975–1977, 2003.

[139] C. Gregg, J. Zhang, B. Weissbourd, S. Luo, G.P. Schroth, D. Haig, and C. Dulac. High resolution analysis of parent-of-origin Allelic expression in the mouse brain: A high resolution approach to analyze imprinting. *Science*, 329(5992):643–648, 2010.

[140] D.A. Griffiths. Maximum likelihood estimation for the beta-binomial distribution and an application to the household distribution of the total number of cases of a a disease. *Biometrics*, 29(4): 637–648, 1973.

[141] P.S. Gromski, Y. Xu, H.L. Kotze, E. Correa, D.I. Ellis, E.G. Armitage, M.L. Turner, and R. Goodacre. Influence of missing values substitutes on multivariate analysis of metabolomics data. *Metabolites*, 4(2):433–452, 2014.

[142] HAPO Study Cooperative Research Group. The hyperglycemia and adverse pregnancy outcome (hapo) study. *International Journal of Gynaecology and Obstetrics*, 78(1):69–77, 2002.

[143] J. Gullberg, P. Jonsson, A. Nordström, M. Sjöström, and T. Moritz. Design of experiments: an efficient strategy to identify factors influencing extraction and derivatization of *Arabidopsis thaliana* samples in metabolomic studies with gas chromatography/mass spectrometry. *Analytical Biochemistry*, 331(2):283–295, 2004.

[144] H. Guo, P. Zhu, X. Wu, X. Li, L. Wen, and F. Tang. Single-cell methylome landscapes of mouse embryonic stem cells and early embryos analyzed using reduced representation bisulfite sequencing. *Genome Research*, 23(12):2126–2135, 2013.

[145] M.J. Ha, V. Baladandayuthapani, and K.-A. Do. Dingo: Differential network analysis in genomics. *Bioinformatics*, 13(21):3413–3420, 2015.

[146] R. Hager, J.M. Cheverud, and J.B. Wolf. Maternal effects as the cause of parent-of-origin effects that mimic genomic imprinting. *Genetics*, 178:1755–1762, 2008.

[147] D. Haig. Genomic imprinting and kinship: How good is the evidence? *Annual Review of Genetics*, 38:553–585, 2004.

[148] David Haig. Genetic conflicts in human pregnancy. *The Quarterly Review of Biology*, 68(4):495–532, 1993.

[149] J.M. Halket, A. Przyborowska, S.E. Stein, W.G. Mallard, S. Down, and R.A. Chalmers. Deconvolution gas chromatography/mass spectrometry of urinary organic acids: potential for pattern recognition and automated identification of metabolic disorders. *Rapid Communications in Mass Spectrometry*, 13:279–284, 1999.

[150] C. Han, J. Park, and S. Lin. *BCurve: Bayesian Curve Credible Bands Approach for Detection of Differentially Methylated Regions. In Epigenome-Wide Association Studies. Guan, W. (Ed.).* Springer, 2021.

[151] C. Han, H. Tang, S. Lou, Y. Gao, M. H. Cho, and S. Lin. Evaluation of recent statistical methods for detecting differential methylation using bs-seq data. *OBM Genetics*, in press, 2(4):041, 2018.

[152] C. Han. *Statistical models and computational methods for studying DNA differential methylation and 3D genome structure.* PhD thesis, Ohio State University, 2021.

[153] C. Han, Q. Xie, and S. Lin. Are dropout imputation methods for scRNA-seq effective for scHi-C data? *Briefings in Bioinformatics*, 2020. bbaa289.

[154] J.-D.J. Han, N. Bertin, Hao. T., D.S. Goldberg, G.F. Berriz, L.V. Zhang, D. Dupuy, A.J.M. Walhout, M.E. Cusick, F.P. Roth, and M. Vidal. Evidence for dynamically organized modularity in the yeast protein-protein interaction network. *Nature*, 430:88–93, 2004.

[155] J.D. Han, D. Dupuy, N. Bertin, M.E. Cusick, and M. Vidal. Effect of sampling on topology predictions of protein-protein interaction networks. *Nature Biotechnology*, 23(7):839–844, 2005.

[156] M. Han, Y.Q. Hu, and S. Lin. Joint detection of association, imprinting and maternal effects using all children and their parents. *European Journal of Human Genetics*, 21(12):1449–1456, 2013.

[157] K.D. Hansen, B. Langmead, and R.A. Irizarry. BSmooth: from whole genome bisulfite sequencing reads to differentially methylated regions. *Genome Biology*, 13(10):R83, 2012.

[158] J. Hao, W. Astle, M. De Iorio, and T.M. Ebbels. Batman–an r package for the automated quantification of metabolites from nuclear magnetic resonance spectra using a bayesian model. *Bioinformatics*, 28:2088–2090, 2012.

[159] J. Hao, M. Liebeke, W. Astle, M. De Iorio, J.G. Bundy, and T.M. Ebbels. Bayesian deconvolution and quantification of metabolites in complex 1d nmr spectra using batman. *Nature Protocols*, 9:1416–1427, 2014.

[160] S. Harney, J. Newton, A. Milicic, M.A. Brown, and B.P. Wordsworth. Non-inherited maternal hla alleles are associated with rheumatoid arthritis. *Rheumatolog*, 42:171–174, 2003.

[161] S. Hassani, H. Martens, M. Qannari, M. Hanafi, G.I. Borge, and A. Kohler. Analysis of –omics data: Graphical interpretation- and validation tools in multi–block methods. *Chemometrics and Intelligent Laboratory Systems*, 104:140–153, 2010.

[162] J. Hastings, P. de Matos, A. Dekker, M. Ennis, B. Harsha, N. Kale, V. Muthukrishnan, G. Owen, S. Turner, M. Williams, and C. Steinbeck. The chebi reference database and ontology for biologically relevant chemistry: enhancements for 2013. *Nucleic Acids Research*, 41(D1):D456–D463, 2013.

[163] T.R. Hazbun, L. Malmström, S. Anderson, B.J. Graczyk, B. Fox, M. Riffle, B.A. Sundin, J.D. Aranda, W.H. McDonald, C.H. Chiu, B.E. Snydsman, P. Bradley, E.G. Muller, S. Fields, D. Baker, J.R. Yates 3rd, and T.N. Davis. Assigning function to yeast proteins by integration of technologies. *Molecular Cell*, 12(6):1353–1365, 2003.

[164] C. He, M.Q. Zhang, and X. Wang. Micc: an r package for identifying chromatin interactions from chia-pet data. *Bioinformatics*, 31:3832–3834, 2015.

[165] F. He, J.Y. Zhou, Y.Q. Hu, F. Sun, J. Yang, S. Lin, and W.K. Fung. Detection of parent-of- origin effects for quantitative traits in complete and incomplete nuclear families with multiple children. *American Journal of Epidemiology*, 174:226–233, 2011.

[166] K. Hebestreit, M. Dugas, and H.-U. Klein. Detection of significantly differentially methylated regions in targeted bisulfite sequencing data. *Bioinformatics*, 29(13):1647–1653, 2013.

[167] W.H.M. Heijne, R.-J.A.N. Lamers, P.J. van Bladeren, J.P. Groten, J.H. van Nesselrooij, and B. van Ommen. Profiles of metabolites and gene expression in rats with chemically induced hepatic necrosis. *Toxicologic Pathology*, 33:425–433, 2005.

[168] B. Hendrich and A. Bird. Identification and characterization of a family of mammalian methyl-CpG binding proteins. *Molecular and Cellular Biology*, 18(11):6538–6547, 1998.

[169] J.G. Herman, J.R. Graff, S. Myohanen, B.D. Nelkin, and S.B. Baylin. Methylation-specific PCR: a novel PCR assay for methylation status of CpG islands. *Proceedings of the National Academy of Sciences*, 93(18):9821–9826, 1996.

[170] H. Hermjakob, L. Montecchi-Palazzi, G. Bader, J. Wojcik, L. Salwinski, A. Ceol, S. Moore, S. Orchard, U. Sarkans, C. von Mering, B. Roechert, S. Poux, E. Jung, H. Mersch, P. Kersey, M. Lappe, Y. Li, R. Zeng, D. Rana, M. Nikolski, H. Husi, C. Brun, S.G.N. Shanker, K .and Grant, C. Sander, P. Bork, W. Zhu, A. Pandey, A. Brazma, B. Jacq, M. Vidal, D. Sherman, P. Legrain, G. Cesareni, I. Xenarios, B. Eisenberg, D.and Steipe, C. Hogue, and R. Apweiler. The hupo psi's molecular interaction format—a community standard for the representation of protein interaction data. *Nature Biotechnology*, 22:177–183, 2004.

[171] H. Hermjakob, L. Montecchi-Palazzi, C. Lewington, S. Mudali, S. Kerrien, S. Orchard, M. Vingron, B. Roechert, P. Roepstorff, A. Valencia, H. Margalit, J. Armstrong, A. Bairoch, G. Cesareni, D. Sherman, and R. Apweiler. Intact: an open source molecular interaction database. *Nucleic Acids Research*, 32(Database issue):D452–D455, 2004.

[172] J. Hildebrand and M. E. Dekker. Mechanisms and functions of chromosome compartmentalization. *Trends in Biochemical Sciences*, 45(5):385–396, 2020.

[173] K. Hiller, J. Hangebrauk, C. Jäger, J. Spura, K. Schreiber, and D. Schomburg. Metabolitedetector: comprehensive analysis tool for targeted and nontargeted gc/ms based metabolome analysis. *Analytical Chemistry*, 81(9):3429–3439, 2009.

[174] Y. Ho, A. Gruhler, A. Heilbut, G.D. Bader, L. Moore, S.L. Adams, A. Millar, P. Taylor, K. Bennett, K. Boutilier, L. Yang, C. Wolting, I. Donaldson, S. Schandorff, J. Shewnarane, M. Vo, J. Taggart, M. Goudreault, B. Muskat, C. Alfarano, D. Dewar, Z. Lin, K. Michalickova, A.R. Willems, H. Sassi, P.A. Nielsen, K.J. Rasmussen, J.R. Andersen, L.E. Johansen, L.H. Hansen, H. Jespersen, A. Podtelejnikov, E. Nielsen, J. Crawford, V. Poulsen, B.D. Sørensen, J. Matthiesen, R.C. Hendrickson, F. Gleeson, T. Pawson, M.F. Moran, D. Durocher, M. Mann, C.W. Hogue, D. Figeys, and M. Tyers. Systematic identification of protein complexes in saccharomyces cerevisiae by mass spectrometry. *Nature*, 415(6868):180–183, 2002.

[175] R.D. Hodge, T.E. Bakken, J.A. Miller, K.A. Smith, E.R. Barkan, L.T. Graybuck, J.L. Close, B. Long, N. Johansen, O. Penn, et al. Conserved cell types with divergent features in human versus mouse cortex. *Nature*, 573(7772):61–68, 2019.

[176] E. Holmes, R.L. Loo, J. Stamler, M. Bictash, I.K.S. Yap, Q. Chan, T. Ebbels, M. De Iorio, I.J. Brown, K.A. Veselkov, M.L. Daviglus, H. Kesteloot, H. Ueshima, L. Zhao, J.K. Nicholson, and P. Elliott. Human metabolic phenotype diversity and its association with diet and blood pressure. *Nature*, 453(7193):396–400, 2008.

[177] H. Hong, S. Jiang, H. Li, G. Du, Y. Sun, H. Tao, C. Quan, C. Zhao, R. Li, W. Li, et al. Deephic: A generative adversarial network for enhancing hi-c data resolution. *PLoS Computational Biology*, 16(2):e1007287, 2020.

[178] H. Horai, M. Arita, S. Kanaya, Y. Nihei, T. Ikeda, K. Suwa, Y. Ojima, K. Tanaka, S. Tanaka, K. Aoshima, Y. Oda, Y. Kakazu, M. Kusano, T. Tohge, F. Matsuda, Y. Sawada, M.Y. Hirai, H. Nakanishi, K. Ikeda, N. Akimoto, T. Maoka, H. Takahashi, T. Ara, N. Sakurai, H. Suzuki, D. Shibata, S. Neumann, T. Iida, K. Tanaka, K. Funatsu, F. Matsuura, T. Soga, R. Taguchi, K. Saito, and T. Nishioka. Massbank: a public repository for sharing mass spectral data for life sciences. *Journal of Mass Spectrometry*, 45(7):703–714, 2010.

[179] Y. Hou, H. Guo, C. Cao, X. Li, B. Hu, P. Zhu, X. Wu, L. Wen, F. Tang, Y. Huang, and J. Peng. Single-cell triple omics sequencing reveals genetic, epigenetic, and transcriptomic heterogeneity in hepatocellular carcinomas. *Cell Research*, 26(3):304–319, 2016.

[180] O. Hrydziuszko and M. Viant. Missing values in mass spectrometry based metabolomics: an undervalued step in the data processing pipeline. *Metabolomics*, 8:S161–S174, 2012.

[181] D. Hsu, S.M. Kakade, and T. Zhang. Robust matrix decomposition with sparse corruptions. *IEEE Transactions on Information Theory*, 57(11):7221–7234, 2011.

[182] M. Hu, K. Deng, Z. Qin, J. Dixon, S. Selvaraj, J. Fang, B. Ren, and J.S. Liu. Bayesian Inference of Spatial Organizations of Chromosomes. *PLOS Computational Biology*, 9(1):e1002893+, January 2013.

[183] H. Huang, B.M. Jednyak, and J.S. Bader. Where have all the interactions gone? estimating the coverage of two-hybrid protein interaction maps. *PLoS Computational Biology*, 3(11):2155–2174, 2007.

[184] H.-C. Huang, Y. Niu, and L.-X. Qin. Differential expression analysis for rna-seq: An overview of statistical methods and computational software. *Cancer Informatics*, 14:57–67, 2015.

[185] K. Huang and G. Fan. Dna methylation in cell differentiation and reprogramming: an emerging systematic view. *Regenerative Medicine*, 5(4): 531–544, 2010.

[186] M. Huang, J. Wang, E. Torre, H. Dueck, S. Shaffer, R. Bonasio, J.I. Murray, A. Raj, M. Li, and N.R. Zhang. Saver: gene expression recovery for single-cell rna sequencing. *Nature Methods*, 15(7):539–542, 2018.

[187] T.H. Huang, M.R. Perry, and D.E. Laux. Methylation profiling of cpg islands in human breast cancer cells. *Human Molecular Genetics*, 8:459–470, 1999.

[188] W. Huber, A. von Heydebreck, H. Sültmann, A. Poustka, and M. Vingron. Variance stabilization applies to microarray data calibration and to the quantification of differential expression. *Bioinformatics*, 18 (Suppl) 1:S96–S104, 2002.

[189] K.M. Huffman, S.H. Shah, R.D. Stevens, J.R. Bain, M. Muehlbauer, C.A. Slentz, C.J. Tanner, M. Kuchibhatia, J.A. Houmard, C.B. Newgard, and W.E. Kraus. Relationships between circulating metabolic intermediates and insulin action in overweight to obese, inactive men and women. *Diabetes Care*, 32(9):1678–1683, 2009.

[190] G. Hughes, C. Cruickshank-Quinn, R. Reisdorph, S. Lutz, I. Petrache, N. Reisdorph, R. Bowler, and K. Kechris. Msprep–summarization, normalization and diagnostics for processing of mass spectrometry-based metabolomic data. *Bioinformatics*, 30(1):133–134, 2014.

[191] J. Hughes, N. Roberts, S. Mcgowan, D. Hay, E. Giannoulatou, M. Lynch, M. DeGobbi, S. Taylor, R. Gibbons, and D. Higgs. Analysis of hundreds of cis-regulatory landscapes at high resolution in a single, high-throughput experiment. *Nature Genetics*, 46:205–212, 2014.

[192] P. Hugo, L. Krijger, and W. de Laat. Regulation of disease-associated gene expression in the 3D genome. *Nature Reviews Molecular Cell Biology*, 17:771–782, 2016.

[193] T. Hui, Q. Cao, J. Wegrzyn-Woltosz, K. O'Neill, C. A. Hammond, D. J.H.F. Knapp, E. Laks, M. Moksa, S. Aparicio, C.J. Eaves, A. Karsan, and M. Hirst. High-resolution single-cell dna methylation measurements reveal epigenetically distinct hematopoietic stem cell subpopulations. *Stem Cell Reports*, 11(2):578–592, 2018.

[194] T. Illig, C. Gieger, G. Zhai, W. Romisch-Margl, R. Wang-Sattler, C. Prehn, E. Altmaier, G. Kastenmuller, B.S. Kato, H.-W. Mewes, T. Meitinger, M. Hrabe de Angelis, F. Kronenberg, N. Soranzo, H.-E. Wichmann, T.D. Spector, J. Adamski, and K. Suhre. A genome-wide perspective of genetic variation in human metabolism. *Nature Genetics*, 42(2):137–141, 2010.

[195] R.A. Irizarry, C. Ladd-Acosta, B. Carvalho, H. Wu, S.A. Brandenburg, J.A. Jeddeloh, Bo Wen, and Andrew P. Feinberg. Comprehensive high-throughput arrays for relative methylation (CHARM). *Genome Research*, 18(5):780–790, 2008.

[196] T. Ito, T. Chiba, R. Ozawa, M. Yoshida, M. Hattori, and Y. Sakaki. A comprehensive two-hybrid analysis to explore the yeast protein interactome. *Proceedings of the National Academy of Sciences*, 98(8):4569–4574, 2001.

[197] T. Ito, K. Tashiro, S. Muta, R. Ozawa, T. Chiba, M. Nishizawa, K. Yamamoto, S. Kuhara, and Y. Sakaki. Toward a protein-protein interaction map of the budding yeast: A comprehensive system to examine two-hybrid interactions in all possible combinations between the yeast proteins. *Proceedings of the National Academy of Sciences*, 97(3):1143–1147, 2000.

[198] Robert A. Jacobs and Michael I. Jordan. Adaptive mixtures of local experts, 1991.

[199] A. Jauhiainen, O. Nerman, G. Michailidis, and R. Jörnsten. Transcriptional and metabolic data integration and modeling for identification of active pathways. *Biostatistics*, 13(4):748–761, 2012.

[200] L.E. Jensen, A.J. Etheredge, K.S. Brown, and A.S. Mitchell, L.E. Mitchell and A.S. Whitehead. Maternal genotype for the monocyte chemoattractant protein 1 a(-2518)g promotor polymorphism is associated with the risk of spina bifida in offspring. *American Journal of Medical Genetics Part A*, 140A: 1114–1118, 2006.

[201] H. Jeong, S.P. Mason, A.-L. Barabasi, and Z.N. Olivai. Lethality and centrality in protein networks. *Nature*, 411:41–42, 2001.

[202] L. Jia, B.P. Berman, U. Jariwala, X. Yan, J.P. Cogan, A. Walters, T. Chen, G. Buchanan, B. Frenkel, and G.A. Coetzee. Genomic androgen receptor-occupied regions with different functions, defined by histone acetylation, coregulators and transcriptional capacity. *PLoS ONE*, 3(11), 2008.

[203] D. Jiang, C. Tang, and A. Zhang. Cluster analysis for gene expression data: A survey. *IEEE Transactions on Knowledge and Data Engineering*, 16:1370–1386, 2004.

[204] H. Jiang and W.H. Wong. Statistical inferences for isoform expression in rna-seq. *Bioinformatics*, 25(8):1026–1032, 2009.

[205] H. Jiang and B. Matija Peterlin. Differential Chromatin Looping Regulates CD4 Expression in Immature Thymocytes. *Molecular and Cellular Biology*, 28(3):907–912, 2008.

[206] K. Jin, L. Ou-Yang, X.-M. Zhao, H. Yan, and X.-F. Zhang. scTSSR: gene expression recovery for single-cell RNA sequencing using two-side sparse self-representation. *Bioinformatics*, 36(10):3131–3138, 2020.

[207] S. John, N. Shephard, G. Liu, E. Zeggini, Cao. M., W. Chen, N. Vasavda, T. Mills, A. Barton, A. Hinks, S. Eyre, K.W. Jones, W. Ollier, A. Silman, N. Gibson, J. Worthington, and G.C. Kennedy. Whole-genome scan, in a complex disease, using 11,245 single-nucleotide poly- morphisms: comparison with microsatellites. *American Journal of Human Genetics*, 75:54–64, 2004.

[208] D.S. Johnson, A. Mortazavi, R.M. Myers, and B. Wold. Genome-wide mapping of in vivo protein-dna interactions. *Science*, 316(5830):1497–1502, 2007.

[209] W.E. Johnson, C. Li, and A. Rabinovic. Adjusting batch effects in microarray expression data using empirical bayes methods. *Biostatistics*, 8(1):118–127, 2007.

[210] P.L. Jones, G.J. Veenstra, P.A. Wade, D. Vermaak, S.U. Kass, N. Landsberger, J. Strouboulis, and A.P. Wolffe. Methylated DNA and MeCP2 recruit histone deacetylase to repress transcription. *Nature Genetics*, 19(2):187–191, 1998.

[211] C. Jourdan, A.-K. Petersen, C. Gieger, A. Doring, T. Illig, R. Wang-Sattler, C. Meisinger, A. Peters, J. Adamski, C. Prehn, K. Suhre, E. Altmaier, G. Kastenmuller, W. Romisch-Margl, F.J. Theis, J. Krumsiek, H.-E. Wichmann, and J. Linseisen. Body fat free mass is associated with the serum metabolite profile in a population-based study. *PLoS ONE*, 7(6):e40009, 2012.

[212] M.P. Joy, A. Brock, D.E. Ingber, and S. Huang. High-betweenness proteins in the yeast protein interaction network. *Journal of Biomedicine and Biotechnology*, 2005(2):96–103, 2005.

[213] S. Jozefczuk, S. Klie, G. Catchpole, J. Szymanski, A. Cuadros-Inostroza, D. Steinhauser, J. Selbig, and L. Willmitzer. Metabolomic and transcriptomic stress response of escherichia coli. *Molecular Systems Biology*, 6:364, 2010.

[214] W.G. Kaelin Jr. The concept of synthetic lethality in the context of anticancer therapy. *Nature Reviews Cancer*, 5(9):689–698, 2005.

[215] R. Kalhor, H. Tjong, N. Jayathilaka, F. Alber, and L. Chen. Genome architectures revealed by tethered chromosome conformation capture and population-based modeling. *Nature Biotechnology*, 30(1):90–98, 2012.

[216] A. Kamburov, R. Cavill, T.M. Ebbels, R. Herwig, and H.C. Keun. Integrated pathway-level analysis of transcriptomics and metabolomics data with impala. *Bioinformatics*, 27(20):2917–2918, 2011.

[217] M.A. Kamleh, T.M. Ebbels, K. Spagou, P. Masson, and E.J. Want. Optimizing the use of quality control samples for signal drift corrction in large-scale urine metabolic profiling studies. *Analytical Chemistry*, 84(6):2670–2677, 2012.

[218] M. Kanehisa and S. Goto. Kegg: kyoto encyclopedia of genes and genomes. *Nucleic Acids Research*, 28(1):27–30, 2000.

[219] M. Kanehisa, Y. Sato, M. Kawashima, M. Furumichi, and M. Tanabe. Kegg as a reference resource for gene and protein annotation. *Nucleic Acids Research*, Database issue:doi:10.1093/nar/gkv1070, 2015.

[220] W.C. Kao, K. Stevens, and Y.S. Song. Bayescall: a model-based base-calling algorithm for high-throughput short-read sequencing. *Genome Research*, 19(10):1884–1895, 2009.

[221] C.-A. Kapourani and G. Sanguinetti. Melissa: Bayesian clustering and imputation of single-cell methylomes. *Genome Biology*, 20(1):61, 2019.

[222] A. Karnovsky, T. Weymouth, T. Hull, V.G. Tarcea, G. Scardoni, C. Laudanna, M.A. Sartor, K.A. Stringer, H.V. Jagadish, C. Burant, B. Athey, and G.S. Omenn. Metscape 2 bioinformatics tool for the analysis and visualization of metabolomics and gene expression data. *Bioinformatics*, 28(3):373–380, 2012.

[223] M. Katajamaa, J. Miettinen, and M. Orešič. Mzmine: toolbox for processing and visualizion of mass spectrometry based molecular profile data. *Bioinformatics*, 22(5):634–636, 2006.

[224] M. Katajamaa and M. Orešič. Processing methods for differential analysis of lc/ms profile data. *BMC Bioinformatics*, 6:179, 2005.

[225] Y. Katz, E.T. Wang, E.M. Airoldi, and C.B. Burge. Analysis and design of rna sequencing experiments for identifying isoform regulation. *Nature Methods*, 7(12):1009–1015, 2010.

[226] N. Kessler, H. Neuweger, A. Bonte, G. Langenkämper, K. Niehaus, T.W. Nattkemper, and A. Goesmann. Meltdb 2.0-advances of the metabolomics software system. *Bioinformatics*, 29(19):2452–2459, 2013.

[227] J. Kettunen, T. Tukiainen, A.-P. Sarin, A. Ortega-Alonso, E. Tikkanen, L.-P. Lyytikainen, A.J. Kangas, P. Soininen, P. Wurtz, K. Silander, D.M. Dick, R.J. Rose, M.J. Savolainen, J. Viikari, M. Kahonen, T. Lehtimaki, K.H. Pietilainen, M. Inouye, M.I. McCarthy, A. Jula, J. Eriksson, O.T.

Raitakari, V. Salomaa, J. Kaprio, M.-R. Jarvelin, L. Peltonen, M. Perola, N.B. Freimer, M. Ala-Korpela, A. Palotie, and S. Ripatti. Genomewide association study identifies multiple loci influencing human serum metabolite levels. *Nature Genetics*, 44(3):269–276, 2012.

[228] S.M. Kielbasa, R. Wan, K. Sato, P. Horton, and M.C. Frith. Adaptive seeds tame genomic sequence comparison. *Genome Research*, 21:487–493, 2011.

[229] H.K. Kim and R. Verpoorte. Sample preparation for plant metabolomics. *Phytochemical Analysis*, 21:4–13, 2010.

[230] H.-J. Kim, G.G. Yardımcı, G. Bonora, V. Ramani, J. Liu, R. Qiu, C. Lee, J. Hesson, C.B. Ware, J. Shendure, Z. Duan, and W.S. Noble. Capturing cell type-specific chromatin compartment patterns by applying topic modeling to single-cell hi-c data. *PLoS Computational Biology*, 16(9):e1008173–e1008173, 2020.

[231] S. Kim, P.A. Thiessen, E.E. Bolton, J. Chen, G. Fu, A. Gindulyte, L. Han, J. He, S. He, B.A. Shoemaker, J. Wang, B. Yu, J. Zhang, and S.H. Bryant. Pubchem substance and compound databases. *Nucleic Acids Research*, 44(D1):C1202–D1213, 2016.

[232] T. Kind, G. Wohlgemuth, Y. Lee do, Y. Lu, M. Palazoglu, S. Shabaz, and O. Fiehn. Fiehnlib: mass spectral and retention index libraries for metabolomics based on quadrupole and time-of-flight gas chromatography/mass spectrometry. *Analytical Chemistry*, 81(24):10038–10048, 2009.

[233] D.A. Kleinjan and V. van Heyningen. Long-range control of gene expression: emerging mechanisms and disruption in disease. *American Journal of Human Genetics*, 76(1):8–32, 2005.

[234] A. Kong, V. Steinthorsdottir, G. Masson, G. Thorleifsson, P. Sulem, S. Besenbacher, A. Jonasdottir, A. Sigurdsson, K. Th Kristinsson, A. Jonasdottir, M.L. Frigge, A. Gylfason, P.I. Olason, S.A. Gudjonsson, S. Sverrisson, S.N. Stacey, B. Sigurgeirsson, K.R. Benediktsdottir, H. Sigurdsson, T. Jonsson, R. Benediktsson, J.H. Olafsson, O. Th Johannsson, A.B. Hreidarsson, G. Sigurdsson, A.C. Ferguson-Smith, D.F. Gudbjartsson, U. Thorsteinsdottir, and K. Stefansson. Parental origin of sequence variants associated with complex diseases. *Nature*, 462(7275):868–874, 2009.

[235] Y. Koren, R. Bell, and C. Volinsky. Matrix factorization techniques for recommender systems. *Computer*, 42(8):30–37, 2009.

[236] E. Korpelainen, J. Tuimala, P. Somervuo, M. Huss, and G. Wong. *RNA-seq Data Analysis – A Practical Approach*. Chapman and Hall/CRC, Oxfordshire, UK, 2015.

[237] J. Kriseman, C. Busick, S. Szelinger, and V. Dinu. Bing: biomedical informatics pipeline for next generation sequencing. *Journal of Biomedical Informatics*, 43(3):428–434, 2010.

[238] N.J. Krogan, G. Cagney, H. Yu, G. Zhong, X. Guo, A. Ignatchenko, J. Li, S. Pu, N. Datta, A.P. Tikuisis, T. Punna, J.M. Peregrín-Alvarez, M. Shales, X. Zhang, M. Davey, M.D. Robinson, A. Paccanaro, J.E. Bray, A. Sheung, B. Beattie, D.P. Richards, V. Canadien, A. Lalev, F. Mena, P. Wong, A. Starostine, M.M. Canete, J. Vlasblom, S. Wu, C. Orsi, S.R. Collins, S. Chandran, R. Haw, J.J. Rilstone, K. Gandi, N.J. Thompson, G. Musso, P. St Onge, S. Ghanny, M.H. Lam, G. Butland, A.M. Altaf-Ul, S. Kanaya, A. Shilatifard, E. O'Shea, J.S. Weissman, C.J. Ingles, T.R. Hughes, J. Parkinson, M. Gerstein, S.J. Wodak, A. Emili, and Greenblatt J.F. Global landscape of protein complexes in the yeast saccharomyces cerevisiae. *Nature*, 440(7084):637–643, 2006.

[239] N.J. Krogan, W.T. Peng, G. Cagney, M.D. Robinson, R. Haw, G. Zhong, X. Guo, X. Zhang, V. Canadien, D.P. Richards, B.K. Beattie, A. Lalev, W. Zhang, A.P. Davierwala, S. Mnaimneh, A. Starostine, A.P. Tikuisis, J. Grigull, N. Datta, J.E. Bray, T.R. Hughes, A. Emili, and J.F. Greenblatt. High-definition macromolecular composition of yeast rna-processing complexes. *Molecular Cell*, 13(2):225–239, 2004.

[240] J. Krumsiek, K. Suhre, A.M. Evans, M.W. Mitchell, R.P. Mohney, M.V. Milburn, B. Wägele, W. Römisch-Margl, T. Illig, J. Adamski, C. Gieger, F.J. Theis, and G. Kastenmüller. Mining the unknown: A systems approach to metabolite identification combining genetic and metabolic information. *PLoS Genetics*, 8(10):e1003005, 2012.

[241] J. Krumsiek, K. Suhre, T. Illig, J. Adamski, and F.J. Theis. Gaussian graphical modeling reconstructs pathway reactions from high-throughput metabolomics data. *BMC Systems Biology*, 5:21, 2011.

[242] B. Langmead. Aligning short sequencing reads with bowtie. *Curr. Protoc. Bioinform.*, 32:11–17, 2010.

[243] M. Lappe and L. Holm. Unraveling protein interaction networks with near-optimal efficiency. *Nature Biotechnology*, 22:98–103, 2004.

[244] C.A. Lareau and M.J. Aryee. diffloop: a computational framework for identifying and analyzing differential dna loops from sequencing data. *Bioinformatics*, 34:672–674, 2018.

[245] C.A. Lareau and M.J. Aryee. hichipper: a preprocessing pipeline for calling dna loops from hichip data. *Nature Methods*, 15:155–156, 2018.

[246] H.A. Lawson, J.M. Cheverud, and J.B. Wolf. Genomic imprinting and parent-of-origin effects on complex traits. *Nature Reviews Genetics*, 14:609–617, 2013.

[247] C. Lee. Coimmunoprecipitation assay. *Methods in Molecular Biology*, 362:401–406, 2007.

[248] D.-S. Lee, C. Luo, J. Zhou, S. Chandran, A. Rivkin, A. Bartlett, J.R. Nery, C. Fitzpatrick, C. O'Connor, J.R. Dixon, and J.R. Ecker. Simultaneous profiling of 3d genome structure and dna methylation in single human cells. *Nature Methods*, 16(10):999–1006, 2019.

[249] J.T. Leek and J.D. Storey. Capturing heterogeneity in gene expression studies by surrogate variable analysis. *PLoS Genetics*, 3(9):e161, 2007.

[250] A. Lesne, J. Riposo, P. Roger, A. Cournac, and J. Mozziconacci. 3d genome reconstruction from chromosomal contacts. *Nature Methods*, advance online publication, 2014.

[251] E. Li, T.H. Bestor, and R. Jaenisch. Targeted mutation of the dna methyltransferase gene results in embryonic lethality. *Cell*, 69(6):915–926, 1992.

[252] G. Li, M.J. Fullwood, H. Xu, F.H. Mulawadi, S. Velkov, V. Vega, P.N. Ariyaratne, Y.B. Mohamed, H.S. Ooi, C. Tennakoon, C.L. Wei, Y. Ruan, and W.K. Sung. ChIA-PET tool for comprehensive chromatin interaction analysis with paired-end tag sequencing. *Genome Biology*, 11(2):R22, 2010.

[253] G. Li, X. Ruan, R.K. Auerbach, K.S. Sandhu, M. Zheng, P. Wang, H.M. Poh, Y. Goh, J. Lim, J. Zhang, H.S. Sim, S.Q. Peh, F.H. Mulawadi, C.T. Ong, Y.L. Orlov, S. Hong, Z. Zhang, S. Landt, D. Raha, G. Euskirchen, C.L. Wei, W. Ge, H. Wang, C. Davis, K.I. F.-Aylor, A. Mortazavi, M. Gerstein, T. Gingeras, B. Wold, Y. Sun, M.J. Fullwood, E. Cheung, E. Liu, W.K. Sung, M. Snyder, and Y. Ruan. Extensive promoter-centered chromatin interactions provide a topological basis for transcription regulation. *Cell*, 148(1-2):84–98, 2012.

[254] H. Li. Aligning sequence reads, clone sequences and assembly contigs with bwa-mem. *arXiv preprint arXiv:13033997*, 2013.

[255] H. Li and R. Durbin. Fast and accurate short read alignment with burrows-wheeler transform. *Bioinformatics*, 25(14):1754–1760, 2009.

[256] H. Li, J. Ruan, and R. Durbin. Mapping short dna sequencing reads and calling variants using mapping quality scores. *Genome Research*, 18(11):1851–1858, 2008.

[257] R. Li, Y. Li, K. Kristiansen, and J. Wang. Soap: Short oligonucleotide alignment program. *Bioinformatics*, 24:713–714, 2008.

[258] W. V. Li and J. J. Li. An accurate and robust imputation method scimpute for single-cell rna-seq data. *Nature Communications*, 9(1):1–9, 2018.

[259] X. Li, X. Lu, J. Tian, P. Gao, H. Kong, and Xu. G. Application of fuzzy c-means clustering in data analysis of metabolomics. *Analytical Chemistry*, 81(11):4468–4475, 2009.

[260] X. Li, Z. An, and Z. Zhang. Comparison of computational methods for 3d genome analysis at single-cell hi-c level. Computational modeling of three-dimensional genome structure. *Methods*, 181-182:52–61, 2020.

[261] Z. Li and C. Chan. Integrating gene expression and metabolic profiles. *Journal of Biological Chemistry*, 279:27124–27137, 2004.

[262] E. L.-Aiden, N.L. van Berkum, L. Williams, M. Imakaev, T. Ragoczy, A. Telling, I. Amit, B.R. Lajoie, P.J. Sabo, M.O. Dorschner, R. Sandstrom, B. Bernstein, M.A. Bender, M. Groudine, A. Gnirke, J. Stamatoyannopoulos, L.A. Mirny, E.S. Lander, and J. Dekker. Comprehensive mapping of long-range interactions reveals folding principles of the human genome. *Science*, 326(5950):289–293, 2009.

[263] M. Lienhard, C. Grimm, M. Morkel, R. Herwig, and L. Chavez. MEDIPS: Genome-wide differential coverage analysis of sequencing data derived from DNA enrichment experiments. *Bioinformatics*, 30(2):284–286, 2014.

[264] D. H.K. Lim and E. R. Maher. Human imprinting syndromes. *Epigenomics*, 1(2):347–369, 2009. PMID: 22122706.

[265] S. Lim, Y. Lu, C.Y. Cho, I. Sung, J. Kim, Y. Kim, S. Park, and S. Kim. A review on compound-protein interaction prediction methods: Data, format, representation and model. *Computational and Structural Biotechnology Journal*, 19:1541–1556, 2021.

[266] S. Lin. Assessing the Effects of Imprinting and Maternal Genotypes on Complex Genetic Traits. *In Risk Assessment and Evaluation of Predictions. Lee, Mlt and Gail, M and Pfeiffer, R and Satten, G and Cai, T and Gandy, a (Eds.)* (Springer Lecture Notes in Statistics 210). Springer, New York, 2013.

[267] S.M. Lin, L. Zhu, A.Q. Winter, M. Sasinowski, and W. Kibbe. What is mzxml good for? *Expert Review of Proteomics*, 2(6):839–845, 2005.

[268] R. Lister and J.R. Ecker. Finding the fifth base: Genome-wide sequencing of cytosine methylation. *Genome Research*, 19(6): 959–966, 2009.

[269] R. Lister, R.C. O'Malley, J. Tonti-Filippini, B.D. Gregory, C.C. Berry, A.H. Millar, and J.R. Ecker. Highly Integrated Single-Base Resolution Maps of the Epigenome in Arabidopsis. *Cell*, 133(3):523–536, 2008.

[270] R. Lister, M. Pelizzola, R.H. Dowen, R.D. Hawkins, G. Hon, J. Tonti-Filippini, J.R. Nery, L. Lee, Z. Ye, Q.-M.M. Ngo, L. Edsall, J. Antosiewicz-Bourget, R. Stewart, V. Ruotti, A.H. Millar, J.A. Thomson, B. Ren, and J.R. Ecker. Human DNA methylomes at base resolution show widespread epigenomic differences. *Nature*, 462(7271):315–322, 2009.

[271] R.J.A. Little. Testing the equality of two independent binomial proportions. *The American Statistician*, 43:283–288, 2016.

[272] H. Liu, X. Liu, S. Zhang, J. Lv, S. Li, S. Shang, S. Jia, Y. Wei, F. Wang, J. Su, Q. Wu, and Y. Zhang. Systematic identification and annotation of human methylation marks based on bisulfite sequencing methylomes reveals distinct roles of cell type-specific hypomethylation in the regulation of cell identity genes. *Nucleic Acids Research*, 44:75–94, 2015.

[273] H. Liu, J. Sun, J. Guan, J. Zheng, and S. Zhou. Improving compound–protein interaction prediction by building up highly credible negative samples. *Bioinformatics*, 31:i221–i229, 2015.

[274] S. Lomvardas, G. Barnea, D.J. Pisapia, M. Mendelsohn, J. Kirkland, and R. Axel. Interchromosomal interactions and olfactory receptor choice. *Cell*, 126(2):403–413, 2006.

[275] S. Lou. *Bayesian Analysis for Significant Interactions of Chromatins and Simulation Algorithm, Ph.D Dissertation, The Ohio State University*. OhioLink, 2021.

[276] K. Luck, D.-K. Kim, L. Lambourne, et al. A reference map of the human binary protein interactome. *Nature*, 580(7803):402–408, 2020.

[277] G. Lunter and M. Goodson. Stampy: A statistical algorithm for sensitive and fast mapping of illumina sequence reads. *Genome Research*, 21:936–939, 2011.

[278] H. Luo and S. Lin. Evaluation of classical statistical methods for analyzing bs-seq data. *OBM Genetics*, 2(4):053, 2018.

[279] K.-S. Lynn, M.-L. Cheng, Y.-R. Chen, C. Hsu, A. Chen, T.M. Lih, H.-Y. Chang, C.-J. Huang, M.-S. Chiao, W.-H. Pan, T.-Y. Sung, and W.-L. Hsu. Metabolite identification for mass spectrometry-based metabolomics using multiple types of correlated ion information. *Analytical Chemistry*, 87:2143–2151, 2015.

[280] B.P. Madakashira and K.C. Sadler. Dna methylation, nuclear organization, and cancer. *Frontiers in Genetics*, 8:76, 2017.

[281] M.A. Mahdavi and Y.-H. Lin. False positive reduction in protein-protein interaction predictions using gene ontology annotations. *BMC Bioinformatics*, 8:262, 2007.

[282] M. Maia, F. Monteiro, M. Sebastiana, A.P. Marques, A.E.N. Ferreira, A.P. Freire, C. Cordeiro, A. Figueiredo, and M.S. Silva. Metabolite extraction for high-throughput fticr-ms-based metabolomics of grapevine leaves. *EuPA Open Proteomics*, 12:4–9, 2016.

[283] T. Mallia, A. Grech, A. Hili, J.C.-Agius, and N.P. Pace. Genetic determinants of low birth weight. *Minerva Ginecologica*, 69(6):631–643, 2017.

[284] Y. Mao, L. Kuo, S.-W. Chen, C.J. Heckman, and M.C. Jiang. The essential and downstream common proteins of amyotrophic lateral sclerosis: A protein-protein interaction network analysis. *PLoS ONE*, 12(3):e0172246, 2017.

[285] T. Massingham and N. Goldman. All your base: a fast and accurate probabilistic approach to base calling. *Genome Biology*, 13:R13, 2012.

[286] D.J. McCarthy, Y. Chen, and G.K. Smyth. Differential expression analysis of multifactor rna-seq experiments with respect to biological variation. *Nucleic Acids Research*, 40(10):4288–4297, 2012.

[287] A. Meissner, A. Gnirke, G.W. Bell, B. Ramsahoye, E.S. Lander, and R. Jaenisch. Reduced representation bisulfite sequencing for comparative high-resolution DNA methylation analysis. *Nucleic Acids Research*, 33(18):5868–5877, 2005.

[288] A.M. Mezlini, E.J. Smith, M. Fiume, O. Buske, G.L. Savich, S. Shah, S. Aparicio, D.Y. Chiang, A. Goldenberg, and M. Brudno. ireckon: simultaneous isoform discovery and abundance estimation from rna-seq data. *Genome Research*, 23(3):519–529, 2013.

[289] J.T. Lee and and M.S. Bartolomei. X-Inactivation, Imprinting, and Long Noncoding RNAs in Health and Disease. Cell, 152(6):1308–1323, 2013.

[290] D.S. Millington and R.D. Stevens. Acylcarnitines: analysis in plasma and whole blood using tandem mass spectrometry. *Methods in Molecular Biology*, 708:55–72, 2011.

[291] X. Ming, B. Zhu, and Z. Zhang. Simultaneously measuring the methylation of parent and daughter strands of replicated dna at the single-molecule level by hammer-seq. *Nature Protocols*, 16(4):2131–2157, 2021.

[292] T.P. Minka. Estimating a dirichlet distribution. *Microsoft Technical Report*, 2003.

[293] R. Mitra, R. Gill, S. Datta, and S. Datta. Statistical analyses of next generation sequencing data: an overview. In S. Datta and D. Nettleton, editors, *Statistical Analysis of Next Generation Sequencing Data*. Springer, New York, 2014.

[294] B.J. Molyneaux, P. Arlotta, J.R.L. Menezes, and J.D. Macklis. Neuronal subtype specification in the cerebral cortex. *Nature Reviews Neuroscience*, 8(6):427–437, 2007.

[295] A. Mongia, D. Sengupta, and A. Majumdar. Mcimpute: Matrix completion based imputation for single cell rna-seq data. *Frontiers in Genetics*, 10:9, 2019.

[296] L.M. Montero, J. Filipski, P. Gil, J. Capel, J.M. Martínez-Zapater, and J. Salinas. The distribution of 5-methylcytosine in the nuclear genome of plants. *Nucleic Acids Research*, 20(12):3207–3210, 1992.

[297] V.K. Mootha, C.M. Cecilia M Lindgren, K.-F. Eriksson, A. Subramanian, S. Sihag, J Lehar, P. Puigserver, E. Carlsson, M. Ridderstråle, E. Laurila, N. Houstis, M.J. Daly, N. Patterson, J.P. Mesirov, T.R. Golub, P. Tamayo, B. Spiegelman, E.S. Lander, J.N. Hirschhorn, D. Altshuler, and L.C. Groop. Pgc-1alpha-responsive genes involved in oxidative phosphorylation are coordinately downregulated in human diabetes. *Nature Genetics*, 34:267–273, 2003.

[298] S. Moran, C. Arribas, and M. Esteller. Validation of a dna methylation microarray for 850,000 cpg sites of the human genome enriched in enhancer sequences. *Epigenomics*, 8:389–399, 2016.

[299] A. Morgat, E. Coissac, E. Coudert, K.B. Axelsen, G. Keller, A. Bairoch, A. Bridge, L. Bougueleret, I. Xenarios, and Viari A. Unipathway: a resource for the exploration and annotation of metabolic pathways. *Nucleic Acids Research*, 40(Database issue):D761–D769, 2012.

[300] I.M. Morison, C.J. Paton, and S.D. Cleverley. The imprinted gene and parent-of-origin effect database. *Nucleic Acids Research*, 29:275–276, 2001.

[301] A. Mortazavi, B.A. Williams, K. McCue, L. Schaeffer, and B. Wold. Mapping and quantifying mammalian transcriptomes by rna-seq. *Nature Methods*, 5(7):621–628, 2008.

[302] L.H. Moulton and N.A. Halsey. A mixture model with detection limites for regression analysis of antibody response to vaccine. *Biometrics*, 51(4):1570–1578, 1995.

[303] J.F. Moxley, M.C. Jewett, M.R. Antoniewicz, S.G. Villas-Boas, H. Alper, L. Wheeler, R.T. andTong, A.G. Hinnebusche, T. Ideker, J. Nielsen, and G. Stephanopoulosa. Linking high-resolution metabolic flux phenotypes and transcriptional regulation in yeast modulated by the global regulator gcn4p. *Proceedings of the National Academy of Sciences*, 106(16):6477–6482, 2009.

[304] M.R. Mumbach, A.J. Rubin, R.A. Flynn, C. Dai, P.A. Khavari, W.J. Greenleaf, and H.Y. Chang. Hichip: efficient and sensitive analysis of protein-directed genome architecture. *Nature Methods*, 13:919–922, 2016.

[305] A. Murrell, S. Heeson, and W. Reik. Interaction between differentially methylated regions partitions the imprinted genes Igf2 and H19 into parent-specific chromatin loops. *Nature Genetics*, 36(8):889–893, 2004.

[306] T. Nagano, Y. Lubling, T.J. Stevens, S. Schoenfelder, E. Yaffe, W. Dean, E.D. Laue, A. Tanay, and P. Fraser. Single-cell Hi-C reveals cell-to-cell variability in chromosome structure. *Nature*, 502(7469):59–64, 2013.

[307] T. Nagano, Y. Lubling, C. Várnai, C. Dudley, W. Leung, Y. Baran, N.M. Cohen, S. Wingett, P. Fraser, and A. Tanay. Cell-cycle dynamics of chromosomal organization at single-cell resolution. *Nature*, 547(7661):61–67, 2017.

[308] C.B. Newgard, J. An, J.R. Bain, M. Muehlbauer, R.D. Stevens, L.F. Lien, A.M. Haqq, S.H. Shah, M. Arlotto, C.A. Slentz, J. Rochon, D. Gallup, O. Ilkayeva, B.R. Wenner, W.S. Yancy Jr., H. Eisenson, G. Musante, R.S. Surwit, D.S. Millington, M.D. Butler, and L.P. Svetkey. A branched-chain amino acid-related metabolic signature that differentiates obese and lean humans and contributes to insulin resistance. *Cell Metabolism*, 9(4):311–326, 2009.

[309] M. Newman and M. Girvan. Finding and evaluating community structure in networks. *Physical Review E*, 69:026113, 2004.

[310] M.E.J. Newman. Finding community structure using the eigenvectors of matrices. *Physical Review E*, 74:036104, 2006.

[311] Y. Ni, Y. Qiu, W. Jiang, K. Suttlemyre, M. Su, W. Zhang, W. Jia, and X. Du. Adap-gc 2.0: deconvolution of coeluting metabolites from gc/tof-ms data for metabolomics studies. *Analytical Chemistry*, 84(15):6619–6629, 2012.

[312] J.K. Nicholson, J.C. Lindon, and E. Holmes. 'metabonomics': Understanding the metabolic responses of living systems to pathophysiological stimuli via multivariate statistical analysis of biological nmr spectroscopic data. *Xenobiotica*, 29:1181–1189, 1998.

[313] L. Niu, G. Li, and S. Lin. Statistical models for detecting differential chromatin interactions mediated by a protein. *PLoS One*, 9(5):e97560, 2014.

[314] L. Niu and S. Lin. A Bayesian mixture model for chromatin interaction data. *Statistical Applications in Genetics and Molecular Biology*, 14(1):53–64, 2015.

[315] M. Nodzenski, M.J. Muehlbauer, J.R. Bain, A.C. Reisetter, W.L. Lowe Jr., and D.M. Scholtens. Metabomxtr: an r package for mixture-model analysis of non-targeted metabolomics data. *Bioinformatics*, 30(22):3287–3288, 2014.

[316] J. Noh. Exact scaling properties of a hierarchical network model. *Physical Review E*, 67(The 4), 2003.

[317] D. Nousome, P.J. Lupo, M.F. Okcu, and M.E. Scheurer. Maternal and offspring xenobiotic metabolism haplotypes and the risk of childhood acute lymphoblastic leukemia. *Leukemia Research*, 37:531–535, 2013.

[318] C. Ober. Hla and pregnancy: the paradox of the fetal allograft. *American Journal of Human Genetics*, 62:1–5, 1998.

[319] S.G. Oliver, M.L. Winson, D.B. Kell, and F. Baganz. Systematic functional analysis of the yeast genome. *Trends in Biotechnology*, 16:373–378, 1998.

[320] O. Oluwadare, M. Highsmith, and J. Cheng. An overview of methods for reconstructing 3-d chromosome and genome structures from hi-c data. *Biological Procedures Online*, 21(1):7, 2019.

[321] S. Orchard, M. Ammari, B. Aranda, L. Breuza, L. Briganti, F. Broackes-Carter, N.H. Campbell, G. Chavali, C. Chen, N. del Toro, M. Duesbury, M. Dumousseau, E. Galeota, U. Hinz, M. Iannuccelli, S. Jagannathan, R. Jimenez, J. Khadake, A. Lagreid, L. Licata, R.C. Lovering, B. Meldal, A.N. Melidoni, M. Milagros, D. Peluso, L. Perfetto, P. Porras, A. Raghunath, S. Ricard-Blum, B. Roechert, A. Stutz, M. Tognolli, K. van Roey, G. Cesareni, and Hermjakob H. The mintact project–intact as a common curation platform for 11 molecular interaction databases. *Nucleic Acids Research*, 42(Database issue):D358–D363, 2014.

[322] R. Oughtred, J. Rust, C. Chang, B.J. Breitkreutz, C. Stark, A. Willems, L. Boucher, G. Leung, N. Kolas, F. Zhang, S. Dolma, J. Coulombe-Huntington, A Chatr-Aryamontri, K. Dolinski, and M. Tyers. The biogrid database: A comprehensive biomedical resource of curated protein, genetic, and chemical interactions. *Protein Science*, 30(1):178–200, 2021.

[323] M. Padi and J. Quackenbush. Integrating transcriptional and protein interaction networks to prioritize condition-specific master regulators. *BMC Systems Biology*, 9:80, 2015.

[324] A. Paksa and J. Rajagopal. The epigenetic basis of cellular plasticity. *Current Opinion in Cell Biology*, 49:116–122, 2017.

[325] C.G.S. Palmer, H.-J. Hsieh, E.F. Reed, J. Lonnqvist, L. Peltonen, J.A. Wood-Ward, and J.S. Sinsheimer. Hla-b maternal-fetal genotype matching increases risk of schizophrenia. *American Journal of Human Genetics*, 79:710–715, 2006.

[326] J. Park and S. Lin. *Statistical Inference on Three-Dimensional Structure of Genome by Truncated Poisson Architecture Model. In Ordered Data Analysis, Modeling, and Health Research Methods - In Honor of H. N. Nagaraja's 60th Birthday. Choudhary, P., Nagaraja, C., Ng, T. (Eds.) (Springer Proceedings in Mathematics and Statistics)*. Springer, New York, 2015.

[327] J. Park and S. Lin. Evaluation and comparison of methods for recapitulation of 3d spatial chromatin structures. *Briefings in Bioinformatics*, page in press, 2017.

[328] J. Park and S. Lin. A random effect model for reconstruction of spatial chromatin structure. *Biometrics*, 73:52–62, 2017.

[329] J. Park and S. Lin. Detection of Differentially Methylated Regions Using Bayesian Curve Credible Bands. *Statistics for Biosciences*, 10:20–40, 2018.

[330] J. Park and S. Lin. Evaluation and comparison of methods for recapitulation of 3d spatial chromatin structures. *Briefings in Bioinformatics*, 20:1205–1214, 2019.

[331] J. Park and S. Lin. Impact of data resolution on three-dimensional structure inference methods. *BMC Bioinformatics*, 17(1):70, 2016.

[332] Y. Park, M.E. Figueroa, L.S. Rozek, and M.A. Sartor. MethylSig: a whole genome DNA methylation analysis pipeline. *Bioinformatics*, 30(17):2414–2422, 2014.

[333] Y. Park and H. Wu. Differential methylation analysis for bs-seq data under general experimental design. *Bioinformatics*, 32:1446–1453, 2016.

[334] H.M. Parsons, D.R. Ekman, T.W. Collette, and M.R. Viant. Spectral relative standard deviation: a practical benchmark in metabolomics. *Analyst*, 134(3):478–485, 2009.

[335] R.K. Patel and Jain. M. Ngs qc toolkit: A toolkit for quality control of next generation sequencing data. *PLoS One*, 7:e30619, 2012.

[336] J. Paulsen, E.A. Rodland, L. Holden, M. Holden, and E. Hovig. A statistical model of ChIA-PET data for accurate detection of chromatin 3D interactions. *Nucleic Acids Research*, 42(18):1–11, 2014.

[337] P.G. Pedrioli, J.K. Eng, R. Hubley, M. Vogelzang, E.W. Deutsch, B. Raught, B. Pratt, E. Nilsson, R.H. Angeletti, R. Apweiler, K. Cheung, C.E. Costello, H. Hermjakob, S. Huang, R.K. Julian, E. Kapp, M.E. McComb, S.G. Oliver, G. Omenn, N.W. Paton, R. Simpson, R. Smith, C.F. Taylor, W. Zhu, and R. Aebersold. A common open representation of mass spectrometry data and its application to proteomics research. *Nature Biotechnology*, 22(11):1459–1466, 2004.

[338] H.E. Pence and A. Williams. Chemspider: an online chemical information resource. *Journal of Chemical Education*, 87(11):1123–1124, 2010.

[339] T. Peng, Q. Zhu, P. Yin, and K. Tan. Scrabble: single-cell rna-seq imputation constrained by bulk rna-seq data. *Genome Biology*, 20(1):88, 2019.

[340] J. Peters. The role of genomic imprinting in biology and disease: an expanding view. *Nature Reviews Genetics*, 15(8):517–530, 2014.

[341] D.H. Phanstiel, A.P. Boyle, N. Heidari, and M.P. Snyder. Mango: a biascorrecting chia-pet analysis pipeline. *Bioinformatics*, 31:3092–3098, 2015.

[342] B. Pir, P. andKırdar, Z.İ. Hayes, A. andÖnsan, K.Ö. Ülgen, and S.G. Oliver. Integrative investigation of metabolic and transcriptomic data. *BMC Bioinformatics*, 7:203, 2006.

[343] L. Ploughman and M. Boehnke. Estimating the power of a proposed linkage study for a complex genetic trait. *American Journal of Human Genetics*, 44:543–551, 1989.

[344] T. Pluskal, S. Castillo, A. Villar-Briones, and M. Orešič. Mzmine 2: Modular framework for processing, visualizing, and analyzing mass spectrometry-based molecular profile data. *BMC Bioinformatics*, 11:395, 2010.

[345] D. Polioudakis, L. de la Torre-Ubieta, J. Langerman, A.G. Elkins, Xu Shi, Jason L. Stein, Celine K. Vuong, Susanne Nichterwitz, Melinda

Gevorgian, Carli K. Opland, et al. A single-cell transcriptomic atlas of human neocortical development during mid-gestation. *Neuron*, 103(5):785–801, 2019.

[346] P. Pons and M. Latapy. Computing communities in large networks using random walks. *Journal of Graph Algorithms and Applications*, 10(2):284–293, 2004.

[347] S.B. Pounds, C.L. Gao, and H. Zhang. Empirical bayesian selection of hypothesis testing procedures for analysis of sequence count expression data. *Statistical Applications in Genetics and Molecular Biology*, 11(5):Article 7, 2012.

[348] K.T.S. Prasad, R. Goel, K. Kandasamy, S. Keerthikumar, S. Kumar, S. MAthivanan, D. Telikicherla, R. Raju, B. Shafreen, A. Venugopal, L. Balakrishnan, A. Marimuthu, S. Banerjee, D.S. Somanathan, A. Sebastian, S. Rani, S. Ray, C.J. Harrys Kishore, S. Kanth, M. Ahmed, M.K. Kashyap, R. Mohmood, Y.L. Ramachandra, V. Krishna, B.A. Rahiman, S. Mohan, P. Ranganathan, S. Ramabadran, R. Chaerkady, and A. Pandey. Human protein reference database–2009 update. *Nucleic Acids Research*, 37(Database issue):D767–D772, 2009.

[349] O. Puig, F. Caspary, G. Riguat, B. Rutz, E. Bouveret, E. Bragado-Nilsson, M. Wilm, and B. Séraphin. The tandem affinity purification (tap) method: a general procedure of protein complex purification. *Methods*, 24(3):218–229, 2001.

[350] Z. Qin, B. Li, K.N. Conneely, H. Wu, M. Hu, D. Ayyala, Y. Park, V.X. Jin, F. Zhang, H. Zhang, L. Li, and S. Lin. Statistical challenges in analyzing methylation and long-range chromosomal interaction data. *Statistics in Biosciences*, 8(2):284–309, 2016.

[351] U.N. Raghavan, R. Albert, and S. Kumara. Near linear time algorithm to detect community structures in large-scale networks. *Physical Review E*, 76(3):036106, 2007.

[352] V. Ramani, X. Deng, R. Qiu, K.L. Gunderson, F.J. Steemers, C.M. Disteche, W.S. Noble, Z. Duan, and J. Shendure. Massively multiplex single-cell hi-c. *Nature Methods*, 14(3):263–266, 2017.

[353] S.S.P. Rao, M.H. Huntley, N.C. Durand, and E.K. Stamenova. A 3D map of the human genome at kilobase resolution reveals principles of chromatin looping. *Cell*, 159(7):1665–1680, 2014.

[354] J. Reichardt and S. Bornholdt. Statistical mechanics of community detection. *Physical Review E*, 74:016110, 2006.

[355] W. Reik. Stability and flexibility of epigenetic gene regulation in mammalian development. *Nature*, 447(7143):425–432, 2007.

[356] A.C. Reisetter, M.J. Muehlbauer, J.R. Bain, M. Nodzenski, R.D. Stevens, O. Ilkayeva, B.E. Metzger, C.B. Newgard, W.L. Lowe Jr., and D.M. Scholtens. Mixture model normalization for non-targeted gas chromatography/mass spectrometry metabolomics data. *BMC Bioinformatics*, 18(1):84, 2017.

[357] G. Renaud, M. Kircher, U. Stenzel, and J. Kelso. freeibis: An efficient base-caller with calibrated quality scores for illumina sequencers. *Bioinformatics*, 29(9):1208–1209, 2013.

[358] E.-A.P. Renyi. On the strength of connectedness of a random graph. *Acta Mathematica Academiae Scientiarum Hungaricae*, 12:261–267, 1961.

[359] R.K. Rew and G.P. Davis. Netcdf: An interface for scientific data access. *IEEE Computer Graphics and Applications*, 10(4):76–82, 1990.

[360] E.P. Rhee, J.E. Ho, M.-H. Chen, S. Dongxiao, S. Cheng, M.G. Larson, A. Ghorbani, X. Shi, Helenius I.T., C.J. O'Donnell, A.L. Souza, A. Deik, K.A. Pierce, K. Bullock, G.A. Walford, R.S. Vasan, J.C. Florez, C. Clish, J.-R.R. Yeh, T.J. Wang, and R.E. Greszten. A genome-wide association study of the human metabolome in a community-based cohort. *Cell Metabolism*, 18(1):130–143, 2013.

[361] B. Richardson. Primer: epigenetics of autoimmunity. *Nature Clinical Practice Rheumatology*, 3:521–527, 2007.

[362] J.S. Ried, S.-Y. Shin, J. Krumsiek, T. Illig, F.J. Theis, T.D. Spector, J. Adamski, H.-E. Wichmann, K. Strauch, N. Soranzo, K. Suhre, and C. Gieger. Novel genetic associations with serum level metabolites identified by phenotype set enrichment analyses. *Human Molecular Genetics*, 23(21):5847–5857, 2014.

[363] L.D. Roberts, A.L. Souza, R.E. Gerzsten, and C.B. Clish. Targeted metabolomics. *Current Protocols in Molecular Biology*, 98:30.2.1–30.2.24, 2012.

[364] M. Robinson, D. McCarthy, and G. Smyth. edger: a bioconductor package for differential expression analysis of digital gene expression data. *Bioinformatics*, 26:139–140, 2010.

[365] M.D. Robinson, A. Kahraman, C.W. Law, H. Lindsay, M. Nowicka, L.M. Weber, M.D. Robinson, A. Kahraman, and C.W. Law. Bioinformatics and computational biology statistical methods for detecting differentially methylated loci and regions MINI REVIEW : Statistical methods for detecting differentially methylated loci and regions. *Frontiers in Genetics*, 5:324, 2014.

[366] M.D. Robinson, D.J. McCarthy, and G.K. Smyth. edgeR: A Bioconductor package for differential expression analysis of digital gene expression data. *Bioinformatics*, 26(1):139–140, 2009.

[367] M.D. Robinson and G.K. Smyth. Small-sample estimation of negative binomial dispersion, with applications to sage data. *Biostatistics*, 9:321–332, 2008.

[368] P.W. Rose, A. Prlić, A. Altunkaya, C. Bi, A.R. Bradley, C.H. Christie, L.D. Costanzo, J.M. Duarte, S. Dutta, Z. Feng, R.K. Green, D.S. Goodsell, B. Hudson, T. Kalro, R. Lowe, E. Peisach, C. Randle, A.S. Rose, C. Shao, Y.P. Tao, Y. Valasatava, M. Voigt, J.D. Westbrook, J. Woo, H. Yang, J.Y. Young, C. Zardecki, H.M. Berman, and S.K. Burley. The rcsb protein data bank: integrative view of protein, gene and 3d structural information. *Nucleic Acids Research*, 45(D1):D271–D281, 2017.

[369] M. Rosenthal, D. Bryner, F. Huffer, S. Evans, A. Srivastava, and N. Neretti. Bayesian estimation of three-dimensional chromosomal structure from single-cell hi-c data. *Journal of Computational Biology: A Journal of Computational Molecular Cell Biology*, 26(11):1191–1202, 2019.

[370] J. Rougemont, A. Amzallag, C. Iseli, L. Farinelli, I. Xenarios, and F. Naef. Probabilistic base calling of solexa sequencing data. *BMC Bioinformatics*, 9:431, 2008.

[371] M. Rousseau, J. Fraser, M. Ferraiuolo, J. Dostie, and Mathieu Blanchette. Three-dimensional modeling of chromatin structure from interaction frequency data using Markov chain Monte Carlo sampling. *BMC Bioinformatics*, 12(1):414+, 2011.

[372] L. Salwinski, C.S. Miller, A.J. Smith, F.K. Pettit, J.U. Bowie, and D. Eisenberg. The database of interacting proteins: 2004 update. *Nucleic Acids Research*, 32(Database Issue):D449–D451, 2004.

[373] J. Salzman, H. Jiang, and W.H. Wong. Statistical modeling of rna-seq data. *Statistical Science*, 26(1):62–83, 2011.

[374] R.C. Samaco, A. Hogart, and J.M. LaSalle. Epigenetic overlap in autism-spectrum neurodevel- opmental disorders: Mecp2 deficiency causes reduced expression of ube3a and gabrb3. *Human Molecular Genetics*, 14:483–492, 2005.

[375] A. Sanyal, B.R. Lajoie, G. Jain, and J. Dekker. The long-range interaction landscape of gene promoters. *Nature*, 489(7414):109–113, 2012.

[376] R. Sanz-Pamplona, A. Berenguer, X. Sole, D. Cordero, M. Crous-Bou, J. Serra-Musach, E. Guinó, M.Á. Pujana, and V. Moreno. Tools for protein-protein interaction network analysis in cancer research. *Clinical and Translational Oncology*, 14(1):3–14, 2012.

[377] F. Savorani, G. Tomasi, and S.B. Engelsen. icoshift: A versatile tool for the rapid alignment of 1d nmr spectra. *Journal of Magnetic Resonance*, 202(2):190–202, 2010.

[378] M.H. Schaefer, L. Serrano, and M.A. Andrade-Navarro. Correcting for the study bias associated with protein–protein interaction measurements reveals differences between protein degree distributions from different cancer types. *Frontiers in Genetics*, 6:260, 2015.

[379] M.H. Schaefer, L. Serrano, and M.A. Andrade-Navarro. Recording negative results of protein–protein interaction assays: an easy way to deal with the biases and errors of interactomic data sets. *Frontiers in Genetics*, 6:260, 2015.

[380] K. Schneeberger, J. Hagmann, S. Ossowski, N. Warthmann, S. Gesing, O. Kohlbacher, and D. Weigel. Simultaneous alignment of short reads against multiple genomes. *Genome Biology*, 10:R98, 2009.

[381] S. Schoenfelder and P. Fraser. Long-range enhancer–promoter contacts in gene expression control. *Nature Reviews Genetics*, 20:437–455, 2019.

[382] D. Scholtens, T. Chiang, W. Huber, and R. Gentleman. Estimating node degree in bait-prey graphs. *Bioinformatics*, 24(2):218–224, 2008.

[383] D. Scholtens and B. Spencer. Node sampling for protein complex estimation in bait-prey graphs. *Statistical Applications in Genetics and Molecular Biology*, 14(4):391–411, 2015.

[384] D. Scholtens, M. Vidal, and R. Gentleman. Local modeling of global interactome netowkrs. *Bioinformatics*, 21(17):3548–3557, 2005.

[385] D.M. Scholtens, J.R. Bain, A.C. Reisetter, M. Muehlbauer, M. Nodzenski, R.D. Stevens, O. Ilkayeva, L.P. Lowe, B.E. Metzger, C.B. Newgard, and W.L. Lowe Jr. Metabolic networks and metabolites underlie associations between maternal glucose during pregnancy and newborn size at birth. *Diabetes*, 65(7):2039–2050, 2016.

[386] D.M. Scholtens, M. Muehlbauer, N.R. Daya, R.D. Stevens, A.R. Dyer, L.P. Lowe, B.E. Metzger, C.B. Newgard, J.R. Bain, and W.L. Lowe Jr. Metabolomics reveals broad-scale metabolic perturbations in hyperglycemic mothers during pregnancy. *Diabetes Care*, 37(1):158–166, 2014.

[387] A.S. Schwartz, J. Yu, K.R. Gardenour, R. Finley Jr, and T. Ideker. Cost-effective strategies for completing the interactome. *Nature Methods*, 6:55–61, 2009.

[388] G. Schwarz. Estimating the dimension of a model. *Annals of Statistics*, 6:461–464, 1978.

[389] M.R. Segal and H.L. Bengtsson. Reconstruction of 3D genome architecture via a two-stage algorithm. *BMC Bioinformatics*, 16(1):373, 2015.

[390] C.A. Sellick, D. Knight, A.S. Croxford, A.R. Maqsood, G.M. Stephens, R. Goodacre, and A.J. Dickson. Evaluation of extraction processes for intracellular metabolite profiling of mammalian cells: Matching extraction approaches to cell type and metabolite targets. *Metabolomics*, 6:427–438, 2010.

[391] D. Serre, B.H. Lee, and A.H. Ting. Mbd-isolated genome sequencing provides a high-throughput and comprehensive survey ofdnamethylation in the human genome. *Nucleic Acids Research*, 38:391–399, 2010.

[392] A. Shafi, C. Mitrea, T. Nguyen, and S. Draghici. A survey of the approaches for identifying differential methylation using bisulfite sequencing data. *Briefings in Bioinformatics*, 19:737–753, 2017.

[393] J.S. Shah, G.N. Brock, and S.N. Rai. Metabolomics data analysis and missing value issues with application to infarcted mouse hearts. *BMC Bioinformatics*, 16(Suppl 15):P16, 2015.

[394] S.H. Shah, W.E. Kraus, and C.B. Newgard. Metabolomic profiling for identification of novel biomarkers and mechanisms related to common cardiovascular diseases: form and function. *Circulation*, 126(9):1110–1120, 2012.

[395] N. Shahaf, I. Rogachev, U. Heinig, S. Meir, S. Malitsky, M. Battat, H. Wyner, S. Zheng, R. Wehrens, and A. Aharoni. The weizmass spectral library for high-confidence metabolite identification. *Nature Communications*, 7:12423, 2016.

[396] P. Shannon, A. Markiel, O. Ozier, N.S. Baliga, J.T. Wang, D. Ramage, N. Amin, B. Schwikowski, and T. Ideker. Cytoscape: a software environment for integrated models of biomolecular interaction networks. *Genome Research*, 13(11):2498–2504, 2003.

[397] R. Sharan, S. Suthram, R.M. Kelley, T. Kuhn, S. McCuine, P. Uetz, T. Sittler, R.M. Karp, and T. Ideker. Conserved patterns of protein interaction in multiple species. *Proceedings of the National Academy of Sciences*, 102(6):1974–1979, 2005.

[398] S. Shete and C.I. Amos. Testing for genetic linkage in families by a variance-components approach in the presence of genomic imprinting. *American Journal of Human Genetics*, 70:751–757, 2002.

[399] M. Shi, D.M. Umbach, S.H. Vermeulen, and C.R. Weinberg. Making the most of case-mother/control-mother studies. *American Journal of Epidemiology*, 168(5):541–547, 2008.

[400] M. Siddappa, S.A. Wani, M.D. Long, and et al. Identification of transcription factor co-regulators that drive prostate cancer progression. *Scientific Reports*, 10(1):20332, 2020.

[401] M. Simonis, P. Klous, E. Splinter, Y. Moshkin, R. Willemsen, E. de Wit, B. van Steensel, and W. de Laat. Nuclear organization of active and inactive chromatin domains uncovered by chromosome conformation capture-on-chip (4C). *Nature Genetics*, 38(11):1348–1354, 2006.

[402] J.S. Sinsheimer, C.G.S. Palmer, and J. Arthur Woodward. Detecting genotype combinations that increase risk for disease: The maternal-fetal genotype incompatibility test. *Genetic Epidemiology*, 24(1):1–13, 2003.

[403] S.A. Smallwood, H.J. Lee, C. Angermueller, F. Krueger, H. Saadeh, J. Peat, S.R. Andrews, O. Stegle, W. Reik, and G. Kelsey. Single-cell genome-wide bisulfite sequencing for assessing epigenetic heterogeneity. *Nat Methods*, 11:217–220, 2014.

[404] C.A. Smith, G. O'Maille, E.J. Want, C. Qin, S.A. Trauger, T.R. Brandon, D.E. Custodio, R. Abagyan, and G. Siuzdak. Metlin: a metabolite mass spectral database. *Therapeutic Drug Monitoring*, 27:747–841, 2005.

[405] C.A. Smith, E.J. Want, G. O'Maille, R. Abagyan, and G. Siuzdak. Xcms: Processing mass spectrometry data for metabolite profiling using nonlinear peak alignment, matching, and identification. *Analytical Chemistry*, 78(3):779–787, 2006.

[406] Z.D. Smith, M.M. Chan, T.S. Mikkelsen, H. Gu, A. Gnirke, A. Regev, and A. Meissner. A unique regulatory phase of dna methylation in the early mammalian embryo. *Nature*, 484:339–344, 2012.

[407] Z.D. Smith and A. Meissner. Dna methylation: roles in mammalian development. *Nature Reviews Genetics*, 14:2014–220, 2013.

[408] F. Spencer, M. Lagarde, A. Geloen, and M. Record. What is lipidomics? *European Journal of Lipid Science and Technology*, 105:481–482, 2003.

[409] C.G. Spilianakis and R.A. Flavell. Long-range intrachromosomal interactions in the T helper type 2 cytokine locus. *Nature Immunology*, 5(10):1017–1027, 2004.

[410] G.L. Splansky, D. Corey, Q. Yang, L.D. Atwood, L.A. Cupples, E.J. Benjamin, and et al. The third generation cohort of the national heart, lung, and blood institutes framingham heart study: design, recruitment, and initial examination. *American Journal of Epidemiology*, 165:1328–1335, 2007.

[411] S. Srivastava and L. Chen. A two-parameter generalized poisson model to improve the analysis of rna-seq data. *Nucleic Acids Research*, 38(17):e170, 2010.

[412] C. Stark, B.-J. Breitkreutz, T. Reguly, L. Boucher, A. Breitkreutz, and M. Tyers. Biogrid: a general repository for interaction datasets. *Nucleic Acids Research*, 34(Suppl 1):D535–D539, 2005.

[413] S.E. Stein. An integrated method for spectrum extraction and compound identification from gc/ms data. *Journal of the American Society of Mass Spectrometry*, 10:770–781, 1999.

[414] T.J. Stevens, D. Lando, S. Basu, L.P. Atkinson, Y. Cao, S.F. Lee, M. Leeb, K.J. Wohlfahrt, W. Boucher, A. O'Shaughnessy-Kirwan, J. Cramard, A.J. Faure, M. Ralser, E. Blanco, L. Morey, M. Sansó, M.G.S. Palayret, B. Lehner, L. Di Croce, A. Wutz, B. Hendrich, D. Klenerman, and E.D. Laue. 3d structures of individual mammalian genomes studied by single-cell hi-c. *Nature*, 544(7648):59–64, 2017.

[415] T. Strachan and A.P. Read. *Human Molecular Genetics*. Wiley, New York, 1999.

[416] M. Sturm, A. Bertsch, C. Gröpl, A. Hildebrandt, R. Hussong, N. Lange, E. Pfeifer, O. Schulz-Trieglaff, A. Zerck, and O. Reinert, K. Kohlbacher. Openms – an open-source software framework for mass spectrometry. *BMC Bioinformatics*, 9(1):1–11, 2008.

[417] J. Su, H. Yan, Y. Wei, H. Liu, H. Liu, F. Wang, J. Lv, Q. Wu, and Y. Zhang. Cpg mps: identification of cpg methylation patterns of genomic regions from high-throughput bisulfite sequencing data. *Nucleic Acids Research*, 41:e4, 2012.

[418] A. Subramanian, P. Pablo Tamayo, V.K. Mootha, S. Mukherjee, B.L. Ebert, M.A. Gillette, A. Paulovich, S.L. Pomeroy, T.R. Golub, E.S. Lander, and J.P. Mesirova. Gene set enrichment analysis: A knowledge-based approach for interpreting genome-wide expression profiles. *Proceedings of the National Academy of Sciences*, 102(43):15545–15550, 2005.

[419] M. Sugimoto, M. Kawakami, M. Robert, T. Soga, and M. Tomita. Bioinformatics tools for mass spectroscopy-based metabolomic data processing and analysis. *Current Bioinformatics*, 7(1):96–108, 2012.

[420] K. Suhre, S.-Y. Shin, A.-K. Petersen, R.P. Mohney, D. Meredith, B. Wagele, E. Altmaier, CARDIoGRAM, P. Deloukas, J. Erdmann, E. Grundberg, C.J. Hammond, M. Hrabe de Angelis, G. Kastenmuller, A. Kottgen, F. Kronenberg, M. Mangino, C. Meisinger, T. Meitinger, H.-W. Mewes, M.V. Milburn, C. Prehn, J. Raffler, J.S. Ried, W. Romisch-Margl, N.J. Samani, K.S. Small, H.-E. Wichmann, G. Zhai, T. Illig, T.D. Spector, J. Adamski, N. Soranzo, and C. Gieger. Human metabolic individuality in biomedical and pharmaceutical research. *Nature 2011*, 7362:54–60, 477.

[421] D. Sun, Y. Xi, B. Rodriguez, H. Park, P. Tong, M. Meong, M. Goodell, and W. Li. MOABS: model based analysis of bisulfite sequencing data. *Genome Biology*, 15(2):R38, 2014.

[422] S. Sun and X. Yu. Hmm-fisher: identifying differential methylation using a hidden markov model and fisher's exact test. *Statistical Applications in Genetics and Molecular Biology*, 15:55–67, 2016.

[423] M. Sysi-Aho, M. Katajamaa, L. Yetukuri, and M. Orešič. Normalization method for metabolomics data using optimal selection of multiple internal standards. *BMC Bioinformatics*, 8:93, 2007.

[424] O. Taiwo, G.A. Wilson, T. Morris, S. Seisenberger, W. Reik, D. Pearce, S. Beck, and L.M. Butcher. Methylome analysis using MeDIP-seq with low DNA concentrations. *Nature Protocols*, 7(4):617–636, 2012.

[425] K.M. Tan, A. Petersen, and D. Witten. Classification of rna-seq data. In S. Datta and D. Nettleton, editors, *Statistical Analysis of Next Generation Sequencing Data*. Springer, New York, 2014.

[426] X. Tang, Y. Huang, J. Lei, H. Luo, and X. Zhu. The single-cell sequencing: new developments and medical applications. *Cell & Bioscience*, 9(1):1–9, 2019.

[427] H. Tanizawa, O. Iwasaki, A. Tanaka, J.R. Capizzi, P. Wickramasinghe, M. Lee, Z. Fu, and K.-i. Noma. Mapping of long-range associations throughout the fission yeast genome reveals global genome organization linked to transcriptional regulation. *Nucleic Acids Research*, 38(22):8164–8177, 2010.

[428] B. Tasic, Z. Yao, L.T. Graybuck, K.A. Smith, T.N. Nguyen, D. Bertagnolli, J. Goldy, E. Garren, M.N. Economo, S. Viswanathan, et al. Shared and distinct transcriptomic cell types across neocortical areas. *Nature*, 563(7729):72–78, 2018.

[429] R. Tautenhan, K. Cho, W. Uritboonthai, Z. Zhu, G.J. Patti, and G. Siuz-dak. An accelerated workflow for untargeted metabolomics using the metlin database. *Nature Biotechnology*, 30(9):826–827, 2012.

[430] B.H. Ter Kuile and H.V. Westerhoff. Transcriptome meets metabolome: hierarchical and metabolic regulation of the glycolytic pathway. *FEBS Letters*, 500:169–171, 2001.

[431] E.E. Thompson, Y. Sun, D. Nicolae, and C. Ober. Shades of gray: a comparison of linkage disequilibrium between hutterites and europeans. *Genetic Epidemiology*, 34:133–139, 2010.

[432] Y. Tian, T.J. Morris, A.P. Webster, Z. Yang, S. Beck, A. Feber, and A.E. Teschendorff. Champ: updated methylation analysis pipeline for illumina beadchips. *Bioinformatics*, 33:3982–3984, 2017.

[433] R. Tibshirani, T. Hastie, B. Narasimhan, and G. Chu. Diagnosis of multiple cancer types by shrunken centroids of gene expression. *Proceedings of the National Academy of Sciences*, 99(10):6567–6572, 2002.

[434] R. Tibshirani, T. Hastie, B. Narasimhan, and G. Chu. Class prediction by nearest shrunken centroids, with applications to dna microarrays. *Statistical Science*, 18(1):104–117, 2003.

[435] R. Tirado-Magallanes, K. Rebbani, R. Lim, S. Pradhan, and T. Benoukraf. Whole genome dna methylation: beyond genes silencing. *Oncotarget*, 8:5629–5637, 2017.

[436] B. Tolhuis, R.J. Palstra, E. Splinter, F. Grosveld, and W. de Laat. Looping and interaction between hypersensitive sites in the active beta-globin locus. *Molecular Cell*, 10(6):1453–1465, 2002.

[437] G. Tomasi, F. van den Berg, and C. Andersson. Correlation optimized warping and dynamic time warping as preprocessing methods for chromatographic data. *Journal of Chemometrics*, 18(5):231–241, 2004.

[438] A.H. Tong, Drees B., G. Nardelli, Bader G.D., B. Brannetti, L. Castagnoli, M. Evangelista, S. Ferracuti, B. Nelson, S. Paoluzi, M. Quondam, A. Zucconi, C.W. Hogue, S. Fields, C. Boone, and G. Cesareni. A combined experimental and computational strategy to define protein interaction networks for peptide recognition modules. *Science*, 295(5553):321–324, 2002.

[439] S. Tornow and H.-W. Mewes. Functional modules by relating protein interaction networks and gene expression. *Nucleic Acids Research*, 31(21):6283–6289, 2003.

[440] C. Trapnell, L. Pachter, and S.L. Salzberg. Tophat: discovering splice junctions with rna-seq. *Bioinformatics*, 25(9):1105–1111, 2009.

[441] R.C. Tripathi, R.C. Gupta, and J. Gurland. Estimation of parameters in the beta binomial model. *Annals of the Institute of Statistical Mathematics*, 46:317–331, 1994.

[442] S. Tripathi, S. Moutari, M. Dehmer, and F. Emmert-Streib. Comparison of module detection algorithms in protein networks and investigation of the biological meaning of predicted modules. *BMC Bioinformatics*, 17:129, 2016.

[443] K. Tu, H. Yu, and Y.-X. Li. Combining gene expression profiles and protein–protein interaction data to infer gene functions. *Journal of Biotechnology*, 124(3):475–485, 2006.

[444] G. Tucker, P.-R. Loh, and B. Berger. A sampling framework for incorporating quantitative mass spectrometry data in protein interaction analysis. *BMC Bioinformatics*, 14:299, 2013.

[445] S. Tulipani, R. Llorach, M. Urpi-Sarda, and C.A. Andres-Lacueva. Comparative analysis of sample preparation methods to handle the complexity of the blood fluid metabolome: When less is more. *Analytical Chemistry*, 85(1):341–348, 2013.

[446] P. Uetz, L. Giot, G. Cagney, T.A. Mansfield, R.S. Judson, J.R. Knight, D. Lockshon, V. Narayan, M. Srinivasan, P. Pochart, A. Qureshi-Emili, Y. Li, B. Godwin, D. Conover, T. Kalbfleisch, G. Vijayadamodar, M. Yang, M. Johnston, S. Fields, and J.M. Rothberg. A comprehensive analysis of protein–protein interactions in saccharomyces cerevisiae. *Nature*, 403:623–627, 2000.

[447] K. Uppal, Q.A. Soltow, F.H. Strobel, W.S. Pittard, K.M. Gernert, T. Yu, and D.P. Jones. xmsanalyzer: automated pipeline for improved feature detection and downstream analysis of large-scale, non-targeted metabolomics data. *BMC Bioinformatics*, 14:15, 2013.

[448] G.J.G. Upton. A comparison of alternative tests for the 2 x 2 comparative trial. *Journal of the Royal Statistical Society Series A*, 145:86–105, 2016.

[449] E. Urbanczyk-Wochniak, A. Luedemann, J. Kopka, J. Selbig, U. Roessner-Tunali, L. Willmitzer, and A.R. Fernie. Parallel analysis of transcript and metabolic profiles: a new approach in systems biology. *EMBO Reports*, 4(10):989–993, 2003.

[450] O. Ursu, N. Boley, M. Taranova, Y.X.R. Wang, G.G. Yardimci, W.S. Noble, and A. Kundaje. Genomedisco: A concordance score for chromosome conformation capture experiments using random walks on contact map graphs. *Bioinformatics*, 34(16):2701–2707, 2018.

[451] R. van de Putte, H. E.K. de Walle, K.J.M. van Hooijdonk, I. de Blaauw, C.L.M. Marcelis, A. van Heijst, J.C. Giltay, K.Y. Renkema, P.M.A. Broens, E. Brosens, C.E.J. Sloots, J.E.H. Bergman, N. Roeleveld, and I. A.L.M. van Rooij. Maternal risk associated with the vacterl association: A case-control study. *Birth Defects Research*, 112(18):1495–1504, 2020.

[452] E.J. van den Oord. The use of mixture models to perform quantitative tests for linkage disequilibrium, maternal effects, and parent-of-origin effects with incomplete subject-parent triads. *Behavior Genetics*, 30:335–343, 2000.

[453] L. van der Maaten and G. Hinton. Visualizing data using t-sne. *Journal of Machine Learning Research*, 9(86):2579–2605, 2008.

[454] D. van Dijk, J. Nainys, R. Sharma, P. Kathail, A.J. Carr, K.R. Moon, L. Mazutis, G. Wolf, S. Krishnaswamy, and D. Pe'er. Magic: A diffusion-based imputation method reveals gene-gene interactions in single-cell rna-sequencing data. *BioRxiv*, page 111591, 2017.

[455] V. Vapnik. The nature of statistical learning theory. In V. Vapnik, editor, *The Nature of Statistical Learning Theory*. Springer, New York, 2000.

[456] N. Varoquaux, F. Ay, W.S. Noble, and Jean-Philippe Vert. A statistical approach for inferring the 3d structure of the genome. *Bioinformatics*, 30(12):26–33, 2014.

[457] S.H. Vermeulen, M. Shi, C.R. Weinberg, and D.M. Umbach. A hybrid design: Case-parent triads supplemented by control-mother dyads. *Genetic Epidemiology*, 33(2):136–144, 2009.

[458] D. Vernimmen, M. De Gobbi, J.A. Sloane-Stanley, W.G. Wood, and D.R. Higgs. Long-range chromosomal interactions regulate the timing of the transition between poised and active gene expression. *The EMBO Journal*, 26(8):2041–2051, April 2007.

[459] R. Vershynin. Introduction to the non-asymptotic analysis of random matrices, 2011.

[460] K.A. Veselkov, J.C. Lindon, T.M.D. Ebbels, D. Crockford, V.V. Volynkin, Holmes, D.B. Davies, and J.K. Nicholson. Recursive segment-wise peak alignment of biological 1h nmr spectra for improved metabolic biomarker discovery. *Analytical Chemistry*, 81(1):56–66, 2009.

[461] K.A. Veselkov, L.K. Vingara, P. Masson, S.L. Robinette, E. Want, J.V. Li, R.H. Barton, C. Boursier-Neyret, B. Walther, T.M. Ebbels, I. Pelczer, E. Holmes, J.C. Lindon, and J.K. Nicholson. Optimized preprocessing of ultra-performance liquid chromatography/mass spectrometry urinary metabolic profiles for improved information recovery. *Analytical Chemistry*, 83(15):5864–5972, 2011.

[462] C. von Mering, L.J. Jensen, B. Snel, S.D. Hooper, M. Krupp, M. Foglierini, N. Jouffre, M.A. Huynen, and Bork P. String: known and predicted protein-protein associations, integrated and transferred across organisms. *Nucleic Acids Research*, 33(Database issue):D433–D437, 2005.

[463] T. Vu and K. Laukens. Getting your peaks in line: a review of alignment methods for nmr spectral data. *Metabolites*, 3:259–276, 2013.

[464] J. Wang, et al. Double restriction-enzyme digestion improves the coverage and accuracy of genome-wide cpg methylation profiling by reduced representation bisulfite sequencing. *BMC Genomics*, 14:11, 2013.

[465] L. Wang, Z. Feng, X. Wang, X. Wang, and X. Zhang. Degseq: an r package for identifying differentially expressed genes from rna-seq data. *Bioinformatics*, 26:136–138, 2010.

[466] Q. Wang, W. Li, Y. Zhang, X. Yuan, K. Xu, J. Yu, Z. Chen, R. Beroukhim, H. Wang, M. Lupien, T. Wu, M. M. Regan, C.A. Meyer, J.S. Carroll, A. Kumar K. Manrai, O.A. Jänne, S.P. Balk, R. Mehra, B. Han, A.M. Chinnaiyan, M.A. Rubin, L. True, M. Fiorentino, C. Fiore, M. Loda, P.W. Kantoff, X. S. Liu, and M. Brown. Androgen receptor regulates a distinct transcription program in androgen-independent prostate cancer. *Cell*, 138(2):245–256, 2009.

[467] S.-Y. Wang, C.-H. Kuo, and Y.J. Tseng. Batch normalizer: A fast total abundance regression calibration method to simultaneously adjust batch and injection order effects in liquid chromatography/time-of-flight mass spectrometry-based metabolomics data and comparison with current calibration methods. *Analytical Chemistry*, 85:1037–1046, 2013.

[468] W. Wang, H. Zhou, H. Lin, S. Roy, T.A. Shaler, L.R. Hill, S. Norton, P. Kumar, M. Anderle, and C.H. Becker. Quantification of proteins and metabolites by mass spectrometry without isotopic labeling or spiked standards. *Analytical Chemistry*, 75:4818–4826, 2003.

[469] X. Wang. Approximating bayesian inference by weighted likelihood. *Canadian Journal of Statistics*, 34(2):279–298, 2006.

[470] X. Wang. *Next-Generation Sequencing Data Analysis*. Chapman & Hall/CRC Oxfordshire, UK, 2016.

[471] X. Wang, Q. Sun, S.D. McGrath, E.R. Mardis, P.D. Soloway, and A.G. Clark. Transcriptome- wide identification of novel imprinted genes in neonatal mouse brain. *PLoS One*, 3:e3839, 2008.

[472] S. Wasserman and K. Faust. *Social Network Analysis*. Cambridge University Press, 1994.

[473] C.L. Waterman, R.A. Currie, L.A. Cottrell, J. Dow, J. Wright, C.J. Waterfield, and J.L. Griffin. An integrated functional genomic study of acute phenobarbital exposure in the rat. *BMC Genomics*, 11:9, 2010.

[474] D.J. Watts and S.H. Strogatz. Collective dynamics of 'small-world' networks. *Nature*, 393:440–442, 1998.

[475] A. Wawrzyniak, A. Kosnowska, S. Macioszek, R. Bartoszewski, and M.J. Markuszewski. New plasma preparation approach to enrich metabolome coverage in untargeted metabolomics: plasma protein bound hydrophobic metabolite release with proteinase k. *Scientific Reports*, 8:9541, 2018.

[476] M. Weber, J.J. Davies, D. Wittig, E.J. Oakeley, M. Haase, W.L. Lam, and D. Schubeler. Chromosome- wide and promoter-specific analyses identify sites of differential dna methylation in normal and transformed human cells. *Nature Genetics*, 37:853–862, 2005.

[477] C.R. Weinberg. Methods for detection of parent-of-origin effects in genetic studies of case-parents triads. *American Journal of Human Genetics*, 65(1):229–235, 1999.

[478] C.R. Weinberg, A.J. Wilcox, and R.T. Lie. A log-linear approach to case-parent-triad data: Assessing effects of disease genes that act either directly or through maternal effects and that may be subjected to parental imprinting. *American Journal of Human Genetics*, 62:969–978, 1998.

[479] C.R. Weinberg and D.M. Umbach. A hybrid design for studying genetic influences on risk of diseases with onset early in life. *American Journal of Human Genetics*, 77:627–636, 2005.

[480] Y. Wen, F. Chen, Q. Zhang, Y. Zhuang, and Z. Li. Detection of differentially methylated regions in whole genome bisulfite sequencing data using local getis-ord statistics. *Bioinformatics*, 32:3396–3404, 2016.

[481] M.R. Wenk. The emerging field of lipidomics. *Nature Reviews Drug Discovery*, 4:594–610, 2005.

[482] A.G. West and P. Fraser. Remote control of gene transcription. *Human Molecular Genetics*, 14 Spec No 1, April 2005.

[483] A.J. Wilcox, C.R. Weinberg, and R.T. Lie. Distinguishing the effects of maternal and offspring genes through studies of case-parent-triads. *American Journal of Epidemiology*, 148(9):893–901, 1998.

[484] D.S. Wishart. Quantitative metabolomics using nmr. *Trends in Analytical Chemistry*, 27:228–237, 2008.

[485] D.S. Wishart, T. Jewison, A.C. Guo, M. Wilson, C. Knox, Y. Liu, Y. Djoumbou, R. Mandal, F. Aziat, E. Dong, S. Bouatra, I. Sinelnikov, D. Arndt, J. Xia, P. Liu, F. Yallou, T. Bjorndahl, R. Perez-Pineiro, R. Eisner, F. Allen, V. Neveu, R. Greiner, and A. Scalbert. Hmdb 3.0–the human metabolome database in 2013. *Nucleic Acids Research*, 41(Database issue):D801–D807, 2013.

[486] D.S. Wishart, C. Knox, A.C. Guo, R. Eisner, N. Young, B. Gautam, D.D. Hau, N. Psychogios, E. Dong, S. Bouatra, R. Mandal, I. Sinel-nikov, J. Xia, L. Jia, J.A. Cruz, E. Lim, C.A. Sobsey, S. Shrivastava, P. Huang, P. Liu, L. Fang, J. Peng, R. Fradette, D. Cheng, D. Tzur, M. Clements, A. Lewis, A. De Souza, A. Zuniga, M. Dawe, Y. Xiong, D. Clive, R. Greiner, A. Nazyrova, R. Shaykhutdinov, L. Li, H.J. Vogel, and Forsythe I. Hmdb: a knowledgebase for the human metabolome. *Nucleic Acids Research*, 37(Database issue):D603–D610, 2009.

[487] D.S. Wishart, R. Mandal, A. Stanislaus, and M. Ramirez-Gaona. Cancer metabolomics and the human metabolome database. *Metabolites*, 6(1):E10, 2016.

[488] D.M. Witten. Classification and clustering of sequencing data using a poisson model. *AAnnals of Applied Statistics*, 5(4):2493–2518, 2011.

[489] P.J. Wittkopp, B.K. Haerum, and A.G. Clark. Parent-of-origin effects on mrna expression in drosophila melanogaster not caused by genomic imprinting. *Genetics*, 173:1817–1821, 2006.

[490] S. Wold, K. Esbensen, and P. Geladi. Principle component analysis. *Chemometrics and Intelligent Laboratory Systems*, 2:37–52, 1987.

[491] D. Wu, C. Zhang, Y. Shen, K.P. Nephew, and Q. Wang. Androgen receptor-driven chromatin looping in prostate cancer. *Trends in Endocrinology and Metabolism*, 22(12):474–480, 2011.

[492] G. Wu, X. Feng, and L. Stein. A human functional protein interaction network and its application to cancer data analysis. *Genome Biology*, 11(5):R53, 2010.

[493] J. Xia, I. Sinelnikov, B. Han, and D.S. Wishart. Metaboanalyst 3.0 - making metabolomics more meaningful. *Nucleic Acids Research*, 43(W1):W251–W257, 2015.

[494] J. Xia and D.S. Wishart. Metpa: a web-based metabolomics tool for pathway analysis and visualization. *Bioinformatics*, 26(18):2342–2344, 2010.

[495] Q. Xie, C. Han, V. Jin, and S. Lin. HiCImpute: A bayesian hierarchical model for identifying structural zeros and enhancing single cell Hi-C data. *PLoS Computational Biology*, 2022. doi.org/10.1371/journal.pcbi.1010129

[496] Q. Xie and S. Lin. ScHiCSRS: A self-representation smoothing method with gaussian mixture model for digitally enhancing single-cell Hi-C data. *bioRxiv*, 2022.

[497] Y. Xing, T. Yu, Y.N. Wu, M. Roy, J. Kim, and C. Lee. An expectation-maximization algorithm for probabilistic reconstructions of full-length isoforms from splice graphs. *Nucleic Acids Research*, 34(10):3150–3160, 2006.

[498] K. Xiong and J. Ma. Revealing hi-c subcompartments by imputing inter-chromosomal chromatin interactions. *Nature Communications*, 10(1):5069, 2019.

[499] E. Yaffe and A. Tanay. Probabilistic modeling of Hi-C contact maps eliminates systematic biases to characterize global chromosomal architecture. *Nature Genetics*, 43(11):1059–1065, 2011.

[500] P. Yan, D. Frankhouser, M. Murphy, H.-H. Tam, B. Rodriguez, J. Curfman, M. Trimarchi, S. Geyer, Y.-Z. Wu, S.P. Whitman, K. Metzeler, a. Walker, R. Klisovic, S. Jacob, M.R. Grever, J.C. Byrd, C.D. Bloomfield, R. Garzon, W. Blum, M.a. Caligiuri, R. Bundschuh, and G. Marcucci. Genome-wide methylation profiling in decitabine-treated patients with acute myeloid leukemia. *Blood*, 120(12):2466–2474, 2012.

[501] C. Yang, Z. He, and W. Yu. Comparison of public peak detection algorithms for maldi mass spectrometry data analysis. *BMC Bioinformatics*, 10:4, 2009.

[502] J Yang. *Likelihood approaches for detecting imprinting and maternal effects in family-based association studies.* PhD thesis, The Ohio State University, June 2010.

[503] J. Yang and S. Lin. Detection of imprinting and heterogeneous maternal effects on high blood pressure using Framingham Heart Study data. *BMC Proceedings*, 3 (Suppl 7):S125, 2009.

[504] J. Yang and S. Lin. Likelihood Approach for Detecting Imprinting and In Utero Maternal Effects Using General Pedigrees from Prospective Family-Based Association Studies. *Biometrics*, 68(2):477–485, 2012.

[505] J. Yang and S. Lin. Robust partial likelihood approach for detecting imprinting and maternal effects using case-control families. *Annals of Applied Statistics*, 7(1):249–268, 2013.

[506] T. Yang, F. Zhang, G.G. Yardımcı, F. Song, R.C. Hardison, W.S. Noble, F. Yue, and Q. Li. Hicrep: assessing the reproducibility of hi-c data using a stratum-adjusted correlation coefficient. *Genome Research*, 27(11):1939–1949, 2017.

[507] Y. Yang, R. Huh, H.W. Culpepper, Y. Lin, M.I. Love, and Y. Li. SAFE-clustering: Single-cell Aggregated (from Ensemble) clustering for single-cell RNA-seq data. *Bioinformatics*, 35(8):1269–1277, 2018.

[508] S.H. Yook, Z.N. Oltvai, and A.-L. Barabasi. Functional and topological characterization of protein interaction networks. *Proteomics*, 4(4):928–942, 2004.

[509] H. Yu and M. Gerstein. Genomic analysis of the hierarchical structure of regulatory networks. *Proceedings of the National Academy of Sciences*, 103:14724–14731, 2006.

[510] J. Yu, J. Yu, R.-S. Mani, Q. Cao, C. J. Brenner, X. Cao, X. Wang, L. Wu, J. Li, M. Hu, Y. Gong, H. Cheng, B. Laxman, A. Vellaichamy, S. Shankar, Y. Li, S.M. Dhanasekaran, R. Morey, T. Barrette, R.J. Lonigro, S. A. Tomlins, S. Varambally, Z.S. Qin, and A.M. Chinnaiyan. An integrated network of androgen receptor, polycomb, and TMPRSS2-ERG gene fusions in prostate cancer progression. *Cancer Cell*, 17(5):443–454, 2010.

[511] M. Yu, A. Abnousi, Y. Zhang, G. Li, L. Lee, Z. Chen, R. Fang, J. Wen, Q. Sun, Y. Li, et al. Snaphic: a computational pipeline to map chromatin contacts from single cell hi-c data. *bioRxiv*, 2020.

[512] T. Yu, Y. Park, J.M. Johnson, and D.P. Jones. aplcms—adaptive processing of high-resolution lc/ms data. *Bioinformatics*, 25(15):1930–1936, 2009.

[513] X. Yu and S. Sun. Hmm-dm: identifying differentially methylated regions using a hidden markov model. *Statistical Applications in Genetics and Molecular Biology*, 15:69–81, 2016.

[514] C. Yu Lin, H. Wu, R.S. Tjeerdema, and M.R. Viant. Evaluation of metabolite extraction strategies from tissue samples using nmr metabolomics. *Metabolomics*, 3:55–67, 2007.

[515] Y. Yuan, C. Norris, Y. Xu, K.W. Tsui, Y. Ji, and H. Liang. Bm-map: An efficient software package for accurately allocating multireads of rna-sequencing data. *BMC Genomics*, 13(Suppl 8):S9, 2012.

[516] V. Zhabotynsky, K. Inoue, T. Magnuson, J.M. Calabrese, and W. Sun. A statistical method for joint estimation of cis-eqtls and parent-of-origin effects under family trio design. *Biometrics*, 75(3):864–874, 2019.

[517] B. Zhang, B.-H. Park, T. Karpinets, and N.F. Samatova. From pull-down data to protein interaction networks and complexes with biological relevance. *Bioinformatics*, 24(1):979–986, 2008.

[518] F. Zhang, A. Khalili, and S. Lin. Imprinting and maternal effect detection using partial likelihood based on discordant sibpair data. *Statistica Sinica*, 29:1915–1937, 2019.

[519] F. Zhang, A. Khalili, and S. Lin. Optimum study design for detecting imprinting and maternal effects based on partial likelihood. *Biometrics*, 72(1):95–105, 2016.

[520] F. Zhang and S. Lin. Nonparametric method for detecting imprinting effect using all members of general pedigrees with missing data. *Journal of Human Genetics*, 59(10):541–548, 2014.

[521] F. Zhang and S. Lin. Incorporating information from markers in ld with test locus for detecting imprinting and maternal effects. *European Journal of Human Genetics*, 28(8):1087–1097, 2020.

[522] Kai Zhang, Hong Zhang, Hagit Hochner, and Jinbo Chen. Covariate adjusted inference of parent-of-origin effects using case–control mother–child paired multilocus genotype data. *Genetic Epidemiology*, 45(8):830–847, 2021.

[523] L. Zhang and S. Zhang. Comparison of computational methods for imputing single-cell rna-sequencing data. *IEEE/ACM Transactions on Computational Biology and Bioinformatics*, 17(2):376–389, 2020.

[524] L. Zhang and S. Zhang. Comparison of computational methods for imputing single-cell rna-sequencing data. *IEEE/ACM Transactions on Computational Biology and Bioinformatics*, 17(2):376–389, 2018.

[525] R. Zhang, T. Zhou, and J. Ma. Multiscale and integrative single-cell hi-c analysis with higashi. *bioRxiv*, 2021.

[526] W. Zhang, Z. Li, N. Wei, H.-J. Wu, and H. Zheng. Detection of differentially methylated cpg sites between tumor samples with uneven tumor purities. *Bioinformatics*, 36:2017–2024, 2020.

[527] X. Zhang, J.M. Asara, J. Adamec, M. Ouzzani, and A.K. Elmagarmid. Data pre-processing in liquid chromatography–mass spectrometry-based proteomics. *Bioinformatics*, 21(12):4054–4059, 2005.

[528] Y. Zhang, E.D. Kolaczyk, and B. Spencer. Estimating node degree distributions under sampling: An inverse problem, with applications to monitoring social networks. *The Annals of Applied Statistics*, 9(1):166–199, 2015.

[529] Y. Zhang, H. Liu, J. Lv, X. Xiao, J. Zhu, X. Liu, J. Su, X. Li, Q. Wu, F. Wang, and Y. Cui. Qdmr: a quantitative method for identification of differentially methylated regions by entropy. *Nucleic Acids Research*, 39:e58, 2011.

[530] Y. Zhang, L. An, J. Xu, B. Zhang, W.J. Zheng, M. Hu, J. Tang, and F. Yue. Enhancing hi-c data resolution with deep convolutional neural network hicplus. *Nature Communications*, 9(1):750, 2018.

[531] Z. Zhang, G. Li, K.-C. Toh, and W.-K. Sung. 3d chromosome modeling with semi-definite programming and hi-c data. *Journal of Computational Biology*, 20(11):831–846, 2013.

[532] Z. Zhang, G. Li, K.-C. Toh, and W.-K. Sung. Inference of spatial organizations of chromosomes using semi-definite embedding approach and hi-c data. In *Proceedings of the 17th International Conference on Research in Computational Molecular Biology*, RECOMB'13, pages 317–332, Berlin, Heidelberg, 2013. Springer-Verlag.

[533] P. Zhao and B. Yu. On model selection consistency of lasso. *Journal of Machine Learning Research*, 7:2541–2563, 2006.

[534] Z. Zhao, G. Tavoosidana, M. Sjölinder, A. Göndör, P. Mariano, S. Wang, C. Kanduri, M. Lezcano, K.S. Sandhu, U. Singh, V. Pant, V. Tiwari, S. Kurukuti, and R. Ohlsson. Circular chromosome conformation capture (4C) uncovers extensive networks of epigenetically regulated intra- and interchromosomal interactions. *Nature Genetics*, 38(11):1341–1347, 2006.

[535] C. Zhen, Y. Wang, L. Han, J. Li, J. Peng, T. Wang, J. Hao, X. Shang, Z. Wei, and J. Peng. A novel framework for single-cell hi-c clustering based on graph-convolution-based imputation and two-phase-based feature extraction. *bioRxiv*, 2021.

[536] X. Zheng, N. Zhang, H-J. Wu, and H. Wu. Estimating and accounting for tumor purity in the analysis of dna methylation data from cancer studies. *Genome Biology*, 18:17, 2017.

[537] Y. Zheng and S. Keleş. Freehi-c: high fidelity hi-c data simulation for benchmarking and data augmentation. *bioRxiv*, 2019.

[538] J. Zhou, Y. Hu, S. Lin, and W.K. Fung. Detection of parent-of-origin effects based on complete and incomplete nuclear families with multiple affected children. *Human Heredity*, 67:1–12, 2009.

[539] J.-Y. Zhou, H.-Q. He, X.-P. You, S.-Z. Li, P.-Y. Chen, and W. K. Fung. A powerful association test for qualitative traits incorporating imprinting effects using general pedigree data. *Journal of Human Genetics*, 60(2):77–83, 2015.

[540] J.-Y. Zhou, W.-G. Mao, D.-L. Li, Y.-Q. Hu, F. Xia, and W.K. Fung. A powerful parent-of-origin effects test for qualitative traits incorporating control children in nuclear families. *Journal of Human Genetics*, 57(8):500–507, 2012.

[541] J. Zhou, J. Ma, Y. Chen, C. Cheng, B. Bao, J. Peng, T.J. Sejnowski, J.R. Dixon, and J.R. Ecker. Robust single-cell hi-c clustering by convolution-and random-walk–based imputation. *Proceedings of the National Academy of Sciences*, page 201901423, 2019.

[542] J.Y. Zhou, J. Ding, W.K. Fund, and S. Lin. Detection of parent-of-origin effects using general pedigree data. *Genetic Epidemiology*, 34:151–158, 2010.

[543] H. Zhu and Z. Wang. Scl: a lattice-based approach to infer 3d chromosome structures from single-cell hi-c data. *Bioinformatics*, 35:3981–3988, 2019.

[544] H. Zhu and Z. Wang. SCL: a lattice-based approach to infer 3D chromosome structures from single-cell Hi-C data. *Bioinformatics*, 35(20):3981–3988, 2019.

[545] H. Zhu and Z. Wang. Scl: a lattice-based approach to infer 3d chromosome structures from single-cell hi-c data. *Bioinformatics*, 35(20):3981–3988, 2019.

[546] D. Zilberman, M. Gehring, R.K. Tran, T. Ballinger, and S. Henikoff. Genome-wide analysis of Arabidopsis thaliana DNA methylation uncovers an interdependence between methylation and transcription. *Nature Genetics*, 39(1):61–69, 2007.

Index